Human Anatomy
Laboratory Manual

with Cat Dissections

The Benjamin/Cummings Series in Human Anatomy and Physiology

R. A. Chase
The Bassett Atlas of Human Anatomy (1989)

S. W. Langjahr and R. D. Brister
Human Anatomy Coloring Atlas, Second Edition (1992)

E. N. Marieb
Human Anatomy and Physiology, Second Edition (1992)

E. N. Marieb
Human Anatomy and Physiology, Study Guide, Second Edition (1992)

E. N. Marieb
Human Anatomy and Physiology Laboratory Manual: Cat Version,
Third Edition (1989)

E. N. Marieb
Human Anatomy and Physiology Laboratory Manual: Fetal Pig Version,
Third Edition (1989)

E. N. Marieb
Human Anatomy and Physiology, Laboratory Manual: Brief Version,
Third Edition (1992)

E. N. Marieb
Essentials of Human Anatomy and Physiology, Third Edition (1991)

E. N. Marieb
The A & P Coloring Workbook: A Complete Study Guide, Third Edition (1991)

E. B. Mason
Human Physiology (1983)

A. P. Spence
Basic Human Anatomy, Third Edition (1991)

R. L. Vines and A. Hinderstein
Human Musculature Videotape (1989)

R. L. Vines and Michaeline Veden
Human Nervous System Videotape (1992)

Human Anatomy Laboratory Manual

with Cat Dissections

Elaine N. Marieb, R.N., Ph.D.

Holyoke Community College

The Benjamin/Cummings Publishing Company, Inc.

Redwood City, California • Menlo Park, California
Reading, Massachusetts • New York • Don Mills, Ontario
Wokingham, U.K. • Amsterdam • Bonn • Sydney
Singapore • Tokyo • Madrid • San Juan

Sponsoring editor: Melinda Adams

Editorial coordinator: Mark Childs

Production coordinator: Andy Marinkovich

Text design: Gary Head

Cover design: Mark Ong

Photo editor: Darcy Lanham

Page layout: Victoria Ann Philp

Compositor: Graphic Typesetting Service

Artists: Raychel Ciemma, Barbara Cousins, Marjorie Garlin, Charles Hoffman, Cathleen Jackson, Jill Leland, Doreen Davis Masterson, Stephanie McCann, Linda McVay, Fran Milner, Elizabeth Morales-Denney, Nadine Sokol, Sara Lee Steigerwald, Carol Verbeeck, Neesh Wallace

Library of Congress Cataloging in Publication Data

Marieb, Elaine Nicpon, 1936–
    Human anatomy laboratory manual with cat
  dissections / Elaine N. Marieb.
        p.   cm.
    Includes index.
    ISBN 0-8053-4050-5
    1. Human anatomy—Laboratory
manuals.   2. Cats—Dissection—
  Laboratory manuals.   I. Title.
  QM34.M33   1992
611′.0078—dc20   91-20592
                                        CIP

  ISBN 0-8053-4050-5
   2 3 4 5 6 7 8 9 10 – VH – 95 94 93 92

The Benjamin/Cummings Publishing Company, Inc.
390 Bridge Parkway
Redwood City, California 94065

Preface

Human Anatomy Laboratory Manual is a self-contained learning aid designed for introductory courses using a systemic approach in teaching human anatomy. It presents a wide range of laboratory experiences for students with a minimal background in the biological sciences, who are pursuing careers in nursing, physical therapy, respiratory therapy, dental hygiene, pharmacology, health and physical education, as well as biology, pre-medical programs, and other allied health professions.

This manual studies anatomy of the human specimen in particular, but the cat is used in the dissection experiments.

ORGANIZATION

The organization and scope of this manual lend themselves to use in the one-term human anatomy course. The variety of anatomical studies enables instructors to gear their courses to specific academic programs, or to their own teaching preferences. Although the textbook *Human Anatomy* (Marieb and Mallatt, Benjamin/Cummings, 1992), provided the impetus for its preparation, the manual is largely based upon exercises developed for use independent of any textbook. It contains all the background discussion and terminology necessary to perform all experiments effectively. This eliminates the need for students to bring a textbook into the laboratory.

The manual is comprehensive and balanced enough to be flexible, and carefully written so that students can successfully complete each of its 31 exercises with little supervision. The manual begins with an exercise on anatomical terminology and an orientation to the human body. The second exercise provides the necessary tools for studying the various body systems. Succeeding exercises on the microscope, the cell, and tissues lay the groundwork for a study of each body system from the cellular to the organ level. Exercises 7 through 31 explore the anatomy of the organ systems in detail.

FEATURES

The following features have been carefully developed to provide students with an interesting and challenging experience in the human anatomy laboratory:

- Learning objectives are clearly written and precede each exercise.
- A materials list precedes each exercise as well. A complete materials list also appears in the instructor's guide.
- Key terminology appears in boldface type, and each term is defined when introduced.
- Illustrations are plentiful, large, and of exceptional quality and clarity. Structures are consistently shown in their natural anatomical positions within the body.
- Exercises are balanced between histological and gross anatomical topics.
- The prologue, "Getting Started—The Scientific Method, Scientific Notation, and Metrics," explains the scientific method, the logical, practical, and reliable way of approaching and solving problems in the laboratory. It also reviews the use of exponents, metric units, and interconversions.
- A dissection icon is used throughout the exercises to clearly designate the dissection instructions.
- All laboratory instructions and procedures have been revised to incorporate the latest precautions as recommended by the Centers for Disease Control (CDC); these are reinforced by the laboratory safety procedures described in the front section of the instructor's guide. These procedures can be easily photocopied and posted in the lab.
- Clinical information is scattered throughout the exercises to point out how systems behave when structural abnormalities occur. It is identifiable by the symbol ⚠ near the margin.
- Tear-out Laboratory Review Study Sections, keyed to the exercises, require students to label diagrams and answer multiple-choice, short-answer,

and essay questions. Every attempt has been made to achieve an acceptable balance between explanatory and recognition questions.

- Fourteen full-color dissection photos of meticulously dissected cats are included in this manual. Identifying muscle in the fast pace of the laboratory session becomes easier when such large, accurate, well-labelled photographs are available.
- The realism of several aspects of anatomy is enhanced by the use of additional full-color photographs and illustrations. Color is used pedagogically; that is, the structures and systems are color-coded to facilitate identification. The following exercises have been embellished with full color:

Exercise 6, "Classification of Tissues," incorporates superbly organized histology tables. These tables present description, location, and function of tissues at a glance. They include color illustrations and photomicrographs for a more complete study of structures. Four additional color photomicrographs can be found in the color foldout section, and twelve more are conveniently located on the inside front and back covers, for easy reference.

Exercise 10, "The Axial Skeleton," features skull art rendered in color-coded sections for ease of identification.

Exercise 15, "Gross Anatomy of the Muscular System," begins with the identification of human muscles, using highly accurate art and accompanying muscles tables. In addition, human anatomy photographs provide the realism lacking when a human specimen is not available in the laboratory. The art is borrowed from the textbook associated with this manual, *Human Anatomy* (Marieb and Mallatt, Benjamin/Cummings, 1992), and the photos were carefully selected from the Stanford Bassett Collection, also featured in the Marieb and Mallatt textbook.

- An Instructor's Guide, complete with directions for lab set-up, comments on chapter exercises, and answers to the lab manual questions, is also available.

ACKNOWLEDGMENTS

I wish to thank the following reviewers for their help and guidance in the preparation of the exercises which make up the majority of the *Human Anatomy Laboratory Manual:* Dean Beckwith (Illinois Central College), Linda Kollett (Massasoit Community College), Donna S. Klaaren (Miami University), Fredericka Kundig (Towson State University), William Matthai (Tarrant County Community College), Evan Oyakawa (California State University, Los Angeles), Carol Spaulding (University of Maryland), Robert Thomas (College of the Redwoods), Glenn Yoshida (Los Angeles Southwest College), and Janice Young, Ph.D (McHenry County College).

I also wish to acknowledge my indebtedness to Melinda Adams, Acquisitions Editor at Benjamin/Cummings, Laura Bonazzoli, Associate Editor, and to Mark Childs, Editorial Assistant, for their assistance. My thanks also go to Andy Marinkovich, and the many others at Benjamin/Cummings who have contributed to this effort, and as always, to my husband, Joseph.

Finally, since preparation for the second edition will begin well in advance of its publication, I invite users of this manual to send me their comments and suggestions for subsequent editions.

Elaine N. Marieb
Department of Biology
Holyoke Community College
303 Homestead Avenue
Holyoke, MA 01040

Contents

Getting Started—
The Scientific Method,
Scientific Notation, and Metrics

Two hundred years ago science was largely a plaything of wealthy patrons, but today's world is dominated by science and its products. Whether or not we believe that such scientific domination is desirable, we all have a responsibility to try to understand the goals and methods of science that have seeded this knowledge explosion.

The biosciences are very special and exciting because they open the doors to an understanding of all the wondrous workings of living things. A course in human anatomy and physiology (a minute subdivision of bioscience) starts you on the road to such insights in relation to your own body. Although some experience in scientific studies is undeniably helpful when beginning a study of anatomy and physiology, perhaps the single most important prerequisite is a pervasive curiosity.

Gaining an understanding of science is a little like becoming acquainted with another person. Even though a written description can provide a good deal of information about the person, you can never really know another unless there is personal contact. And so it is with science—if you are to know it well, you must deal with it intimately.

The laboratory is the setting for "intimate contact" with science. It is where scientists test their ideas (do research), the essential purpose of which is to provide a basis from which predictions about scientific phenomena can be made. Likewise, it will be the site of your "intimate contact" with the subject of human anatomy and physiology as you are introduced to the methods and instruments used in biological research.

For many students, human anatomy and physiology is taken as an introductory-level course; and their scientific background exists, at best, as a dim memory. If this is your predicament, this prologue may be just what you need to fill in a few gaps and to get you started on the right track before your actual laboratory experiences begin. So—let's get to it!

THE SCIENTIFIC METHOD

Science would quickly stagnate if new knowledge were not continually derived from and added to it.

The approach commonly used by scientists when they investigate various aspects of their respective disciplines is called the **scientific method.** This method is *not* a single rigorous technique that must be followed in a lockstep manner; to be perfectly truthful, it is nothing more or less than a logical, practical, and reliable way of approaching and solving problems of every kind—scientific or otherwise. It comprises five major steps.

Step 1: Observation of Phenomena

The crucial first step involves observation of some phenomenon of interest. In other words, before a scientist can investigate anything, he or she must decide on a *problem* or focus for the investigation. In most college laboratory experiments, the problem or focus has been decided for you. However, to illustrate this important step, we will assume that you want to investigate the true nature of apples, particularly green apples. In such a case you would begin your studies by making a number of different observations concerning apples.

Step 2: Statement of the Hypothesis

Once you have decided on a focus of concern, the next step is to design a significant question to be answered. Such a question is usually posed in the form of a **hypothesis,** an unproven conclusion that attempts to explain some phenomenon. (At its crudest level, a hypothesis can be considered to be a "guess" or an intuitive hunch that tentatively explains some observation.) Generally, scientists do not restrict themselves to a single hypothesis; instead, they usually pose several and then test each one systematically.

We will assume that, to accomplish step one, you go to the supermarket and randomly select apples from several bins. When you later eat the apples, you find that the green apples are sour, but the red and yellow apples are sweet. From this observation, you might conclude (*hypothesize*) that "green apples are sour." This statement would represent your current understanding of green apples. You might also reasonably predict that, if you buy more apples, any green ones you buy will be sour. Thus, you would have gone beyond your initial observation that "these" green apples are sour to the prediction that "all" green apples are sour.

Any good hypothesis must meet several criteria. First, *it must be testable*. This characteristic is far more important than its being correct. The tests may prove the hypothesis incorrect; or new information may require that the hypothesis be modified. Clearly the accuracy of a prediction in the green apple example or in any scientific study depends on the accuracy of the initial information on which it is based.

In our example, no great harm will come from an inaccurate prediction—that is, were we to find that some green apples are sweet. However, in some cases human life may depend on the accuracy of the prediction. Take the case of testing drugs for their effectiveness in treating disease. If one set of observations erroneously indicates that the drugs are risky but very effective, such a conclusion could lead to the death of the subsequent drug recipient(s). This illustrates two points: (1) Repeated testing of scientific ideas is important, particularly because scientists working on the same problem do not always agree in their conclusions. The studies on the use of saccharin and amino acid sweeteners are only two examples. (2) Conclusions drawn from scientific tests are only as accurate as the information on which they are based; therefore, careful observation is essential, even at the very outset of a study.

A second criterion is that, even though hypotheses are guesses of a sort, *they must be based on measurable, describable facts. No mysticism can be theorized*. We cannot conjure up, to support our hypothesis, forces that have not been shown to exist. For example, as scientists, we cannot say that the tooth fairy took Johnny's tooth unless we can prove that the tooth fairy exists!

Third, a hypothesis *must not be anthropomorphic*. Human beings tend to anthropomorphize—that is, to relate all experiences to human experience. Because humans are social animals influenced by culture, these two characteristics tend to promote biased thinking. Whereas we could state that bears instinctively protect their young, it would be anthropomorphic to say that bears love their young, because love is a human emotional response. Thus, the initial hypothesis must be stated without interpretation.

Step 3: Data Collection

Once the initial hypothesis has been stated, scientists plan experiments that will provide data (or evidence) to verify or disprove their hypotheses—that is, they *test* their hypotheses. Data are accumulated by making qualitative or quantitative observations of some sort; and the observations are often aided by the use of various types of equipment such as cameras, microscopes, stimulators, or various electronic devices that allow chemical and physiologic measurements to be made.

Observations referred to as **qualitative** are those we are able to make with our senses—that is, by using our vision, hearing, or sense of taste, smell, or touch. The color of an object, its texture, the relationship of one part to another, and its relative size (large versus small) may all be part of a qualitative description. For some quick practice in qualitative observation, compare and contrast* an orange and an apple.

Whereas the differences between an apple and an orange are obvious, this is not always the case in biological observations. Quite often a scientist tries to detect very subtle differences that cannot be determined by qualitative observations; and data must be derived from measurements made using a variety of scientific equipment. Such observations based on precise measurements of one type or another are **quantitative observations.** Examples of quantitative observations include careful measurements of body or organ dimensions such as mass, size, and volume; measurement of volumes of oxygen consumed during metabolic studies; determination of the concentration of glucose and other chemicals in urine; and determination of the differences in blood pressure and pulse under conditions of rest and exercise. An apple and an orange could be compared quantitatively by performing chemical measurements of the relative amounts of sugar and water in a given volume of fruit flesh, by analyzing the pigments and vitamins present in the apple skin and orange peel, and so on.

A valuable part of data gathering is the use of experiments to verify or disprove a hypothesis. An **experiment** is a procedure designed to describe the factors in a given situation that affect one another (that is, to discover cause and effect) under certain conditions.

Two general rules govern experimentation. The first of these rules is that the experiment(s) should be conducted in such a manner that every **variable** (any factor that might affect the outcome of the experiment) is under the control of the experimenter. Thus, the experimenter manipulates the **independent variables** and observes the effects of this manipulation on the **dependent** (or **response**) **variable.** For example, if the goal is to determine the effect of body temperature on breathing rate, the value measured (breathing rate) is called the dependent variable because it "depends on" the value chosen for the independent variable (body temperature). The ideal way to perform such an experiment is to set up and run a series of tests that are all identical, except for one specific factor that is varied.

One specimen (or group of specimens) is used as the **control** against which all other experimental samples are compared. The importance of the control sample cannot be overemphasized. It is essential to know how the system you are investigating works under normal circumstances before you can be sure

---

Compare means to emphasize the similarities between two things, whereas *contrast* means that the differences are to be emphasized.

that the results obtained from experimentation are actually due solely to the manipulation of the independent variable(s). Taking our example one step further, if we wanted to investigate the effects of body temperature (the independent variable) on breathing rate (the dependent variable), we could collect data on the breathing rate of individuals with "normal" body temperature (the implicit control group) and compare these data to breathing-rate measurements obtained from groups of individuals with higher and lower body temperatures. The control group would provide the "normal standard" against which all other samples would be compared relative to the dependent variable.

The second rule governing experimentation is that valid results require that testing be done on large numbers of subjects. It is essential to understand that it is nearly impossible to control all possible variables in biological tests. Indeed, there is a bit of scientific wisdom that mirrors this truth—that is, that laboratory animals, even in the most rigidly controlled and carefully designed experiments, "will do as they damn well please." Thus, stating that the testing of a drug for its pain-killing effects was successful after having tested it on only one postoperative patient would be scientific suicide. Large numbers of patients would have to receive the drug and be monitored for a decrease in postoperative pain before such a statement could have any scientific validity. Then, other researchers would have to be able to uphold those conclusions by running similar experiments. *Repeatability* is an extremely important part of the scientific method, and is the primary basis for acceptance or rejection of many hypotheses.

During experimentation and observation, data must be carefully recorded. Usually, such initial, or raw, data are recorded in tabular (table) form. The table should be labeled to show the variables investigated and the results for each sample. At this point, *accurate recording* of observations is the primary concern. Later, these raw data will be reorganized and manipulated to show more explicitly the outcome of the experimentation.

Some of the observations that you will be asked to make in the anatomy and physiology laboratory will require that a drawing be made. Don't panic! The purpose of making drawings (in addition to providing a record) is to force you to observe things very closely. You need not be an artist (most biological drawings are simple outline drawings), but you do need to be neat and as accurate as possible. (It is advisable to use a 4H pencil to do your drawings, because it is easily erased and doesn't smudge.) Before beginning to draw, examine your specimen closely, studying it as though you were going to have to draw it from memory. For example, when looking at cells you should ask yourself questions such as "What is their shape—the relationship of length and width? How are they joined together? What color are they?" Then decide precisely what you are going to show and how large the drawing must be to show the necessary detail. After making the drawing, add labels

in the margins, and connect them, by straight lines, to the structures being named.

Step 4: Manipulation and Analysis of Data

The form of the final data varies, depending on the nature of the data collected. Usually, the final data represent information converted from the original measured values (raw data) to some other form. This may mean that averaging or some other statistical treatment must be applied, or it may require conversions from one kind of units to another. In other cases, graphs may be needed to display the data.

ELEMENTARY STATISTICAL TREATMENT OF DATA   Only very elementary statistical treatment of data is required in this manual. For example, you will be expected to understand and/or compute an average (mean), percentages, and a range.

Two of these statistics, the average and the range, are useful in describing the *typical* case among a large number of samples evaluated. Let us use a simple example. We will assume that the following heart rates (in beats/min) were recorded during an experiment: 64, 70, 82, 94, 85, 75, 72, 78. If you put these numbers in numerical order, the **range** is easily computed, because the range is the difference between the highest and lowest numbers obtained (highest number minus lowest number). What is the range of the set of numbers just provided?

1. _____* The **average,** or **mean,** is obtained by summing the items and dividing the sum by the number of items. Compute the average for the set of numbers just provided:

2. _____

The word *percent* comes from the Latin meaning "for 100"; thus *percent,* indicated by the percent sign, %, means parts per 100 parts. Thus, if we say that 45% of Americans have type O blood, what we are really saying is that among each group of 100 Americans, 45 (45/100) can be expected to have type O blood.

It is very easy to convert any number (including decimals) to a percent. The rule is to move the decimal point two places to the right and add the percent sign. If no decimal point appears, it is *assumed* to be at the end of the number; and zeros are added to fill any empty spaces. Two examples follow:

$$0.25 = 0.2\!\!\smallsmile\!\!5 = 25\%$$
$$5 = 5\!\!\smallsmile\!\! = 500\%$$

Change the following numbers to percents:

3.   38.2 = _____   5.   1.6 = _____

4.   402 = _____

*Answers are given on page xvi.

Note that although you are being asked here to convert numbers to percents, percents by themselves are meaningless. We always speak in terms of a percentage *of* something.

To change a percent to a whole number (or decimal), remove the percent sign, and move the decimal point two places to the left. Change the following percents to whole numbers or decimals:

6.   36% = _____   8.   25777% = _____

7.   800% = _____   9.   0.05% = _____

MAKING AND READING LINE GRAPHS For some laboratory experiments you will be required to show your data (or part of them) graphically. Simple line graphs allow relationships within the data to be shown interestingly and allow trends (or patterns) in the data to be demonstrated. An advantage of properly drawn graphs is that they save the reader's time because the essential meaning of large numbers of statistical data can be visualized at a glance.

To aid in making accurate graphs, graph paper (or a printed grid in the manual) is used. Line graphs have both horizontal and vertical scales. Each scale should have uniform intervals—that is, each unit measured on the scale should require the same distance along the scale as any other. Variations from this rule may be misleading and result in false interpretations of the data. By convention, the condition that is manipulated (the independent variable) in the experimental series is plotted on the X-axis (the horizontal axis); and the value that we then measure (the dependent variable) is plotted on the Y-axis (the vertical axis). To plot the data, a dot or a small *x* is placed at the precise point where the two variables (measured for each sample) meet; and then a line (this is called the **curve**) is drawn to connect the plotted points.

Sometimes you will see the curve on a line graph extended beyond the last plotted point. This is (supposedly) done to predict "what comes next." When you see this done, be skeptical. The information provided by such a technique is only slightly more accurate than that provided by a crystal ball!

To read a line graph, pick any point on the line, and match it with the information directly below on the horizontal scale and with that directly to the left of it on the vertical scale. Figure P.1 is a graph that illustrates the relationship between breaths per minute (respiratory rate) and body temperature. Answer the following questions about this graph:

10.   What was the respiratory rate at a body temperature of 96°F? _____

11.   Between 98° and 102°F, the respiratory rate increased from _____ to _____ breaths per minute.

12.   Between which two body temperature readings

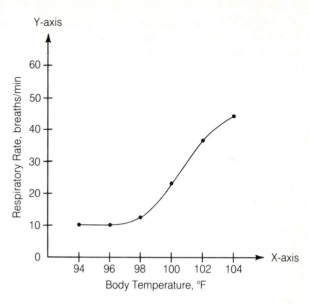

Figure P.1

Example of graphically presented data.

was the increase in breaths per minute greatest?

_____

13.   Are the intervals on each scale uniform? ____

Step 5: Reporting Conclusions of the Study

Drawings, tables, and graphs alone do not suffice as the final presentation of scientific results. The final step requires that you provide a straightforward description of the conclusions drawn from your results. If possible, your findings should be compared to those of other investigators working on the same problem. (For laboratory investigations conducted by students, these comparative figures are provided by classmates.)

It is important to realize that scientific investigations do not always yield the anticipated results. If there are discrepancies between your results and those of others, or what you expected to find based on your class notes or textbook readings, this is the place to try to explain those discrepancies.

Results are often only as good as the observation techniques used. Depending on the type of experiment conducted, several questions may need to be answered. Did you weigh the specimen carefully enough? Did you balance the scale first? Was the subject's blood pressure actually as high as you recorded it, or did you record it hastily (and inaccurately)? If you did record it accurately, is it possible that the subject was emotionally upset about something, which (even though the matter of concern had nothing to do with the experiment) might have given falsely high data for the variable being investigated? Attempting to explain an unexpected result will often teach you more than you would have learned from anticipated results.

When the experiment produces results that are consistent with the hypothesis, then the hypothesis can be said to have reached a higher level of certainty. There is now a greater probability that the hypothesis is correct.

A hypothesis that has been validated by many different investigators is called a **theory.** Theories are useful in two important ways. First, they link sets of data; and second, they make predictions that may lead to additional avenues of investigation. (Okay, we know this with a high degree of certainty; what's next?)

When a theory has been repeatedly verified and appears to have wide applicability in biology, it may assume the status of a **biological principle.** A principle is a statement that applies with a high degree of probability to a range of events. For example, "Living matter is made of cells or cell products" is a principle stated in many biology texts. It is a sound and useful principle, and will continue to be used as such—unless new findings prove it wrong.

We have been through quite a bit of background concerning the scientific method and what its use entails. Because it is important that you remember the phases of the scientific method, they are summarized here:

1. Observation of some phenomenon
2. Statement of a hypothesis (based on the observations)
3. Collection of data (testing the hypothesis with controlled experiments)
4. Manipulation and analysis of the data
5. Reporting of the conclusions of the study

SCIENTIFIC NOTATION AND METRICS

No matter how highly developed our ability to observe, observations have scientific value only if they can be communicated to others. This necessitates the use of scientific notation and a widely accepted system of measurements.

Scientific Notation

Because quantitative measurements often yield very large or very small numbers, you are quite likely to encounter numbers such as 3.5×10^{12} or 10^{-3}. It is important that you understand what this **scientific notation** means.

Scientific notation is dependent on the properties of exponents and on the movement of the decimal point when multiplying or dividing by 10. When you multiply 10 by itself, you get a product that is one followed by zeros. The number of zeros (two) in the product is equal to the number of times you have used 10 as a factor and is shown as an **exponent.** Thus, the following notation

$$\text{base} \rightarrow 10^2 \leftarrow \text{exponent}$$

translates to "the base 10 multiplied by itself (10×10)."

The powers of 10 are represented as follows:

$10^0 = 1$ (Any number followed by a zero exponent is one.)

$10^1 = 10$ ($10 \times 1 = 10$)

$10^2 = 100$ ($10 \times 10 = 100$)

$10^3 = 1000$ ($10 \times 10 \times 10 = 1000$)

$10^4 = 10,000$ ($10 \times 10 \times 10 \times 10 = 10,000$)

As you can see, each time the exponent is increased by one, another zero ($\times 10$) is added to the answer.

When you multiply any number by a power of 10 written with exponents, the decimal point is moved to the right the number of times shown in the exponent. Thus:

$$3.25 \times 10^1 = 3.2\underset{\frown}{5} \qquad = 32.5$$
$$3.25 \times 10^3 = 3.2\underset{\frown}{5} \qquad = 3250$$
$$3.25 \times 10^5 = 3.2\underset{\frown}{5} \qquad = 325,000$$

By using such exponential notation, very large numbers may be written in far simpler form.

Write the following numbers using the proper scientific notation:

14. $140,000 = 1.4 \times$ _____
15. $9,650,000 = 9.65 \times$ _____
16. $852 = 8.52 \times$ _____
17. $10 = 1.0 \times$ _____

Notice that proper scientific notation entails only one number to the left of the decimal point. Thus 1.03×10^3 is correct, but 10.3×10^2 is not.

In the above examples, all of the numbers used were greater than one. Scientific notation can also be used to report numbers less than one. To do this, negative exponents are used. For example, in

$$3.25 \times 10^2$$

the positive exponent means that the decimal point is to be moved two places to the right, and the number designated is 325 ($3.25 \times 10 \times 10$). However, in

$$3.25 \times 10^{-2}$$

the negative exponent means that the number is to be divided by the power of 10 indicated by the exponent and the decimal point is to be moved two places to the left. The number so designated is 0.0325 [$3.25 \div (10 \times 10)$].

Thus, the rule converting scientific notation (using powers of 10) to decimal notation is to move the decimal point the number of places indicated by the exponent. When the exponent is positive (with or without a plus sign), the decimal point is moved to the right. When the exponent is negative (always provided with a minus sign), the decimal point is moved to the left.

For a little more practice, write the following numbers in scientific notation: (18–21)

$0.0000063 = 6.3 \times$ _____ $0.265 = 2.65 \times$ _____

$0.00054 = 5.4 \times$ _____ $0.10 = 1.0 \times$ _____

Metrics

Without measurement, we would be limited to qualitative description. However, with a system of measurement, quantitative description becomes possible.

Anyone can establish a system of measurement. All that is required is a reference point; and, historically, much of our common (the British) system of measurement evolved from units based on objects everyone knew. For example, horses were measured in "hands," and a "fathom" was the distance between outstretched arms. However, the variability in such measurements is immediately apparent—for example, an infant's hand is substantially smaller than that of an adult. Therefore, for precise and repeatable communication of information, the agreed-upon system of measurement used by scientists is the **metric system,** a nonvarying standard of reference.

A major advantage of the metric system is that it is based on units of 10. This allows rapid conversion to workable numbers so that neither very large nor very small figures need be used in calculations. Fractions or multiples of the standard units of length, volume, mass, time, and temperature have been assigned specific names. Table P.1 shows the commonly used units of the metric system, along with the prefixes used to designate fractions and multiples thereof.

To change from smaller units to larger units, you must *divide* by the appropriate factor of 10 (because there are fewer of the larger units). For example, a milliunit (milli = one thousandth), such as a milliliter or millimeter, is one step smaller than a centiunit (centi = one hundredth), such as a centiliter or centimeter. Thus to change milliunits to centiunits, you must divide by 10. On the other hand, when converting from larger units to smaller ones, you must *multiply* by the appropriate factor of 10 (because there will be more of the smaller units). A partial scheme for conversions between the metric units is shown below. (See also Table P.1.)

Students studying a science or preparing for a profession in the health-related fields find that certain of the metric units are encountered and dealt with more frequently than others. Thus, the objectives of the sections that follow are to provide a brief overview of these most-used measurements and to help you gain some measure of confidence in dealing with them.

LENGTH MEASUREMENTS   The metric unit of length is the **meter (m).** In addition to measuring things in meters, you will be expected to measure smaller objects in centimeters or millimeters. Subcellular structures are measured in micrometers.

To help you picture these units of length, some equivalents follow:

One meter (m) is slightly longer than one yard (1 m = 39.37 in.).

One centimeter (cm) is approximately the width of a piece of chalk. (Note: there are 2.54 cm in 1 in.)

One millimeter (mm) is approximately the thickness of the wire of a paper clip or of a mark made by a No. 2 pencil lead.

One micrometer (μm) is extremely tiny and can be measured only microscopically.

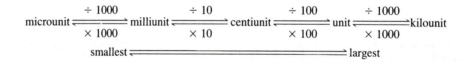

TABLE P.1   Commonly Used Units of the Metric System, and Their Fractions and Multiples

Measurement	Unit	Fraction or Multiple		Prefix	Symbol
Length	Meter (m)	10^6	one million	mega	M
Volume	Liter (l)	10^3	one thousand	kilo	k
Mass	Gram (g)	10^{-1}	one tenth	deci	d
Time*	Second (s)	10^{-2}	one hundredth	centi	c
Temperature	Degree Celsius (°C)	10^{-3}	one thousandth	milli	m
		10^{-6}	one millionth	micro	μ
		10^{-9}	one billionth	nano	n

* The accepted standard for time is the second; and thus hours and minutes are used in scientific, as well as everyday, measurement of time. The prefixes used in the designation of units of length, mass, and volume are also used in specifying units of time. However, because minutes and hours are terms that indicate *multiples* of seconds, the only prefixes generally used are those indicating *fractional portions* of seconds—for example, millisecond and microsecond.

Make the following conversions between metric units of length: (22–25)

352 cm = _____ mm

150 km = _____ m

2000 μm = _____ mm

1 mm = _____ m

VOLUME MEASUREMENTS

The metric unit of volume is the liter (1). A **liter (1)** is slightly more than a quart (1 1 = 1.057 quarts). Liquid products, measured in liters, are becoming more common, and laboratory solutions are often prepared in 1-liter quantities. Liquid volumes measured out for laboratory experiments are usually measured in milliliter (ml) volumes. (The terms *ml* and *cc,* cubic centimeter, are used interchangeably in laboratory and medical settings.)

To help you visualize metric volumes, the equivalents of some common substances follow:

A 12-oz can of soda is just slightly more than 360 ml.
A cup of coffee is approximately 180 ml.
A fluid ounce is 30 ml (cc).
A teaspoon of vanilla is about 5 ml (cc), and many drug injections are given in 5-ml volumes.

Compute the following:

26. How many 5-ml injections can be prepared from 1 liter of a medicine? _____

27. A 450-ml volume of alcohol is _____ 1.

28. The volume of one grape is approximately 0.004 1. What is the volume of the grape in milliliters? _____

MASS MEASUREMENTS

Although many people use the terms *mass* and *weight* interchangeably, this usage is inaccurate. **Mass** is the amount of matter in an object; and an object has a constant mass, regardless of where it is—that is, at sea level, on a mountaintop, or in outer space. However, weight varies with gravitational pull; the greater the gravitational pull, the greater the weight. Thus, our astronauts are said to be weightless* when in outer space, but they still have the same mass as they do on earth.

The metric unit of mass is the **gram (g),** and most objects weighed in the laboratory will be measured in terms of this unit or fractions thereof. Medical dosages are usually prescribed in milligrams (mg) or micrograms (μg); and, in the clinical agency, body weight (particularly of infants) is typically specified in kilograms (kg) (1 kg = 2.2 lb).

The following examples are provided to help you become familiar with the masses of some common objects:

A nickel has a mass of 5 g.
The mass of an average woman (132 lb) is 60 kg.

Make the following conversions:

29. 300 g = _____ mg = _____ μg

30. 4000 μg = _____ mg = _____ g

31. A nurse must administer, to her patient, Mrs. Smith, 5 mg of a drug per kg of body mass. Mrs. Smith weighs 140 lb. How many grams of the drug should the nurse administer to her patient?

TEMPERATURE MEASUREMENTS

In the laboratory and in the clinical agency, temperature is measured both in metric units (degrees Celsius, °C) and in British units (degrees Fahrenheit, °F). Thus it helps to be familiar with both temperature scales.

The temperatures of boiling and freezing water can be used to compare the two scales:

The boiling point of water is 100°C and 212°F.
The freezing point of water is 0°C and 32°F.

As you can see, the range from the freezing point to the boiling point of water on the Celsius scale is 100 degrees, whereas the comparable range on the Fahrenheit scale is 180 degrees. Hence, one degree on the Celsius scale represents a greater change in temperature.

Normal body temperature is approximately 98.6°F and 37°C.

To convert from the Fahrenheit scale to the Celsius scale (or vice versa), the following equation is used:

$$°C = 5(°F - 32)/9$$

Perform the following temperature conversions. (Remember, the items within parentheses must be attended to first.)

32. Convert 38°C to °F: _____

33. Convert 158°F to °C: _____

---

*Astronauts are not *really* weightless. It is just that they and their surroundings are being pulled toward the earth at the same speed; and so, in reference to their environment, they appear to float.

ANSWERS

1. range of 94−64, 30 bpm
2. average 77.5
3. 3820%
4. 40200%
5. 160%
6. 0.36
7. 8
8. 257.77
9. 0.0005
10. 10 breaths/min
11. 12−36 bpm
12. interval between 100−102° (22−36 bpm)
13. yes
14. 10^5
15. 10^6
16. 10^2
17. 10^1
18. 6.3×10^{-6}
19. 5.4×10^{-4}
20. 2.65×10^{-1}
21. 1.0×10^{-1}
22. cm = 3520 mm
23. km = 150,000 m
24. μm = 2 mm
25. mm = 0.001 m
26. 200
27. 0.45 l
28. 4 ml
29. 300 g = $\underline{3 \times 10^5}$ mg = $\underline{3 \times 10^8}$ μg
30. 4000 μg = $\underline{4}$ mg = $\underline{4 \times 10^{-3}}$ g (0.004)
31. 0.3g
32. 100.4°F
33. 70°C

The Language of Anatomy

OBJECTIVES

1. To describe the anatomic position verbally or by demonstration.
2. To use proper anatomic terminology to describe body directions, planes, and surfaces.
3. To name the body cavities and note the important organs in each.

MATERIALS

Human torso model (dissectible)
Human skeleton
Demonstration: sectioned and labeled kidneys (three separate kidneys uncut or cut so that a. entire, b. transverse section, and c. longitudinal sectional views are visible)

Most of us have a natural curiosity about our bodies. This fact is amply demonstrated by infants, who early in life become fascinated with their own waving hands or their mother's nose. The study of the gross anatomy of the human body elaborates on this fascination. Unlike the infant, however, the student of anatomy must learn to identify and observe the dissectible body structures formally. The purpose of any gross-anatomy experience is to examine the three-dimensional relationships of body structures—a goal that can never completely be achieved by using illustrations and models, regardless of their excellence.

When beginning the study of any science, the student is often initially overcome by the jargon unique to the subject. The study of anatomy is no exception. But without this specialized terminology, confusion is inevitable. For example, what do *over, on top of, superficial to, above,* and *behind* mean in reference to the human body? Anatomists have an accepted set of reference terms that are universally understood. These allow body structures to be located and identified with a minimum of words and a high degree of clarity.

This unit presents some of the most important anatomic terminology used to describe the body and introduces you to basic concepts of **gross anatomy,** the study of body structures visible to the naked eye.

ANATOMICAL POSITION

When anatomists or doctors discuss the human body, they refer to specific areas in accordance with a universally accepted standard position called the **anatomical position.** It is essential to understand this position, because much of the body terminology employed in this book refers to this body positioning, regardless of the position the body happens to be in. In the anatomical position the human body is erect, with feet together, head and toes pointed forward, and arms hanging at the sides with palms facing forward (Figure 1.1).

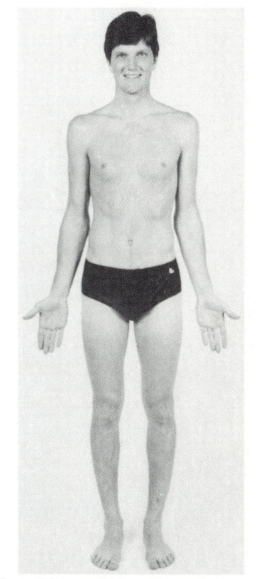

Figure 1.1

Anatomical position.

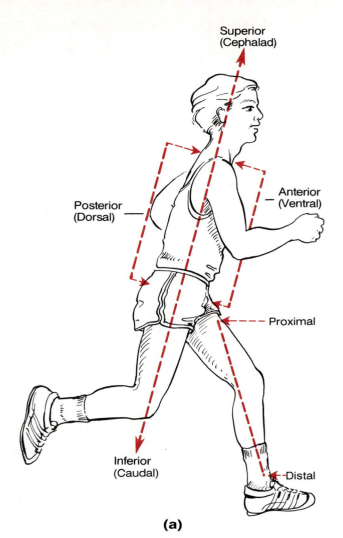

Superior
(Cephalad)

Posterior
(Dorsal)

Anterior
(Ventral)

Proximal

Inferior
(Caudal)

Distal

(a)

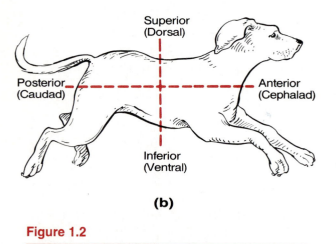

Superior
(Dorsal)

Posterior
(Caudad)

Anterior
(Cephalad)

Inferior
(Ventral)

(b)

Figure 1.2

Anatomic terminology describing body orientation and direction: (a) with reference to a human, (b) with reference to a four-legged animal.

Assume the anatomic position, and note that it is not particularly comfortable. The hands are held unnaturally forward rather than hanging partially cupped toward the thighs.

BODY ORIENTATION AND DIRECTION

Study the terms below, referring to Figure 1.2. Note that certain terms have a different connotation for a four-legged animal than they do for a human.

Superior/inferior (*above/below*): These terms refer to placement of a body structure along the long axis of the body. Superior structures always appear above other structures. For example, the nose is superior to the mouth, and the abdomen is inferior to the chest region.

Anterior/posterior (*front/back*): In humans the most anterior surfaces are those that are most forward—the face, chest, and abdomen. Posterior surfaces are those on the backside of the body—the back and buttocks.

Medial/lateral (*toward the midline/away from the midline or median plane*): The ear is lateral to the eye; the breastbone is medial to the ribs.

The terms of position described above are dependent on an assumption of anatomic position. The next three term pairs, however, are more absolute; that is, their applicability is not relative to a particular body position, and they consistently have the same meaning in all vertebrate animals.

Cephalad/caudal (*toward the head/toward the tail*): In humans these terms are used interchangeably with *superior* and *inferior.* But in four-legged animals they are synonymous with *anterior* and *posterior.*

Dorsal/ventral (*backside/belly side*): These terms are used chiefly in discussing the comparative anatomy of animals, assuming the animal is on all fours. *Dorsum* is a Latin word meaning "back"; thus *dorsal* refers to the backside of the animal's body or of any other structures. For instance, the posterior surface of the leg is its dorsal surface. The term *ventral* derives from the Latin term *venter,* meaning "belly," and thus always refers to the belly side of animals. In humans the terms *ventral* and *dorsal* may be used interchangeably with the terms *anterior* and *posterior,* but in four-legged animals *ventral* and *dorsal* are synonymous with *inferior* and *superior.*

Proximal/distal (*nearer the trunk or attached end/farther from the trunk or point of attachment*): These terms are used primarily to locate various areas of the body limbs. For example, the fingers are distal to the elbow; the knee is proximal to the toes.

Before continuing, use a human torso model, a skeleton, or your own body to specify the relationship between the following structures. Use the correct anatomic terminology:

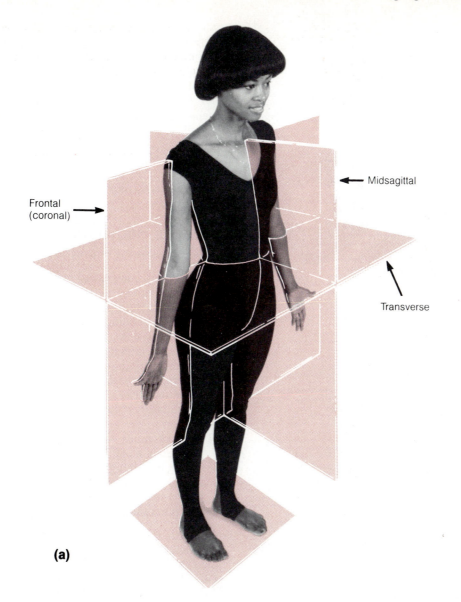

(a)

Figure 1.3

Planes of the body.

The wrist is ___Superior___ to the hand.

The trachea is ___anterior___ to the spine.

The brain is ___Superior___ to the spinal cord.

The kidneys are ___inferior___ to the liver.

The bridge of the nose is ___medial___ to the eyes.

BODY PLANES AND SECTIONS

To observe internal structures, it is often helpful and necessary to make use of a **section**, or cut. When the section is made through the body wall or through an organ, it is made along an imaginary line called a **plane.** Anatomists commonly refer to three planes (Figure 1.3), or sections, which lie at right angles to one another.

Sagittal section: When the cut is made along a longitudinal plane dividing the body into right and left parts, it is referred to as a sagittal section. If it divides the body into equal parts, right down the median plane of the body, it is called a **midsagittal section.**

Frontal section: Sometimes called a **coronal section,** this cut is made along a longitudinal plane that divides the body into anterior and posterior parts.

Transverse section: A cut made along a transverse or horizontal plane, dividing the body into superior and inferior parts, may also be called a **cross section.**

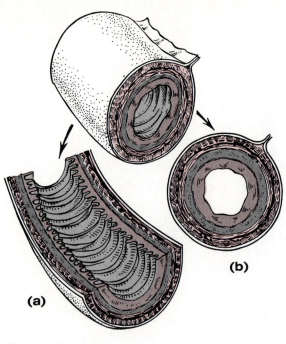

Figure 1.4

Segment of the small intestine: (a) cut longitudinally, (b) cut transversely.

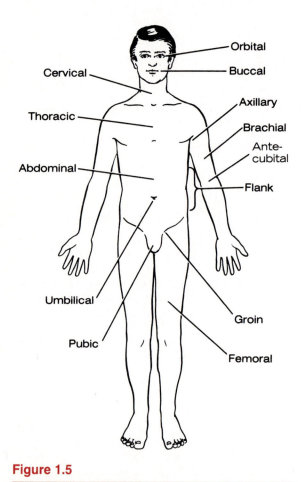

Figure 1.5

Anterior body landmarks.

These terms can also be used to describe the way organs are cut for observation. An organ cut along a longitudinal plane provides quite a different view from one cut along the transverse plane as shown in Figure 1.4.

Go to the demonstration area and observe the transversely and longitudinally cut organ specimens.

SURFACE ANATOMY

Body surfaces provide visible landmarks for study of the body.

Anterior Body Landmarks

Note the following regions in Figure 1.5:

Orbital: pertaining to the eye
Buccal: pertaining to the cheek
Cervical: pertaining to the neck region
Thoracic: pertaining to the chest
Axillary: pertaining to the armpit
Brachial: pertaining to the arm
Umbilical: pertaining to the navel
Abdominal: pertaining to the anterior body trunk inferior to the ribs
Antecubital: pertaining to the anterior surface of the elbow
Groin: pertaining to the area where the thigh meets the body trunk
Femoral: pertaining to the thigh
Flank: pertaining to the lateral surface of the body from the rib cage to the hip
Pubic: pertaining to the genital region

Since the abdominal surface covers a large area, it is helpful to divide it into smaller areas for study. Two medical division schemes exist for this purpose. One scheme divides the abdominal surface (and the abdominopelvic cavity deep to it) into four more or less equal regions called **quadrants;** these quadrants are then named according to their relative position; that is, right upper quadrant, right lower quadrant, left upper quadrant, and left lower quadrant (Figure 1.6(a)). The second scheme divides the abdominal surface into nine separate regions by four planes, as shown in Figure 1.6(b) and described on the next page.

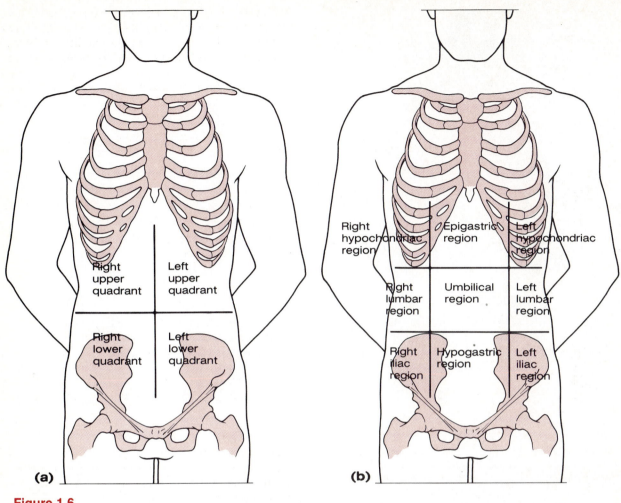

Figure 1.6

Abdominopelvic cavity: (a) four quadrants, (b) nine regions delineated by four planes. The superior horizontal plane is just inferior to the ribs; the inferior horizontal plane is just superior to the hipbones. Vertical planes are just medial to the nipples.

Umbilical region: the centermost region, which includes the umbilicus

Epigastric region: immediately above the umbilical region; overlies most of the stomach

Hypogastric region: immediately below the umbilical region; encompasses the pubic area

Iliac regions: lateral to the hypogastric region and overlying the hip bones

Lumbar regions: between the ribs and the flaring portions of the hip bones

Hypochrondriac regions: flanking the epigastric region and overlying the lower ribs

Locate the anterior body landmarks, including the regions of the abdominal surface, on a torso model and on yourself before continuing.

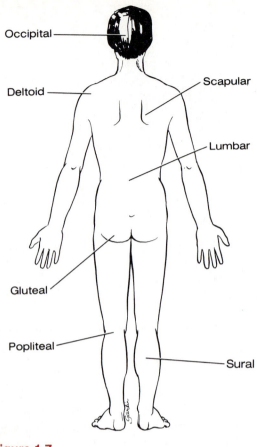

Figure 1.7

Posterior body landmarks.

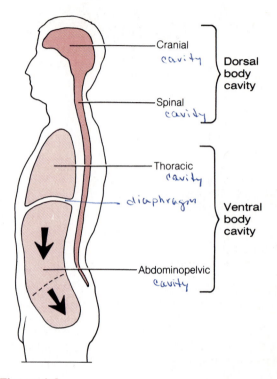

Figure 1.8

Body cavities; the angle of the relationship between the abdominal and pelvic cavities is shown by arrows.

Posterior Body Landmarks

Note the following body surface regions in Figure 1.7:

Scapular: pertaining to the scapula or shoulder blade area

Lumbar: pertaining to the area of the back between the ribs and hips

Gluteal: pertaining to the buttocks or rump

Popliteal: pertaining to the knee region

Sural: pertaining to the posterior surface of the leg

Occipital: pertaining to the posterior surface of the head

Deltoid: pertaining to the curve of the shoulder formed by the large deltoid muscle

BODY CAVITIES

The body has two sets of cavities, which provide quite different degrees of protection to the organs within them (Figure 1.8).

Dorsal Body Cavity

The dorsal body cavity can be subdivided into the **cranial cavity,** in which the brain is enclosed within the rigid skull, and the **spinal cavity,** in which the delicate spinal cord is protected within a bony vertebral column. Because the cord is a continuation of the brain, these cavities are continuous with each other.

Ventral Body Cavity

Like the dorsal cavity, the ventral body cavity is subdivided. The superior **thoracic cavity** is separated from the rest of the ventral cavity by the dome-shaped diaphragm. The heart and lungs, located in the thoracic cavity, are afforded some measure of protection by the bony rib cage. The cavity below the diaphragm is often referred to as the **abdominopelvic cavity,** since there is no further physical separation of the ventral cavity. Some prefer to subdivide the abdominopelvic cavity into a superior **abdominal cavity,** which houses the stomach, intestines, liver, and other organs, and an inferior **pelvic cavity,** containing the reproductive organs, bladder, and rectum. Note in Figure 1.8 that the abdominal and pelvic cavities are not continuous with each other in a straight plane but that the pelvic cavity is tipped away from the perpendicular.

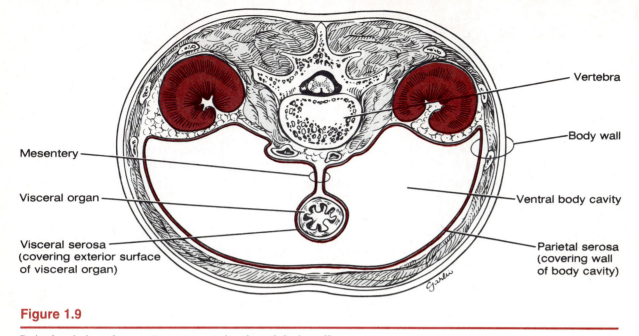

Figure 1.9

Parietal and visceral serosa (transverse section through body wall).

The inner body wall of the ventral cavity is lined with a smooth serous membrane, the **parietal serosa,** which is continuous with a similar membrane, the **visceral serosa,** covering the external surfaces of the organs within the cavity (Figure 1.9). These membranes produce a thin lubricating fluid that allows the organs to slide over one another or to rub against the body wall without friction.

The specific names of the serous membranes depend on the structures they envelop. Thus the serosa lining the abdominal cavity and covering its organs is the **peritoneum,** that enclosing the lungs is the **pleura,** and that around the heart is the **pericardium.**

Organ Systems Overview

OBJECTIVES

1. To name the human organ systems and state the major functions of each.
2. To list two or three organs of each system, and categorize the various organs by organ system.
3. To identify these organs in a dissected rat or on a dissectible human torso model.
4. To identify the correct organ system for each organ when presented with a list of organs (as studied in the laboratory).

MATERIALS

Freshly killed or preserved rat for dissection (one for every two to four students), or dissected human cadaver
Dissecting pans and pins
Scissors
Forceps
Twine
Human torso model (dissectible)

The basic unit or building block of all living things is the **cell.** Cells fall into four different categories according to their functions. Each of these corresponds to one of the four **tissue** types: epithelial, muscular, nervous, and connective. An **organ** is a structure composed of two or more tissue types that performs a specific function for the body. For example, the small intestine, which digests and absorbs nutrients, is composed of all four tissue types. An **organ system** is a group of organs that act together to perform a particular body function. For example, the organs of the digestive system work together to assure that food moving through the digestive system is properly broken down and that the end products are absorbed into the bloodstream to provide nutrients and fuel for all the body's cells. In all, there are 11 organ systems, which are described in Table 2.1. Read through this summary before beginning the rat dissection.

RAT DISSECTION

Now you will have a chance to observe the size, shape, location, and distribution of the organs and organ systems. Many of the external and internal structures of the rat are quite similar in structure and function to those of the human, so a study of the gross anatomy of the rat should help you understand your own physical structure.

The following instructions have been written to complement and direct the student's dissection and observation of a rat, but the descriptions for organ observations from procedure 4 (p. 10) apply as well to superficial observations of a previously dissected human cadaver. In addition, the general instructions for observation of external structures can easily be extrapolated to serve human cadaver observations.

Note that four of the organ systems listed in Table 2.1 will not be studied at this time (integumentary, nervous, skeletal, and muscular), as they require microscopic study or more detailed dissection.

External Structures

1. Obtain a preserved or freshly killed rat (one for every two to four students), a dissecting pan, dissecting pins, scissors, and forceps.

2. Observe the major divisions of the animal's body—head, trunk, and extremities. Compare these divisions to those of humans.

3. Examine the structures of the oral cavity. Identify the teeth and tongue. Observe the extent of the hard palate (the portion underlain by bone) and the soft palate (immediately posterior to the hard palate, with no bony support). Note that the posterior end of the oral cavity leads into the throat, or pharynx. The pharynx is a passageway used by both the digestive and respiratory systems.

Ventral Body Cavity

1. Pin the animal to the wax of the dissecting pan by placing its dorsal side down and securing its extremities to wax as shown in Figure 2.1(a). (If the dissecting pan is not waxed, secure the animal with twine. Make a loop knot around one upper limb, pass the twine under the pan, and secure the opposing limb. Repeat for the lower extremities.)

2. Lift the abdominal skin with a forceps, and cut through it with the scissors (Figure 2.1(b)). Close the scissor blades and insert them under the cut skin. Moving in a superior direction, open and close the blades to loosen the skin from the underlying connective tissue and muscle. Once this skin-freeing procedure has been completed, cut the skin on the

TABLE 2.1   Overview of Organ Systems* of the Body

Organ system	Major component organs	Function
Integumentary (Skin)	Epidermal and dermal regions; cutaneous sense organs and glands	• Protects deeper organs from mechanical, chemical, and bacterial injury, and dessication (drying out) • Excretion of salts and urea • Aids in regulation of body temperature • Produces vitamin D
Skeletal	Bones, cartilages, tendons, ligaments, and joints	• Body support and protection of internal organs • Provides levers for muscular action • Cavities provide a site for blood cell formation
Muscular	Muscles attached to the skeleton	• Primarily function to contract or shorten; in doing so, skeletal muscles allow locomotion (running, walking, etc.), grasping and manipulation of the environment, and facial expression. • Generates heat
Nervous	Brain, spinal cord, nerves, and sensory receptors	• Allows body to detect changes in its internal and external environment and to respond to such information by activating appropriate muscles or glands • Maintains homeostasis of the body
Endocrine	Pituitary, thyroid, parathyroid, adrenal, and pineal glands; ovaries, testes, and pancreas	• Maintains body homeostasis, growth, and development; produces chemical "messengers" (hormones) that travel in the blood to exert their effect(s) on various "target organs" of the body
Cardiovascular	Heart, blood vessels, and blood	• Primarily a transport system that carries blood containing oxygen, carbon dioxide, nutrients, wastes, ions, hormones, and other substances to and from the tissue cells where exchanges are made; blood is propelled through the blood vessels by the pumping action of the heart • White blood cells and antibodies in the blood act to protect the body
Lymphatic	Lymphatic vessels, lymph nodes, spleen, thymus, and scattered collections of lymphoid tissue	• Picks up fluid leaked from the blood vessels and returns it to the blood • Cleanses blood of pathogens and other debris • Houses lymphocytes that act in body immunity
Respiratory	Nasal passages, pharynx, larynx, trachea, bronchi, and lungs	• Keeps the blood continuously supplied with oxygen while removing carbon dioxide
Digestive	Oral cavity, esophagus, stomach, small and large intestines, rectum, and accessory structures (teeth, salivary glands, liver, and pancreas)	• Acts to break down ingested foods to minute particles, which can be absorbed into the blood for delivery to the body cells • Undigested residue removed from the body as feces
Urinary	Kidneys, ureters, bladder, and urethra	• Rids the body of nitrogen-containing wastes (urea, uric acid, and ammonia), which result from the breakdown of proteins and nucleic acids by body cells • Maintains water, ionic, and acid–base balance of blood
Reproductive	Male: testes, scrotum, penis, and duct system, which carries sperm to the body exterior Female: ovaries, uterine tubes, uterus, and vagina	• Provides germ cells (sperm and eggs) for perpetuation of the species • Female uterus houses the developing fetus until birth

*The immune system, composed of a mobile army of cells (lymphocytes and macrophages) that protects the body from antigens (foreign substances) via the immune response, is not included because it is a *functional system* rather than an organ system in the true sense.

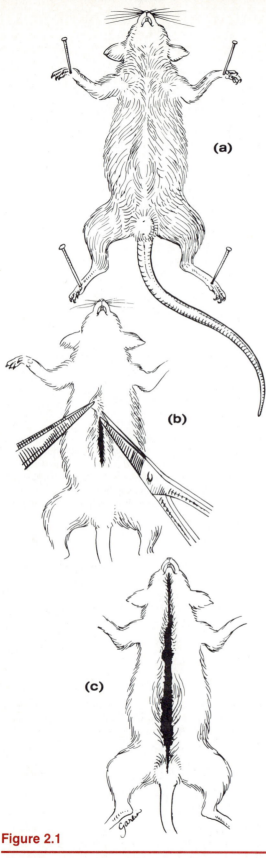

Figure 2.1

Rat dissection: pinning and the initial incision.
(a) Securing the rat to the dissection tray; (b) using scissors to make the incision on the medial line of the abdominal region; (c) completed incision from the pelvic region to the lower jaw.

medial line, from the pubic region to the lower jaw (Figure 2.1(c)). Make a lateral cut about halfway down the ventral surface of each limb. Complete the job of freeing the skin with the scissor tips, and pin the flaps to the tray (Figure 2.2(a)). The underlying tissue that is now exposed is the skeletal musculature of the body wall and limbs. It allows voluntary body movement. Note that the muscles are packaged in sheets of pearly white connective tissue (fascia), which protects the muscles and binds them together.

3.   Carefully cut through the muscles of the abdominal wall in the pubic region, avoiding the underlying organs. Remember, to *dissect* means "to separate"—not mutilate! Now, hold and lift the muscle layer with a forceps and cut through the muscle layer from the pubic region to the bottom of the rib cage. Make two lateral cuts through the rib cage (Figure 2.2(b)). A thin membrane attached to the inferior boundary of the rib cage should be obvious; this is the **diaphragm,** which separates the thoracic and abdominal cavities. Cut the diaphragm away to loosen the rib cage. You can now lift the ribs to view the contents of the thoracic cavity.

4.   Examine the structures of the thoracic cavity, starting with the most superficial structures and working deeper. As you work, refer to Figure 2.3 (p. 12), which shows the superficial organs.

Thymus: an irregular mass of glandular tissue overlying the heart

Push the thymus to the side to view the heart.

Heart: median oval structure enclosed within the pericardium (serous membrane sac)
Lungs: flanking the heart on either side

Now observe the throat region.

Trachea: tubelike "windpipe" running medially down the throat; part of the respiratory system

Follow the trachea into the thoracic cavity; note where it divides. These are the bronchi.

Bronchi: two passageways that plunge laterally into the tissue of the two lungs

Now push the trachea to one side to expose the esophagus.

Esophagus: literally a food chute; the part of the digestive system that transports food from the throat to the stomach

Follow the esophagus through the diaphragm to its junction with the stomach.

Stomach: a C-shaped organ important in food digestion and temporary storage

5.   Examine the superficial structures of the abdominopelvic cavity. Beginning with the stomach, trace the rest of the digestive tract.

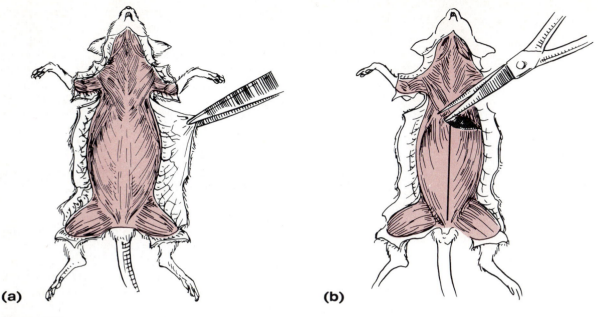

(a)   **(b)**

Figure 2.2

Rat dissection: (a) reflection (folding back) of the skin to expose the underlying muscles, (b) making lateral cuts at the base of the rib cage.

Small intestine: connected to the stomach and ending just before a large saclike cecum

Cecum: the initial portion of the large intestine

Large intestine: a large muscular tube coiled within the abdomen

Follow the course of the large intestine to the rectum, which is partially covered by the urinary bladder.

Rectum: Terminal part of the large intestine; continuous with the anal canal

Anus: the opening of the digestive tract (anal canal) to the exterior

Now lift the small intestine with the forceps to view the mesentery.

Mesentery: an apronlike serous membrane; suspends many digestive organs in the abdominal cavity. Notice that it is heavily invested with blood vessels and, more likely than not, riddled with large fat deposits.

Locate the remaining abdominal structures.

Pancreas: a diffuse gland; rests in the mesentery between the first portion of the small intestine and the stomach

Spleen: a dark red organ curving around the left lateral side of the stomach; considered part of the lymphatic system and often called the red blood cell graveyard

Liver: large and brownish red; the most superior organ in the abdominal cavity, directly beneath the diaphragm

6. To locate the deeper structures of the abdominopelvic cavity, cut through the superior margin of the stomach and the distal end of the large intestine and lay them aside. (Refer to Figure 2.4 as you work.)

Examine the posterior wall of the abdominal cavity to locate the two kidneys.

Kidneys: bean-shaped organs; retroperitoneal (behind the peritoneum)

Adrenal glands: large glands that sit astride the superior margin of each kidney; considered part of the endocrine system

Carefully strip away part of the peritoneum and attempt to follow the course of one of the ureters to the bladder.

Ureter: tube running from the indented region of a kidney to the urinary bladder

Urinary bladder: the sac that serves as a reservoir for urine

7. In the midline of the body cavity lying between the kidneys are the two principal abdominal blood vessels. Identify each.

Inferior vena cava: the large vein that returns blood to the heart from the lower regions of the body

Descending aorta: deep to the inferior vena cava; the largest artery of the body; carries blood away from the heart down the midline of the body

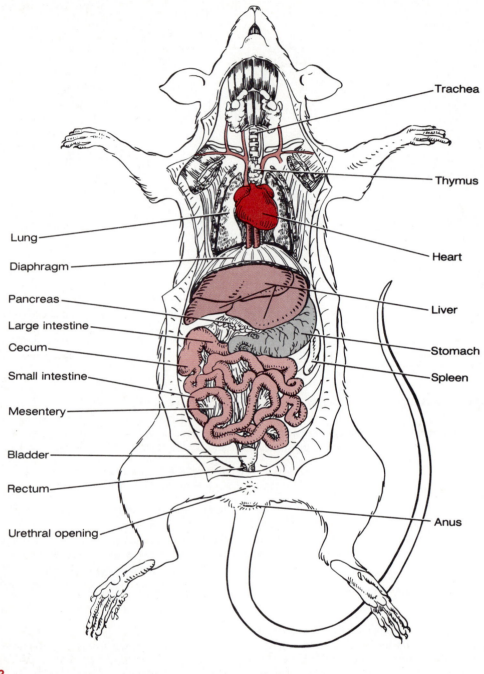

Figure 2.3

Rat dissection: superficial organs of the thoracic and abdominal cavities.

8. Only a cursory examination of reproductive organs will be done. First determine if the animal is a male or female. Observe the ventral body surface beneath the tail. If a saclike scrotum and a single body opening are visible, the animal is a male. If three body openings are present, it is a female. (See Figure 2.4.)

MALE ANIMAL   Make a shallow incision into the

scrotum. Loosen and lift out the oval **testis.** Exert a gentle pull on the testis to identify the slender **vas deferens,** or sperm duct, which carries sperm from the testis superiorly into the abdominal cavity and joins with the urethra. The urethra runs through the penis of the male and carries both urine and sperm out of the body. Identify the **penis,** extending from the bladder to the ventral body wall. Figure 2.4(a)

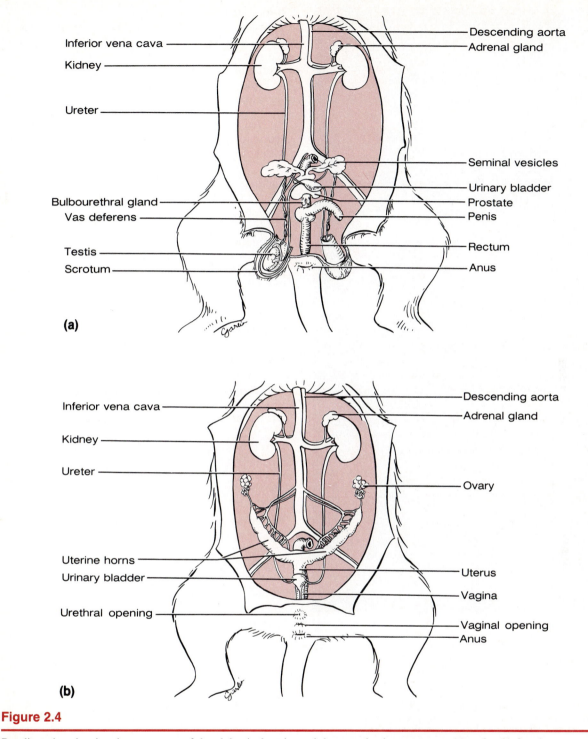

Figure 2.4

Rat dissection showing deeper organs of the abdominal cavity and the reproductive structures: (a) male, (b) female.

indicates other glands of the male reproductive system, but they need not be identified at this time.

FEMALE ANIMAL Inspect the pelvic cavity to identify the Y-shaped **uterus** lying against the dorsal body wall and beneath the bladder (Figure 2.4(b)). Follow one of the uterine horns superiorly to identify an **ovary,** a small oval structure at the end of the uterine horn. The inferior undivided part of the uterus is continuous

with the vagina, which leads to the body exterior. Identify the **vaginal orifice** (external vaginal opening).

9. When you have finished your observations, store or dispose of the rat according to your instructor's directions. Wash the dissecting pan and tools with laboratory detergent, dry them, and return them to the proper storage area.

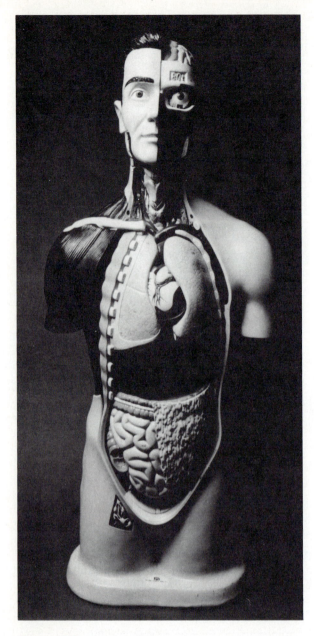

Figure 2.5

Human torso model.

EXAMINING THE HUMAN TORSO MODEL

Examine a human torso model to identify the organs listed below. (Note: If a torso model is not available, Figure 2.5 may be used for this part of the exercise.)

Dorsal cavity: brain, spinal cord
Thoracic cavity: heart, lungs, bronchi, trachea, esophagus, diaphragm, descending aorta, inferior vena cava
Abdominopelvic cavity: liver, stomach, pancreas, spleen, small intestine, large intestine, rectum, kidneys, ureters, bladder, adrenal gland

As you observe these structures, locate the nine abdominopelvic areas studied earlier and determine which organs would be found in each area.

umbilical region _____

epigastric region _____

hypogastric region _____

right iliac region _____

left iliac region _____

right lumbar region _____

left lumbar region _____

right hypochondriac region _____

left hypochondriac region _____

Would you say that the shape and location of the human organs are similar or dissimilar to those of

the rat? _____

Assign each of the organs just identified to one of the organ system categories below.

Digestive: _____

Urinary: _____

Cardiovascular: _____

Reproductive: _____

Respiratory: _____

Lymphatic: _____

Nervous: _____

The Microscope

OBJECTIVES

1. To identify the parts of the microscope and list the function of each.
2. To describe and demonstrate the proper techniques for care of the microscope.
3. To define *total magnification* and *resolution*.
4. To demonstrate the proper focusing technique.
5. To define *parfocal, field,* and *depth of field*.
6. To estimate the size of objects in a field.

MATERIALS

Compound microscope
Millimeter ruler
Prepared slides of the letter *e* or newsprint
Immersion oil
Lens paper
Prepared slide of grid ruled in millimeters (grid slide)
Prepared slide of 3 crossed colored threads
Clean microscope slide and cover slip
Toothpicks (flat-tipped)
Physiologic saline in a dropper bottle
Methylene blue stain (dilute) in a dropper bottle
Forceps
Beaker containing 10% household bleach solution for wet mount disposal
Disposable autoclave bag

With the invention of the microscope, biologists gained a valuable tool to observe and study structures (like cells) that are too small to be seen by the unaided eye. As a result, many of the theories basic to the understanding of biologic sciences have been established. Microscopes range in magnification from the 3 × hand lens to the 1,000,000 × electron microscope. This exercise will familiarize you with the workhorse of microscopes—the compound microscope—and provide you with the necessary instructions for its proper use.

CARE AND STRUCTURE OF THE COMPOUND MICROSCOPE

The compound microscope is a precision instrument and should always be handled with care. At all times you must observe the following rules for its transport, cleaning, use, and storage:

Note to the Instructor: The slides and coverslips used for viewing cheek cells are to be soaked for two hours (or longer) in 10% bleach solution and then drained. The slides, coverslips, and disposable autoclave bag (containing used toothpicks) are to be autoclaved for 15 min at 121°C and 15 pounds pressure to insure sterility. After autoclaving, the disposable autoclave bag may be discarded in any disposal facility and the glassware washed with laboratory detergent and reprepared for use. These instructions apply as well to any blood-stained glassware or disposable items used in other experimental procedures.

● When transporting the microscope, hold it in an upright position with one hand on its arm and the other supporting its base. Avoid jarring the instrument when setting it down.
● Use only special grit-free lens paper to clean the lenses. Clean all lenses before and after use.
● Always begin the focusing process with the lowest-power objective lens in position, changing to the higher-power lenses if necessary.
● *Never* use the coarse adjustment knob with the high-dry or oil immersion lenses.
● A cover slip must always be used with temporary (wetmount) preparations.
● Before putting the microscope in the storage cabinet, remove the slide from the stage, rotate the lowest-power objective lens into position, and replace the dust cover.
● Never remove any parts from the microscope; inform your instructor of any mechanical problems that arise.

 1. Obtain a microscope and bring it to the laboratory bench. (Use the proper carrying technique!) Compare your microscope with the illustration in Figure 3.1 and identify the following microscope parts:

Base: supports the microscope. (Note: Some microscopes are provided with an inclination joint, which allows the instrument to be tilted backward for viewing dry preparations.)

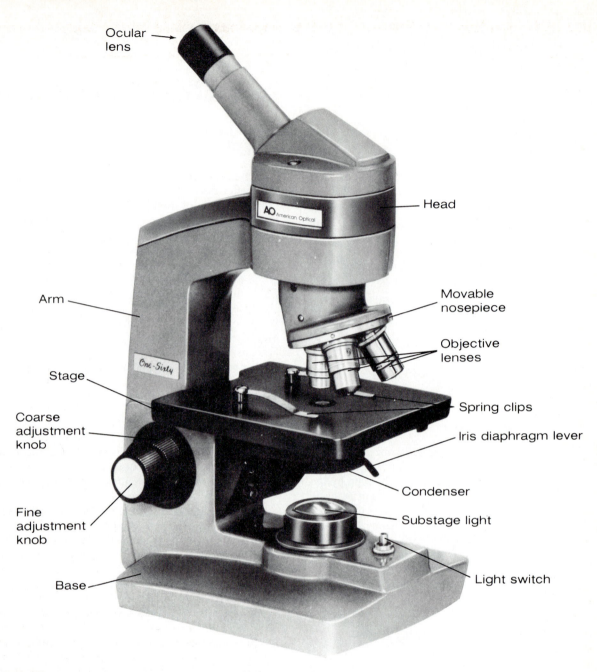

Ocular lens

Head

Arm

Movable nosepiece

Objective lenses

Stage

Spring clips

Coarse adjustment knob

Iris diaphragm lever

Condenser

Fine adjustment knob

Substage light

Base

Light switch

Figure 3.1

Compound microscope and its parts. (Courtesy of AO Scientific Instruments, Division of Warner-Lambert Technologies, Inc.)

Substage light (or *mirror*): located in the base. In microscopes with a substage light source, the light passes directly upward through the microscope. If a mirror is used, light must be reflected from a separate free-standing lamp.

Stage: the platform the slide rests on while being viewed. The stage always has a hole in it to permit light to pass through both it and the specimen. Some microscopes have a stage equipped with *spring clips;* others have a clamp-type *mechanical stage.* Both hold the slide in position

for viewing; in addition, the mechanical stage permits precise movement of the specimen.

Condenser: concentrates the light on the specimen. The condenser may be equipped with a height-adjustment knob that raises and lowers the condenser to vary the delivery of light. Generally, the best position for the condenser is close to the inferior surface of the stage.

Iris diaphragm lever: arm attached to the condenser that regulates the amount of light passing through the condenser. The iris diaphragm permits the best possible contrast when viewing the specimen.

Coarse adjustment knob: used to focus the specimen.

Fine adjustment knob: used for precise focusing once coarse focusing has been completed.

Head or **body tube:** supports the objective lens system (which is mounted on a movable nosepiece), and the ocular lens.

Arm: vertical portion of the microscope connecting the base and head.

Ocular (or *eyepiece*): lens contained at the superior end of the head or body tube. Observations are made through the ocular.

Nosepiece: generally carries three objective lenses.

Objective lenses: adjustable lens system that permits the use of a **low-power lens, a high-dry lens,** or an **oil immersion lens.** The objective lenses have different magnifying and resolving powers.

2. Examine the objectives carefully, noting their relative lengths and the numbers inscribed on their sides. On most microscopes, the low-power (l.p.) objective is the shortest and generally has a magnification of 10× (it increases the apparent size of the object by ten times or ten diameters). The high-dry (h.p.) objective is of intermediate length and has a magnification range from 40× to 50×, depending on the microscope. The oil immersion objective is usually the longest of the objectives and has a magnifying power of 95× to 100×. (Note: some microscopes lack the oil immersion lens but have a very low magnification lens called the **scanning lens,** which is a very short objective with a magnification of 4× to 5×). Record the magnification of each objective lens below. If your microscope has a scanning lens instead of the oil immersion lens, cross out "oil immersion" and substitute "scan" on the chart.

3. Rotate the l.p. objective into position, and turn the coarse adjustment knob about 180 degrees. Note how far the stage (or objective) travels during this adjustment. Move the fine adjustment knob 180 degrees, noting again the distance that the stage (or the objective) moves.

Magnification and Resolution

The microscope is an instrument of magnification. In the compound microscope, magnification is

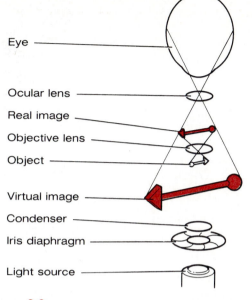

Eye

Ocular lens

Real image

Objective lens

Object

Virtual image

Condenser

Iris diaphragm

Light source

Figure 3.2

Optical system of the compound microscope. Note the real and virtual images.

achieved through the interplay of two lenses—the ocular lens and the objective lens. The objective lens magnifies the specimen to produce a **real image** that is projected to the ocular. This real image is magnified by the ocular lens to produce the **virtual image** seen by your eye (Figure 3.2).

The **total magnification** of any specimen being viewed is equal to the power of the ocular lens multiplied by the power of the objective lens used. For example, if the ocular lens magnifies 10× and the objective lens being used magnifies 45×, the total magnification is 450×. Determine the total magnification you may achieve with each of the objectives on your microscope and record the figures on the chart below.

The compound light microscope has certain limitations. Although the level of magnification is almost limitless, the **resolution** (or resolving power), the ability to discriminate two close objects as separate, is not. The human eye can resolve objects about 100 μm apart, but the compound microscope has a res-

	Low power		High power		Oil immersion	
Magnification of the objective lenses	X		X		X	
Total magnification	X		X		X	
Detail observed						
Field size (diameter)	mm	μm	mm	μm	mm	μm
Working distance		mm		mm		mm

olution of 0.2 μm under ideal conditions. Objects closer than 0.2 μm are seen as a single fused image.

Resolving power (RP) is determined by the amount and physical properties of the visible light that enters the microscope. In general, the greater the amount of light delivered to the objective lens, the greater the resolution. The size of the objective lens aperture decreases with increasing magnification, allowing less light to enter the objective; thus you will probably find it necessary to increase the light intensity at the higher magnifications.

VIEWING OBJECTS THROUGH THE MICROSCOPE

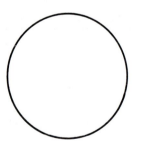
1. Obtain a millimeter ruler, a prepared slide of the letter *e* or newsprint, a dropper bottle of immersion oil, and some lens paper. Secure the slide on the stage so that the letter *e* is centered over the hole, and switch on the light source. (If the light source is not built into the base, use the curved surface of the mirror to reflect the light up into the microscope.) The condenser should be in its highest position.

2. With the l.p. objective in position over the stage, use the coarse adjustment knob to bring the objective and stage as close together as possible.

3. Looking through the ocular, adjust the light for comfort. Now use the coarse adjustment knob to focus slowly away from the *e* until it is as clearly focused as possible. Complete the focusing with the fine adjustment knob.

4. Sketch the letter in the circle just as it appears in the **field** (the area you see through the microscope).

What is the total magnification? _____ ×

How far is the bottom of the objective from the specimen? In other words, what is the **working distance?**

_____ mm

(Use a millimeter ruler to make this measurement and record it in the chart on page 17. How has the apparent orientation of the *e* changed top to bottom,

right to left, and so on? _____

_____

5. Move the slide slowly away from you on the stage as you look through the ocular. In what direc-

tion does the image move? _____

Move the slide to the left. In what direction does the

image move? _____

At first this change in orientation will confuse you, but with practice you will learn to move the slide in the desired direction with no problem.

6. Without touching the focusing knobs, increase the magnification by rotating the high-dry objective into position over the stage. Using the fine adjustment only, sharpen the focus.* What new details

become clear? _____

_____

What is the total magnification now? _____ ×

As best you can, measure the distance between the objective and the slide (the working distance) and record it on the chart.

Why should the coarse focusing knob *not* be used when focusing with the higher-powered objective

lenses? _____

_____

Is the image larger or smaller? _____

Approximately how much of the letter is visible now?

_____

Is the field larger or smaller? _____

Why is it necessary to center your object (or the portion of the slide you wish to view) before changing to a higher power?

_____

*Today most good laboratory microscopes are **parfocal;** that is, the slide should be in focus (or nearly so) at the higher magnifications once you have properly focused in l.p. If you are unable to swing the objective into position without raising the objective, your microscope is not parfocal. Consult your instructor.

Move the iris diaphragm lever while observing the field. What happens? _____

_____

Is it more desirable to increase *or* decrease the light when changing to a higher magnification? _____

Why? _____

_____

7.   Without touching the focusing knob, rotate the high-dry lens out of position so that the area of the slide over the opening in the stage is unobstructed. Place a drop of immersion oil over the *e* on the slide and rotate the immersion lens into position. Adjust the fine focus and the light for the best possible resolution. Is the field again decreased in size? _____

What is the total magnification with the immersion lens? _____ × Is the working distance less or greater than it was when the high-dry lens was focused? _____

Compare your observations on the relative working distances of the objective lenses with the illustration in Figure 3.3. Explain why it is desirable to begin the focusing process in l.p. _____

_____

_____

8.   Rotate the immersion lens slightly to the side and remove the slide. Clean the oil immersion lens carefully with lens paper and then clean the slide in the same manner.

DETERMINING THE SIZE OF THE MICROSCOPE FIELD

By this time you should know that the size of the microscope field decreases with increasing magnification. For future microscope work, it will be useful to determine the diameter of each of the microscope fields. This information will allow you to make a fairly accurate estimate of the size of the objects you view in any field. For example, if you have calculated the field diameter to be 4 mm and the object extends across half this diameter, you can estimate the size of the object to be approximately 2 mm.

   Microscopic specimens are usually measured in micrometers and millimeters, both units of the met-

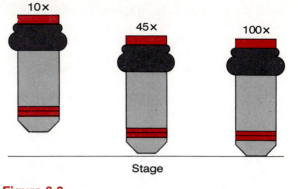

Figure 3.3

Relative working distance of the 10×, 45×, and 100× objectives.

ric system. You can get an idea of the relationship and meaning of these units from Table 3.1.

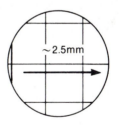

1.   Return the letter *e* slide and obtain a grid slide, a slide prepared with graph paper ruled in millimeters. Each of the squares in the grid is 1 mm on each side. Focus in low power.

2.   Move the slide so that one grid line touches the edge of the field on one side, and count the number of squares you can see across the diameter of the l.p. field. If you can see only part of a square, as in the accompanying diagram, estimate the part of a millimeter that the partial square represents.

~2.5mm

TABLE 3.1   Comparison of Metric Units of Length

Metric unit	Abbreviation	Equivalent
Meter	m	(about 39.3 in.)
Centimeter	cm	10^{-2} m
Millimeter	mm	10^{-3} m
Micrometer (or micron)	μm (μ)	10^{-6} m
Nanometer (or millimicrometer, or millimicron)	nm (mμ)	10^{-9} m
Angstrom	Å	10^{-10} m

For future reference, record this figure in the appropriate space marked "field size" on the summary chart on page 17. Complete the chart by computing the approximate diameter of the high-dry and immersion fields. Say the diameter of the l.p. field (total magnification of 50×) is 2 mm. You would compute the diameter of a high-dry field with a total magnification of 100× as follows:

$$2\,mm \times 50 = X\,(\text{diameter of h.p. field}) \times 100$$
$$100\,mm = 100X$$
$$1\,mm = X\,(\text{diameter of the h.p. field})$$

The formula is:
Diameter of the l.p. field (mm) × Total magnification of the l.p. field = Diameter of field X × Total magnification of field X

3.   Estimate the size of the following microscopic objects. *Base your calculations on the field sizes you have determined for your microscope.*

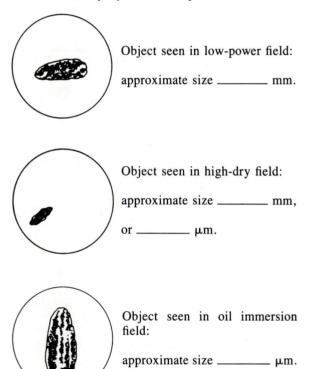

Object seen in low-power field:

approximate size _____ mm.

Object seen in high-dry field:

approximate size _____ mm,

or _____ μm.

Object seen in oil immersion field:

approximate size _____ μm.

4.   If an object viewed with the oil immersion lens looked like the field depicted here, could you determine its approximate size from this view? _____

If not, then how could you determine it?

_____

_____

_____

PERCEIVING DEPTH

Any specimen mounted on a slide has depth as well as length and width; it is rare indeed to view a tissue slide with just one layer of cells. Normally you can see two or three cell thicknesses. Therefore, it is important to learn how to determine relative depth with your microscope.*

   1.   Return the grid slide and obtain a slide of colored crossed threads. Focusing in l.p., locate a point where the threads cross.

2.   Focus down with the coarse adjustment until the threads are out of focus, then slowly focus upward again, noting which thread comes into clear focus first. This one is the lowest or most inferior thread. (Note: you will see two or even all three threads, so you must be very careful in determining which one comes into focus first.)

Record your observation: _____ thread

over _____

Continue to focus upward from the middle to the superiormost thread. Again record your observation.

_____ thread over _____

Which of the threads is uppermost? _____

Lowest? _____

PREPARING AND OBSERVING A WET MOUNT

1.   Obtain the following: a clean microscope slide and cover slip, a flat-tipped toothpick, a dropper bottle of physiologic saline, a dropper bottle of methylene blue stain, forceps.

*In microscope work the **depth of field** (the depth of the specimen clearly in focus) is greater at lower magnifications.

2. Place a drop of physiologic saline in the center of the slide. Using the flat end of the toothpick, gently scrape the inner lining of your cheek. Agitate the end of the toothpick containing the cheek scrapings in the drop of saline (Figure 3.4(a)). *Immediately* dispose of the used toothpick in the disposable autoclave bag provided at the supplies area.

Add a small drop of the methylene blue stain to the preparation. (These epithelial cells are nearly transparent and thus difficult to see without the stain, which colors the nuclei of the cells and makes them look much darker than the cytoplasm.) Stir again.

3. Hold the cover slip with the forceps so that its inferior edge touches one side of the fluid drop (Figure 3.4(b)), then *carefully* lower the cover slip onto the preparation (Figure 3.4(c)). *Do not just drop the cover slip,* or you will trap large air bubbles under it, which will obscure the cells. *A cover slip should always be used with a wet mount* to prevent soiling the lens if you should misfocus.

4. Place the slide on the stage and locate the cells in l.p. You will probably want to dim the light with the iris diaphragm to provide more contrast for viewing the lightly stained cells. (Furthermore, a wet mount will dry out quickly in bright light, since a bright light source is hot.)

5. Cheek epithelial cells are very thin, six-sided cells. In the cheek, they provide a smooth, tilelike lining, as shown in Figure 3.5.

6. Make a sketch of the epithelial cells that you observed.

Approximately how large are the cheek epithelial

cells? _____ mm

Why do *your* cheek cells look different than those illustrated in Figure 3.5? (Hint: what did you have

to *do* to your cheek to obtain them?) _____

_____

_____

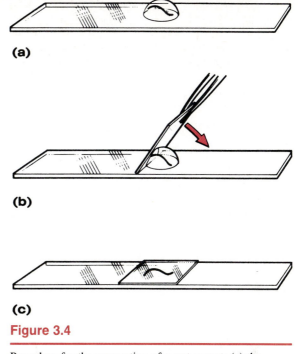

(a)

(b)

(c)

Figure 3.4

Procedure for the preparation of a wet mount: (a) the object is placed in a drop of water on a clean slide, (b) a cover slip is held at a 45° angle with forceps, and (c) it is lowered carefully over the water and the object.

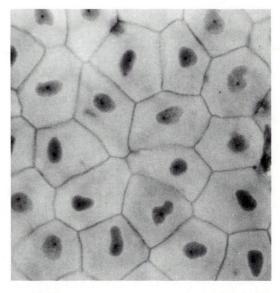

Figure 3.5

Epithelial cells of the cheek cavity (surface view), 160×.

7. Dispose of your wet mount preparation in the beaker of bleach solution and make sure all materials are properly disposed of or returned to the proper laboratory station: clean the microscope lenses and slide, and put the dust cover on the microscope before you return it to the storage cabinet.

The Cell—Anatomy and Division

OBJECTIVES

1. To define *cell, organelle,* and *inclusion.*
2. To identify on a cell model or diagram the following cellular regions and to list the major function of each: nucleus, cytoplasm, and plasma membrane.
3. To identify and list the major functions of the various organelles studied.
4. To compare and contrast specialized cells with the concept of the "generalized cell."
5. To define *interphase, mitosis,* and *cytokinesis.*
6. To list the stages of mitosis and describe the events of each stage.
7. To identify the mitotic phases on projected slides or appropriate diagrams.
8. To explain the importance of mitotic cell division and its product.

MATERIALS

Three-dimensional model of the "composite" animal cell or laboratory chart of cell anatomy
Prepared slides of simple squamous epithelium ($AgNO_3$ stain), teased smooth muscle, human blood cell smear, and sperm
Compound microscope
Clean slide and cover slip
Physiologic saline
Flat-tipped toothpick
Janus green stain and medicine dropper
Filter paper or some other type of porous paper
Forceps
Beaker containing 10% household bleach solution for wet mount disposal
Disposable autoclave bag
Prepared slides of whitefish blastulae
Three-dimensional models of mitotic states

The **cell,** defined as the structural and functional unit of all living things, is a very complex entity. The cells of the human body are highly diverse; and their differences in size, shape, and internal composition reflect their specific roles in the body. Yet cells do have many common anatomical features, and there are some functions that all must perform to maintain life. For example, all cells have the ability to metabolize, to digest foods, and dispose of wastes, to grow and reproduce, to move (mobility), and to respond to a stimulus (irritability). Most of these functions are considered in detail in later exercises. This exercise focuses on structural similarities that typify the "composite," or "generalized," cell and considers only the function of cell reproduction (cell division). Transport mechanisms (the means by which substances cross the plasma membrane) are dealt with separately in Exercise 5.

ANATOMY OF THE COMPOSITE CELL

In general, all cells have three major regions, or parts, that can readily be identified with a light microscope:

Note to the Instructor: See directions for handling of toothpicks and wet mount preparations on p. 15.

the **nucleus,** the **plasma membrane,** and the **cytoplasm.** The nucleus is usually seen as a round or oval structure near the center of the cell. It is surrounded by cytoplasm, which in turn is enclosed by the plasma membrane. Since the advent of the electron microscope, even smaller cell structures—organelles—have been identified. Figure 4.1(a) represents the fine structure of the composite cell as revealed by the electron microscope.

Nucleus

The nucleus can quite accurately be described as the control center of the cell and is necessary for cell reproduction. A cell that has lost or ejected its nucleus (for whatever reason) is literally programmed to die because the nucleus is the site of the "genes," or genetic material—DNA.

When the cell is not dividing, the genetic material is loosely dispersed throughout the nucleus in a form called **chromatin,** which is granular or thread-like. When the cell is in the process of dividing to form daughter cells, the chromatin coils and condenses to form dense, darkly staining rodlike bodies called **chromosomes**—much in the way a stretched spring becomes shorter and thicker when relaxed. (Cell division is discussed later in this exercise.) Notice the appearance of the nucleus carefully—it is somewhat nondescript. When the nucleus appears dark and the chromatin becomes clumped, this is an

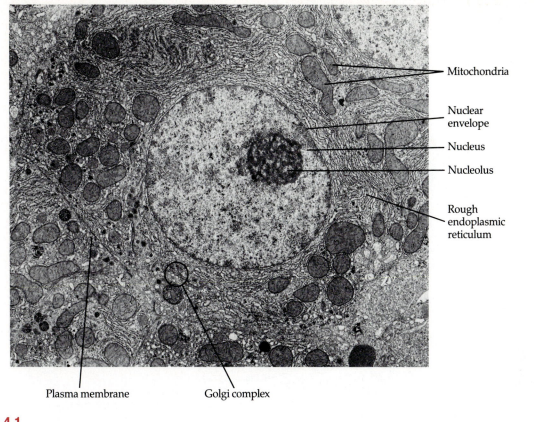

Figure 4.1

Anatomy of the composite animal cell (a) as revealed by transmission electron microscopy.

indication that the cell is dying and undergoing degeneration.

The nucleus also contains one or more small round bodies, called **nucleoli,** composed primarily of proteins and ribonucleic acid (RNA). The nucleoli are believed to be storage sites for RNA and/or assembly sites for ribosomal particles (particularly abundant in the cytoplasm), which are the actual "factories" synthesizing protein.

The nucleus is bound by a double-layered porous membrane, the **nuclear membrane** (or nuclear envelope), which is similar in composition to other cellular membranes.

- **Label** the nuclear membrane, chromatin threads, and a nucleolus in Figure 4.1(b).

Plasma Membrane

The **plasma membrane** separates cell contents from the surrounding environment. It is made up of protein and lipid (fat) and appears to have a bimolecular lipid core that protein molecules float in (see Figure 4.1(c)). Besides providing a protective barrier for the cell, the plasma membrane plays an active role in determining which substances may enter or leave the cell and in what quantity. In some cells the membrane is thrown into minute fingerlike projections or folds called **microvilli,** which greatly increase the surface area of the cell available for absorption or passage of materials.

- **Label** the plasma membrane and microvilli on Figure 4.1(b).

Cytoplasm and Organelles

The cytoplasm consists of the cell contents outside the nucleus. It is the major site of most activities carried out by the cell. Suspended in the cytoplasmic material are many small structures called **organelles** (literally, small organs). The organelles are the metabolic machinery of the cell, and they are highly organized to carry out specific functions for the cell as a whole. The organelles include the ribosomes, endoplasmic reticulum, Golgi apparatus, lysosomes, mitochondria, centrioles, and cytoskeletal elements.

- Each organelle type is described briefly next. Read through this material and then correctly **label** the organelles on Figure 4.1(b).

The **ribosomes** are tiny spherical bodies composed of RNA and protein. They are the actual sites of protein synthesis. They are seen floating free in the cytoplasm or attached to a membranous structure. When they are attached, the whole ribosome–membrane complex is called the granular or rough endoplasmic reticulum.

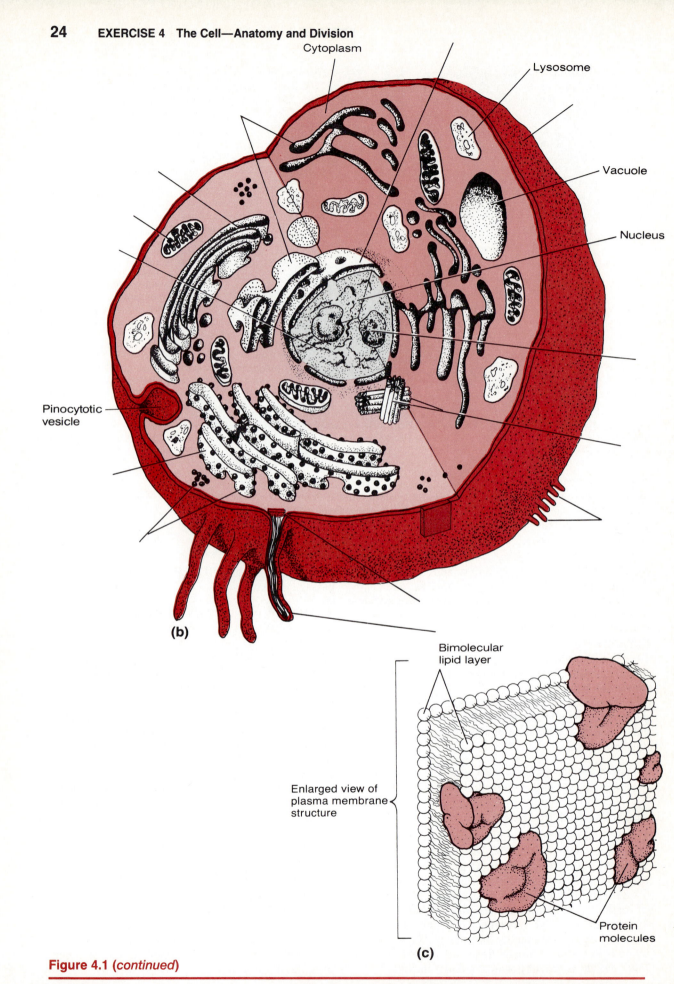

Cytoplasm

Lysosome

Vacuole

Nucleus

Pinocytotic
vesicle

(b)

Bimolecular
lipid layer

Enlarged view of
plasma membrane
structure

Protein
molecules

(c)

Figure 4.1 (*continued*)

Anatomy of the composite animal cell. (b) Diagrammatic view of a composite cell; (c) structural details of the plasma membrane.

The **endoplasmic reticulum** (ER) is a highly folded membranous system of tubules that extends throughout the cytoplasm. The ER has been observed to be continuous with the Golgi apparatus, nuclear membrane, and plasma membrane. Thus it is assumed that the ER provides a system of channels for the transport of cellular substances (primarily proteins) from one part of the cell to another or to the cell exterior. The ER exists in two forms; a particular cell may have both or only one, depending on its specific functions. The granular or **rough ER,** as noted earlier, is studded with ribosomes. Its cisternae proteins deliver them to other areas of the cell; the external face of the rough ER is involved in phospholipid and cholesterol synthesis. The amount of rough ER is closely correlated with the amount of protein a cell manufactures and is especially abundant in cells that produce protein products for export—for example, pancreas cells, which produce digestive enzymes destined for the small intestine. The agranular or **smooth ER** has no protein synthesis-related function but is present in conspicuous amounts in cells that produce steroid-based hormones—for example, the interstitial cells of the testes, which produce testosterone, and in cells that are highly active in lipid metabolism and drug detoxification activities—liver cells, for instance.

The **Golgi apparatus** is a stack of flattened sacs (often accompanied by bulbous ends and small vesicles) and is generally found close to the nucleus. It is now known to have a role in packaging proteins destined for incorporation into cellular membranes or lysosomes, or for export (the proteins are delivered to it by the rough ER) and perhaps in attaching carbohydrate groups to some of them. As the proteins accumulate in the Golgi apparatus, the sacs swell, and little vesicles filled with protein pinch off. Those vesicles containing secretory products travel to the plasma membrane, fuse with it, and eject their contents to the cell exterior.

The **lysosomes,** which appear in various sizes, are membrane-bound sacs containing an array of powerful digestive enzymes. Believed to arise from the packaging activities of the Golgi apparatus, the lysosomes contain enzymes capable of digesting worn-out cell structures and foreign substances that enter the cell through phagocytosis or pinocytosis (see Exercise 5). The lysosomes also bring about some of the changes that occur during menstruation, when the uterine lining is sloughed off. Since they have the capacity of total cell destruction, the lysosomes are often referred to as the "suicide sacs" of the cell.

The **mitochondria** are generally rod-shaped bodies with a double-membrane wall; the inner membrane is thrown into folds, or cristae. Oxidative enzymes on or within the mitochondria catalyze the reactions of the Krebs cycle and the electron transport chain (collectively called oxidative respiration), in which foods are broken down to produce energy. The released energy is captured in the bonds of ATP (adenosine triphosphate) molecules, which then diffuse out of the mitochondria to provide a ready energy supply to power the cell. Every living cell requires a constant supply of ATP for its many activities. Since the mitochondria provide the bulk of this ATP, they are referred to as the powerhouses of the cell.

The paired **centrioles** lie close to the nucleus in all animal cells capable of reproducing themselves. They are rod-shaped bodies that lie at right angles to each other; internally they consist of a system of fine tubules. During cell division, the centrioles direct the formation of the mitotic spindle. Centrioles also form the basis for cell projections called cilia and flagella (described below).

The **cytoskeletal elements** are extremely important in cellular support and movement of substances within the cell. **Microtubules** are basically slender tubules formed of proteins called *tubulins*. Since tubulins can aggregate spontaneously to form microtubules and then disaggregate just as quickly, the microtubules have been difficult to study. Microtubules are the basis of the spindle formed by the centrioles during cell division; they act in the transport of substances down the length of elongated cells (such as neurons), and form part of the internal cytoskeleton, providing rigidity to the soft cellular substance. **Intermediate filaments** are *stable* proteinaceous cytoskeletal elements that are important in resisting mechanical forces acting on cells. **Microfilaments,** elements that are ribbon or cordlike rather than hollow, are formed of contractile proteins. Because of their ability to shorten and then relax to assume a more elongated form, these are important in cell mobility and are very conspicuous in cells that are highly specialized to contract (such as muscle cells). Since the cytoskeletal structures are so labile and minute, they are rarely seen, even in electron micrographs, and are not depicted in Figure 4.1. (The exceptions are the microtubules of the spindle, which are very obvious during cell division (see p. 27), and the microfilaments of skeletal muscle cells (see p. 93).)

In addition to these cell structures, some cells have projections called **flagella,** which propel the cells, or **cilia,** which allow cells to sweep substances along a tract. Label the cilium in Figure 4.1(b).

The cell cytoplasm contains various other substances and structures, including stored foods (glycogen granules, lipid droplets, and other nutrients), pigment granules, crystals of various types, water vacuoles, and ingested foreign materials. But these are not part of the active metabolic machinery of the cell and are therefore called **inclusions.**

Once you have located and labeled all of these structures in Figure 4.1(b), examine the cell model (or cell chart) to reinforce your identifications.

OBSERVING DIFFERENCES AND SIMILARITIES IN CELL STRUCTURE

1. Obtain prepared slides of simple squamous epithelium, sperm, smooth muscle cells (teased), and human blood.

2. Observe each slide under the microscope, carefully noting similarities and differences in the cells. (The oil immersion lens will be needed to observe blood and sperm.) Distinguish the limits of the individual cells, and note the shape and position of the nucleus in each case. When you look at the human blood smear, direct your attention to the red blood cells, the pink-stained cells that are most numerous. Diagram your observations in the circles provided.

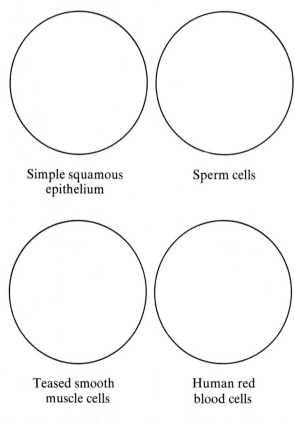

Simple squamous epithelium

Sperm cells

Teased smooth muscle cells

Human red blood cells

3. How do these four cell types differ in shape and size? _____

_____

_____

_____

_____

How might cell shape affect cell function? _____

_____

_____

_____

_____

_____

Which cells have visible projections? _____

_____

How do these projections relate to the function of

this cell? _____

_____

_____

Do any of these cells lack a cell membrane? _____

A nucleus? _____

In the cells with a nucleus, can you discern nucleoli?

_____

Were you able to observe any of the organelles in

these cells? _____ Why or why not?

_____

_____

_____

_____

Many of the organelles are impossible to observe with the light microscope unless they are selectively stained. Selective or differential stains, one of which you will be using here, react to the chemical differences in the organelles in specified ways, thereby allowing you to differentiate among them.

1. Prepare a wet mount of cheek epithelial cells in physiologic saline, as in Exercise 3 (p. 21), but do not add the methylene blue stain. Before continuing, dispose of the used toothpick in the autoclave bag at the supply area.

2. Observe the cells under l.p. to identify the cell membrane, nucleus, and cytoplasm. Keep the light dim to increase contrast.

3. Place a drop of Janus green stain to one side of the cover slip. Place a piece of filter paper or some other porous paper on the other side of the cover slip to remove the excess saline and to facilitate the movement of the Janus green under the cover slip.

4. Observe the cells again under the oil immersion lens. Janus green stains selectively for areas of rapid oxidation. What organelles are reacting with the Janus green? _____

_____

5. Properly dispose of your wet mount in the bleach-containing beaker.

CELL DIVISION: MITOSIS AND CYTOKINESIS

Cell division in all cells other than bacteria consists of a series of events collectively called mitosis and cytokinesis. **Mitosis** is nuclear division; **cytokinesis** is the division of the cytoplasm, which begins after mitosis is nearly complete. Although mitosis is usually accompanied by cytokinesis, in some instances cytoplasmic division does not occur, leading to the formation of binucleate (or multinucleate) cells. This

is relatively common in the human liver.

The process of **mitosis** results in the formation of two daughter nuclei that are genetically identical to the mother nucleus. This distinguishes mitosis from **meiosis,** a specialized type of nuclear division that occurs only in the reproductive organs (testes or ovaries). Meiosis, which yields four daughter nuclei that differ genetically and in composition from the mother nucleus, is used only for the production of eggs and sperm (gametes) for sexual reproduction. The function of cell division, including mitosis and cytokinesis in the body, is to increase the number of cells for growth and repair while maintaining their genetic heritage.

In cells about to divide, an important event precedes cell division. The genetic material (the DNA molecules composing part of the chromatin strands) is duplicated exactly during the portion of the cell cycle called **interphase.** Interphase is *not* part of mitosis; it represents the time when a cell is not actively involved in cell division. Although some people refer to interphase as the cell's resting period, this is an inaccurate description because the cell is quite active in its daily activities and is resting only from cell division. The stages of mitosis diagramed in Figure 4.2 include the following events:

Prophase: At the onset of cell division, the chromatin threads coil and shorten to form densely staining, short, barlike **chromosomes.** By the

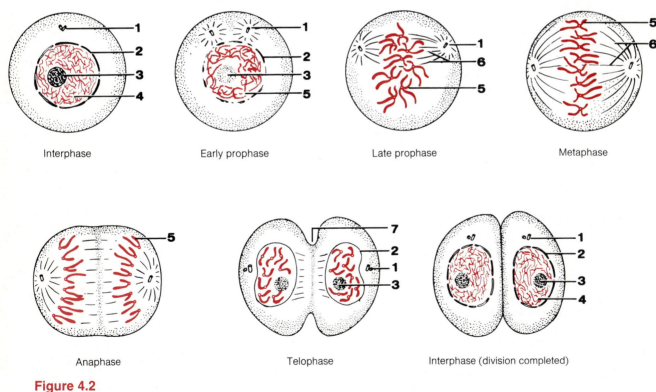

Interphase   Early prophase   Late prophase   Metaphase

Anaphase   Telophase   Interphase (division completed)

Figure 4.2

The stages of mitosis, diagrammatic views. (1 = centriole, 2 = nuclear membrane, 3 = nucleolus, 4 = chromatin, 5 = chromosomes, 6 = spindle, 7 = cleavage furrow.)

middle of prophase the chromosomes appear as double-stranded structures (each strand is a **chromatid**) connected by a small median body called a **centromere.** The centrioles separate and migrate toward opposite poles of the cell, spinning out the mitotic **spindle** between them as they move. The spindle acts as a scaffolding for the attachment and movement of the chromosomes during later mitotic stages. The nuclear membrane and the nucleolus break down and disappear, and the chromosomes randomly attach to the spindle fibers by their centromeres.

Metaphase: A brief stage, during which the chromosomes migrate to the central plane or equator of the spindle and align along that plane in a straight line from the superior to the inferior region of the spindle (lateral view). Viewed from the poles of the cell (end view), the chromosomes appear to be arranged in a "rosette," or circle, around the widest dimension of the spindle.

Anaphase: During anaphase, the centromeres break, and the chromatids (now called chromosomes again) separate from one another and then progress slowly toward opposite ends of the cell. The chromosomes appear to be pulled by their centromere attachment, with their "arms" dangling behind them. Anaphase is complete when poleward movement ceases.

Telophase: During telophase, the events of prophase are essentially reversed. The chromosomes at the poles begin to uncoil and resume the chromatin form, the spindle disappears, a nuclear membrane forms around each chromatin mass, and nucleoli appear in each of the daughter nuclei.

Mitosis is essentially the same in all animal cells, but depending on the type of tissue, it takes from 5 minutes to several hours to complete. In most cells, centriole replication is deferred until interphase of the next cell cycle.

Cytokinesis, or the division of the cytoplasmic mass, begins during telophase. In animal cells, a cleavage furrow begins to form approximately over the equator of the spindle, and eventually splits or pinches the original cytoplasmic mass into two portions. Thus at the end of cell division two daughter cells exist, each smaller in cytoplasmic mass than the mother cell but genetically identical to it. The daughter cells grow and carry out the normal spectrum of metabolic processes until it is their turn to divide.

Cell division is extremely important during the body's growth period. Most cells (excluding nerve cells) undergo mitosis until puberty, when normal body size is achieved and overall body growth ceases. After this time in life, only certain cells routinely carry out cell division—for example, cells subjected to abrasion (epithelium of the skin and lining of the gut). Other cell populations—such as liver cells—

stop dividing but retain this ability should some of them be removed or damaged. Skeletal muscle and nervous tissue completely lose this ability to divide and thus are severely handicapped by injury. Throughout life, the body retains its ability to repair cuts and wounds and to replace some of its aged cells.

 Obtain a prepared slide of whitefish blastulae to study the stages of mitosis. The cells of each blastula (a stage of embryonic development consisting of a hollow ball of cells) are at approximately the same mitotic stage, so it may be necessary to observe more than one blastula to view all the mitotic stages. The exceptionally high rate of mitosis observed in this tissue is typical of embryos, but if occurring in specialized tissues, it can be an indication of cancerous cells, which also have an extraordinarily high mitotic rate. Examine the slide carefully, identifying the four mitotic stages and the process of cytokinesis. Compare your observations with Figure 4.2, and verify your identifications with your instructor. Then sketch your observations of each stage in the circles provided here.

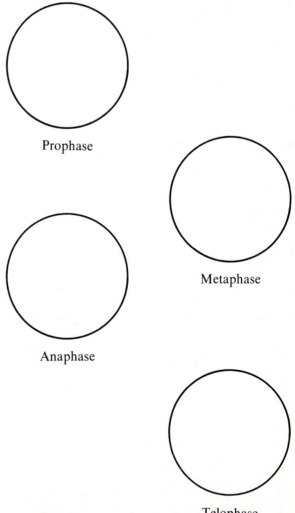

Prophase

Metaphase

Anaphase

Telophase

The Cell—
Transport Mechanisms
and Cell Permeability

OBJECTIVES

1. To define *differential,* or *selective, permeability; diffusion* (*dialysis* and *osmosis*); *Brownian motion; isotonic, hypotonic,* and *hypertonic solutions; active transport; pinocytosis; phagocytosis;* and *permease system.*

2. To explain the processes that account for the movement of substances across the plasma membrane and to note the driving force for each.

3. To determine which way substances will move passively through a selectively permeable membrane (given appropriate information on concentration differences).

MATERIALS

For Passive Transport Experiments:
Clean slides and coverslips
Forceps
Glass stirring rods
15-ml graduated cylinders
Compound microscopes
Hot plate and large beaker for hot water bath
Brownian motion:
Milk in dropper bottles
Diffusion:
Petri plate containing 12 ml of 1.5% agar-agar
Methylene blue dye crystals
Potassium permanganate dye crystals
Millimeter rulers
Four dialysis sacs
Beakers (250 ml)

40% glucose solution
Fine twine
10% NaCl solution, boiled starch solution
Laboratory balance
Benedict's solution in dropper bottle
Test tubes in racks, test tube holder
Wax marker
Silver nitrate ($AgNO_3$) in dropper bottle
Lugol's iodine solution in dropper bottle
Lancets, alcohol swabs
Filter paper
Animal (mammalian) blood (in vials) obtained from an animal hospital or veterinary school—at option of instructor
Physiologic (mammalian) saline solution in dropper bottle
1.5% sodium chloride (NaCl) solution in dropper bottle
Medicine dropper
Basin containing 10% household bleach solution
Disposable autoclave bag
Yeast suspension
Congo red dye in dropper bottle
Filtration:
Ring stand, ring, clamp
Filter paper, funnel
Solution containing a mixture of uncooked starch, powdered charcoal, and copper sulfate ($CuSO_4$)
Active Transport:
Culture of starved ameba (*Amoeba proteus*)
Depression slide
Tetrahymena pyriformis culture
Compound microscope

Because of its molecular composition, the plasma membrane is selective about what passes through it. It allows nutrients to enter the cell but keeps out undesirable substances. By the same token, valuable cell proteins and other substances are kept within the cell, and excreta or wastes pass to the exterior. This property is known as **selective permeability.** Transport through the plasma membrane occurs in two basic ways. In one, the cell must provide energy (ATP) to power the transport process (active trans-

Note to the Instructor: See directions for handling wet mount preparations and disposable supplies on page 15. Exercise. 3.

port); in the other, the transport process is driven by concentration or pressure differences (passive transport).

PASSIVE TRANSPORT

All molecules vibrate randomly (because of their inherent kinetic energy) at all temperatures above absolute zero (about $-460°F$). In general, the smaller the particle, the greater the kinetic energy it possesses and the faster its molecular motion. This random movement may be detected indirectly by

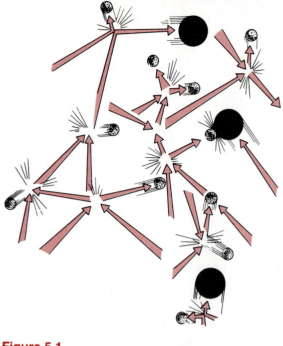

Figure 5.1

Random movement and numerous collisions, causing molecules to become evenly distributed. The small spheres represent water molecules; the large spheres represent glucose molecules.

observing a suspension like milk. The larger particles can be seen moving randomly as they are hit and deflected by the smaller, more rapidly moving particles. The zigzag movement of the larger particles is known as **Brownian motion.**

 1.   Make a wet mount of milk; that is, place a small drop of milk on a slide and cover carefully with a cover slip. Allow the slide to stand on the microscope stage for about 10 minutes before observing.

2.   Observe with high power and then with the oil immersion lens. Keep the light as dim as possible to increase the contrast. As the minute solvent (water) molecules collide with the fat globules of the milk, you can see the larger fat globules ricochet in an erratic manner (Brownian motion).

3.   Place the preparation on a warm hot plate for a few seconds. Observe again. How has the *rate* of

Brownian motion changed? _____

_____

_____

What would you conclude about the effect of increased temperature on the kinetic energy of molecules?

_____

_____

_____

Diffusion

If a **concentration gradient** (difference in concentration) exists, molecules eventually become evenly distributed through random molecular motion (Figure 5.1). **Diffusion** is the movement of molecules from a region of their high concentration to a region of their low concentration; the driving force is the kinetic energy of the molecules themselves.

The diffusion of particles into and out of cells is modified by the plasma membrane, which constitutes a physical barrier. Molecules diffuse passively through the plasma membrane if they are small enough to pass through its pores or if they can dissolve in the lipid portion of the membrane (as in the case of CO_2 and O_2). The diffusion of solutes (particles dissolved in water) through a semipermeable membrane is called **dialysis;** the diffusion of water through a semipermeable membrane is called **osmosis.** Both dialysis and osmosis, examples of diffusion phenomena, involve the movement of a substance from an area of its high concentration to one of its low concentration.

DIFFUSION THROUGH AN AGAR GEL   The relationship between molecular weight and the rate of diffusion can be examined simply by observing the diffusion of the molecules of two different types of dye through an agar gel. The dyes used in this experiment are methylene blue, which has a molecular weight of 320 and is deep blue in color, and potassium permanganate, which has a molecular weight of 158 and is deep purple in color. Although the agar gel appears quite solid, it is primarily (98.5%) water and allows free movement of the diffusing dye molecules through it.

 1.   Obtain a petri dish containing agar gel, a forceps, a millimeter ruler, methylene blue crystals, and potassium permanganate crystals, and bring them to your bench.

2.   Using the forceps, select approximately equal-size crystals of each dye, and place them gently on the agar gel surface, approximately 10 centimeters apart (Figure 5.2).

3.   At 15-minute intervals, use the millimeter ruler to measure the distance the dye has diffused from each crystal. These observations should be continued for 1½ hours, and the results recorded on page 31.

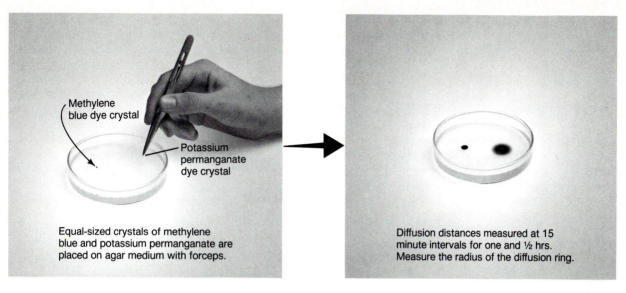

Figure 5.2

Setup for comparing the diffusion rates of molecules of methylene blue and potassium permanganate.

Time (min)	Diffusion of Methylene Blue (mm)	Diffusion of Potassium Permanganate (mm)
15		
30		
45		
60		
75		
90		

Which dye diffused more rapidly? _____

What is the relationship between molecular weight

and rate of molecular movement (diffusion)? _____

_____

Why did the dye molecules move? _____

_____

DIFFUSION THROUGH NONLIVING MEM-BRANES   A diffusion experiment providing information on the passage of water and solutes through semipermeable membranes, which may be applied to the study of transport mechanisms in living membrane-bound cells, is outlined next.

1.   Obtain four dialysis sacs* and four beakers (250 ml). Number the beakers 1 to 4 with a wax marker, and half fill all of them with distilled water except for beaker 2, to which you should add 40% glucose solution.

2.   Prepare the dialysis sacs one at a time by half filling each with 20 ml of the specified liquid, pressing out the air, and tying the neck with fine twine. Before proceeding to the next sac, quickly and carefully blot each sac dry and weigh it with a laboratory balance. Record the weight, and then drop each sac into the corresponding beaker. Be sure the sac is completely covered by the beaker solution, adding more solution if necessary.

● Sac 1: 40% glucose solution. Weight _____ g

● Sac 2: 40% glucose solution. Weight: _____ g

● Sac 3: 10% NaCl solution. Weight: _____ g

● Sac 4: boiled starch solution. Weight: _____ g

Allow sacs to remain undisturbed in the beakers for 1 hour. (Use this time to continue with the rest of the exercise.)

*Dialysis sacs are selectively permeable membranes with pores of a particular size. The selectivity of living membranes depends on more than just pore size, but using the dialysis sacs will allow you to examine selectivity due to this factor.

3. After an hour, get a beaker of water boiling on the hot plate and then quickly blot sac 1 dry and weigh:

_____ g

Has there been any change in weight? _____

Conclusions? _____

_____

_____

Place 5 ml of Benedict's solution in each of two test tubes. Put 4 or 5 drops of the beaker fluid into one test tube and 4 or 5 drops of the sac fluid into the other. Mark the tubes for identification and then place them in a beaker containing boiling water. Boil 2 minutes. Cool slowly. If a green, yellow, or rusty red precipitate forms, the test is positive, meaning that glucose is present. If the solution remains the original blue color, the test is negative.

Was glucose still present in the sac? _____

Was glucose present in the beaker? _____

Conclusions? _____

_____

_____

4. Blot and weigh sac 2: _____ g

Was there an *increase* or *decrease* in weight? _____

With 40% glucose in the sac and 40% glucose in the beaker, would you expect to see any net movements of water (osmosis) or of glucose molecules (dialysis)?

_____

Why or why not? _____

_____

_____

Return the glucose solution in beaker 2 to the *stock* supply bottle before continuing.

5. Blot and weigh sac 3: _____ g

Was there any change in weight? _____

Conclusions? _____

_____

Take a 5 ml sample of beaker 3 solution and put it in a test tube. Add a drop of silver nitrate. The appearance of a white precipitate or cloudiness indicates the presence of AgCl, which is formed by the reaction of $AgNO_3$ with NaCl (sodium chloride).

Results? _____

_____

Conclusions? _____

_____

6. Blot and weigh sac 4: _____ g

Was there any change in weight? _____

Conclusions? _____

_____

Take a 5 ml sample of beaker 4 solution and add a couple of drops of Lugol's iodine solution. The appearance of a black color is a positive test for the presence of starch. Did any starch diffuse from the

sac into the beaker? _____ Explain: ___

_____

_____

7. In which of the test situations did net osmosis

occur? _____

In which of the test situations did net dialysis oc-

cur? _____

What conclusions can you make about the relative size of glucose, starch, NaCl, and water molecules?

_____

_____

_____

_____

With what cell structure can the dialysis sac be com-

pared? _____

_____

DIFFUSION THROUGH LIVING MEMBRANES

To examine permeability properties of cell membranes, conduct the following two experiments.

Experiment 1:   1.   Obtain a clean slide and cover slip, lancets, an alcohol swab, physiologic saline, 1.5% sodium chloride solution, test tubes (3), test tube rack, glass stirring rod, and 15 ml graduated cylinder.

If animal blood is to be used, also obtain a vial of animal blood and a medicine dropper.

2.   (a) Label 3 test tubes A, B, and C, and prepare them as follows:

> A: add 2 ml 1.5% sodium chloride solution
> B: add 2 ml distilled water
> C: add 2 ml physiologic saline

(b) If you are using your own blood, follow step (b)1; if you are using animal blood provided by your instructor, follow the directions in step (b)2.

(b)1 Clean a fingertip with an alcohol swab, puncture it with a lancet, and add five drops of blood to each test tube. Stir each test tube with the glass rod, rinsing between each sample.

(b)2 Using a medicine dropper, add 5 drops of animal blood to each test tube.

(c) Hold each test tube in front of this printed page. *Record* the clarity of print seen through the fluid in each tube, and then proceed to step 3.

Test tube A _____

Test tube B _____

Test tube C _____

3.   (a) Place a very small drop of physiologic saline on a slide. Clean your fingertip once again with an alcohol swab, puncture it with a lancet, and touch a small drop of blood to the saline on the slide *or* add a small drop of animal blood to the saline on the slide. Tilt the slide to mix, cover with a cover slip, and immediately examine the preparation under the h.p. lens. Notice the smooth disklike shape of the red blood cells. Compare to Figure 5.3 (a).

(b) Now add a drop of 1.5% sodium chloride solution to the slide so that it touches one edge of the cover slip, and then carefully observe the red blood cells. What begins to happen to the normally smooth disk shape of the red blood cells as the saline solution diffuses under the cover slip?

_____

_____

This crinkling-up process, called **crenation,** is due to the fact that the 1.5% sodium chloride solution is slightly hypertonic to the cell sap of the red blood cell. A **hypertonic** solution contains more solutes (thus less water) than are present in the cell. Under these circumstances, water tends to leave the cells by osmosis. Compare your observations to Figure 5.3(b).

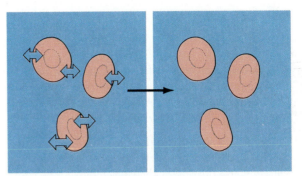

(a) Cells in isotonic solution

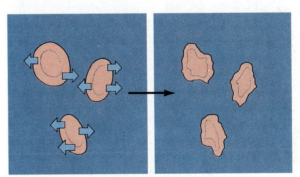

(b) Cells in hypertonic solution

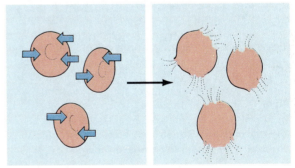

(c) Cells in hypotonic solution

Figure 5.3

Influence of hypertonic and hypotonic solutions on red blood cells. (a) Red blood cells suspended in an isotonic solution, where the cells retain their normal size and shape. (b) Red blood cells suspended in a hypertonic solution. As the cells lose water to the external environment, they shrink and become prickly; this phenomenon is called crenation. (c) Red blood cells suspended in hypotonic solution. Notice their spherical bloated shape, a result of excessive water intake.

(c) Add a drop of distilled water to the edge of the cover slip. Place a piece of filter paper at the opposite edge of the cover slip; it will absorb the saline solution and draw the distilled water across the cells. Watch the red blood cells as they float across the field. Describe the change in their appearance. _____

_____

_____

Distilled water contains *no* solutes (therefore, it is 100% water). Distilled water and *very* dilute solutions (that is, those containing less than 0.9% solutes) have relatively more water than is found inside a living cell. Such solutions are said to be **hypotonic** to the cell. (**Isotonic** solutions have the same solute-solvent ratio as do the cells and therefore cause no visible changes in the cells. See Figure 5.3a.) In a hypotonic solution, the red blood cells first "plump up" (Figure 5.3c) but then they suddenly start to disappear. The red blood cells burst as the water floods into them, leaving "ghosts" in their wake. This phenomenon is called **hemolysis.**

How do your observations of test tube B (step 2) correlate with what you have just observed under the microscope? _____

_____

_____

Dispose of your slides and test tube in the bleach-containing basin. Put your lancets (if used) into the disposable autoclave bag.

Experiment 2:   Make a wet mount of yeast suspension; observe the cells under high power to determine their normal color and structure. Prepare two test tubes by placing 1 ml of the yeast suspension in each. Boil one of the tubes in a water bath for 15 seconds. Remove and add eight drops of Congo red dye to both the boiled and unboiled preparations. Prepare a wet mount from each tube.

Was the dye accepted by the unboiled cells?

_____ By the boiled cells? _____

What are your conclusions about the selectivity of living and nonliving (boiled) cell membranes?

_____

_____

_____

Filtration

Filtration is a physical process by which water and solutes pass through a membrane from an area of higher hydrostatic (fluid) pressure into an area of lower hydrostatic pressure. Like diffusion, it is a passive process. For example, fluids and solutes filter out of the capillaries in the kidneys into the kidney tubules because the blood pressure in the capillaries is greater than the fluid pressure in the tubules. Filtration is not a selective process. The amount of filtrate (fluids and solutes) formed depends almost entirely on the difference in pressure on the two sides of the membrane and on the size of the membrane pores.

1.   Obtain the following equipment: a ring stand, ring, and ring clamp; a piece of filter paper; a beaker; and a solution of uncooked starch, powdered charcoal, and copper sulfate. Attach the ring to the ring stand with the clamp.

2.   Fold the filter paper in half twice, open it onto a cone, and place it in a funnel. Place the funnel in the ring of the ring stand and place a beaker under the funnel. Shake the starch solution, and fill the funnel with it to just below the top of the filter paper. How many drops of filtrate form in the first 10 seconds?

_____ drops

When the funnel is half empty, again count the number of drops formed in 10 seconds and record the count.

_____ drops

3.   After all the fluid has passed through the filter, check the filtrate and paper to see which materials were retained by the paper. (Note: If the filtrate is blue, the copper sulfate passed. Check both the paper and filtrate for black particles to see if the charcoal passed. Finally, add Lugol's iodine to a 2 cc filtrate sample in a test tube. If the sample turns blue/black with the addition of the iodine, starch is present in the filtrate.)

Passed: _____

Retained: _____

What does the filter paper represent? _____

During which counting interval was the filtration rate

greatest? _____

Explain: _____

_____

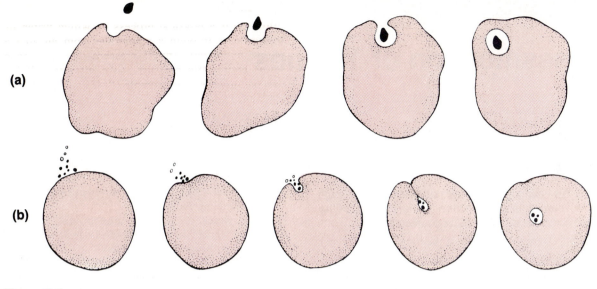

Figure 5.4

Phagocytosis and pinocytosis. (a) In phagocytosis, extensions of the plasma membrane (pseudopodia) flow around the external particle and enclose it within a vacuole. (b) In pinocytosis, dissolved proteins gather on the surface of the cell membrane, causing the membrane to invaginate and to incorporate the protein.

_____

_____

What characteristic of the three solutes determined whether or not they passed through the filter paper?

_____

_____

_____

_____

ACTIVE TRANSPORT

Whenever a cell expends energy to move substances across its boundaries, the process is referred to as an active transport process (ATP). The substances moved by active means are generally unable to pass by diffusion. They may be too large to pass through the pores; they may not be lipid-soluble; or they may have to move against rather than with a concentration gradient.

In one type of active transport, substances move across the membrane by combining with a protein carrier molecule; the process resembles an enzyme–substrate interaction. ATP is required, and in many cases the substances move against concentration or electro-chemical gradients or both. Some of the substances that are moved into the cells by such carriers are amino acids and some sugars; neither is lipid-

soluble, and both are too large to pass through the pores but necessary for cell life. On the other hand, sodium ions (Na^+) are moved out of cells by active transport. There is more Na^+ outside the cell than there is inside, so the Na^+ tends to remain in the cell unless actively transported out. Pinocytosis and phagocytosis also require ATP.

In **pinocytosis** (cell drinking), the cell membrane seems to sink beneath the material to form a small vesicle, which then pinches off into the cell interior (see Figure 5.4). Pinocytosis is most common for taking in liquids containing protein or fat.

In **phagocytosis** (cell eating), parts of the plasma membrane flow around a relatively large or solid material (for example, bacteria, cell debris) and engulf it, enclosing it within a sac. This is a rather uncommon phenomenon in the human body except for certain phagocytic or scavenger cells, such as some white blood cells and macrophages (Figure 5.4).

 1. Obtain a drop of starved _Amoeba proteus culture_ and place it in the well of a depression slide.

2. Locate an ameba under low power and then add a drop of _Tetrahymena pyriformis_ culture (an ameba "meal") to the well. Keep the light as dim as possible, otherwise the ameba will "ball up" and begin to disintegrate.

3. Observe as the ameba phagocytizes the _Tetrahymena_ by forming pseudopods that engulf it. In unicellular organisms like the ameba, phagocytosis is an important food-getting mechanism, but in higher organisms, it is more important as a protective device.

Classification of Tissues

OBJECTIVES

1. To name the four major types of tissues in the human body and the major subcategories of each.
2. To identify the tissue subcategories through microscopic inspection or inspection of an appropriate diagram or projected slide.
3. To state the location of the various tissue types in the body.
4. To state the general functions and structural characteristics of each of the four major tissue types.

MATERIALS

Compound microscope
Prepared slides of simple squamous, simple cuboidal, simple columnar, stratified squamous (nonkeratinizing), pseudostratified ciliated, and transitional epithelium
Prepared slides of adipose, areolar, and dense fibrous connective tissue (tendon); hyaline cartilage; and bone (cross section)
Prepared slides of skeletal, cardiac, and smooth muscle (longitudinal sections)
Prepared slide of nervous tissue (spinal cord smear)

Exercise 4 describes cells as the building blocks of life and the all-inclusive functional units of unicellular organisms like amebae. But in higher organisms cells do not ordinarily operate as isolated, independent entities. In humans and other multicellular organisms, cells depend on one another and cooperate to produce homeostasis in the body.

The most complex animal starts out as a single cell, the fertilized egg, which divides almost endlessly. The trillions of cells that result become specialized for a particular function; some become supportive bone, others the transparent lens of the eye, still others skin cells, and so on. Thus a division of labor exists, with certain groups of cells highly specialized to perform functions that benefit the organism as a whole.

Cell specialization brings about great sophistication of achievement but carries with it certain hazards. When a small specific group of cells is indispensable, any inability to function on its part can paralyze or destroy the entire body. For example, the action of the heart depends on a highly differentiated group of cells in the heart muscle. If they cease functioning, the heart no longer operates efficiently and the whole body suffers or dies from lack of oxygen. The jack-of-all-trades ameba faces no such danger.

Groups of cells that are similar in structure and function are called **tissues.** The four primary tissue types—epithelium, connective tissue, nervous tissue, and muscle—have distinctive structures, patterns, and functions. The four primary tissues are further divided into subcategories.

To perform specific body functions, the tissues are organized into such **organs** as the heart, kidneys, and lungs. Most organs contain several representatives of the primary tissues, and the arrangement of these tissues determines the organ's structure and function. Thus **histology,** the study of tissues, complements a study of gross anatomy and provides the structural basis for a study of organ physiology.

The main objective of this exercise is to familiarize you with the major similarities and dissimilarities of the primary tissues and to teach you to recognize the common tissues you will encounter in units to follow. Because epithelium and some types of connective tissue will not be considered again, they are emphasized more than muscle, nervous tissue, and bone (a connective tissue), which are covered in more depth in later exercises.

EPITHELIAL TISSUE

Epithelial tissue, or **epithelium,** covers surfaces, and since glands almost invariably develop from epithelial membranes, they too are logically classed as epithelium. Epithelium covers the external body surface (as the epidermis), lines its cavities and tubules, and composes the various endocrine (hormone-producing) and exocrine glands of the body.

Epithelial functions include protection, absorption, filtration, and secretion. For example, the epithelium covering the body protects against bacterial and chemical damage; that lining the respiratory tract is ciliated to sweep debris-laden mucus away from the lungs. Epithelium specialized to absorb substances lines the stomach and small intestine. In the kidney tubules, the epithelium both absorbs and filters. Secretion is a specialty of the glands.

Epithelium generally exhibits the following characteristics:

- Cells fit closely together to form membranes, or sheets of cells.
- The membranes always have one free surface.

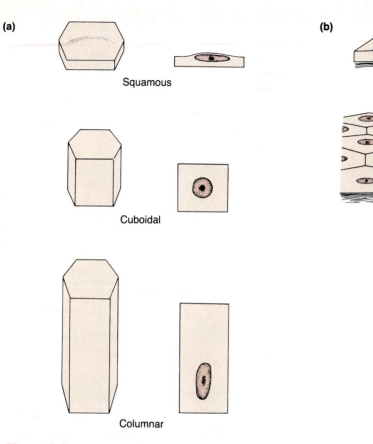

(a)

Squamous

Cuboidal

Columnar

(b)

Simple

Stratified

Figure 6.1

Classification of epithelia. (a) Classification on the basis of cell shape. For each category, a whole cell is shown on the left and a longitudinal section is shown on the right. (b) Classification on the basis of number of cell layers.

● The cells are attached to an adhesive **basement membrane,** a structureless material secreted cooperatively by the adjacent epithelial and connective tissue cells.

● Epithelial tissues have no blood supply of their own (are avascular) but depend on capillaries in the underlying connective tissue for a supply of food and oxygen.

● If well nourished, epithelial cells can easily regenerate themselves.

The covering and lining epithelia are classified according to two criteria—cell shape and arrangement (Figure 6.1). **Squamous** (flattened), **cuboidal** (cubelike), and **columnar** (column-shaped) epithelial cells are the general types based on shape. On the basis of arrangement, there are **simple** epithelia, consisting of one layer of cells attached to the basement membrane, and **stratified** epithelia, consisting of more than one layer of cells. The terms denoting shape and arrangement of the epithelial cells are combined to describe the epithelium fully. *Stratified epithelia are named according to the cells at the surface of the epithelial membrane,* not those resting on the basement membrane.

There are, in addition, two less easily categorized types of epithelia. **Pseudostratified columnar** epithelium is actually a simple epithelium (one layer of cells), but because its cells extend varied distances from the basement membrane, it gives the false appearance of being stratified. **Transitional** epithelium is a rather peculiar stratified squamous epithelium formed of rounded, or "plump," cells with the ability to slide over one another to allow the organ to be stretched. Transitional epithelium is found in urinary system organs subjected to periodic distention, such as the bladder. The superficial cells are flattened (like true squamous cells) when the organ is distended and rounded when the organ is empty.

The most common types of epithelia, their most common locations in the body, and their functions are described in Figure 6.3.

Epithelial cells forming glands are highly specialized to remove materials from the blood and to manufacture them into new materials, which they then secrete. There are two types of glands, as shown in Figure 6.2. The **endocrine glands** lose their surface connection (duct) as they develop; thus they are referred to as ductless glands. Their secretions (all hormones) are extruded directly into the blood or the

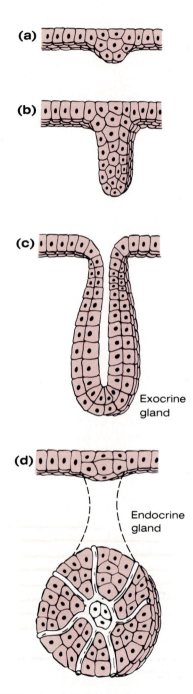

(a)

(b)

(c)

Exocrine gland

(d)

Endocrine gland

Figure 6.2

Formation of endocrine and exocrine glands from epithelial sheets. (a) Epithelial cells grow and push into the underlying tissue. (b) A cord of epithelial cells forms. (c) In an exocrine gland, a lumen (cavity) forms. The inner cells form the duct, the outer cells produce the secretion. (d) In the formation of an endocrine gland, the connecting cells forming the duct atrophy, leaving the secretory cells with no connection to the epithelial surface. However, they do become heavily invested with blood and lymphatic vessels that receive the secretions.

lymphatic vessels that weave through the glands. The **exocrine glands** retain their ducts, and their secretions empty through these ducts to an epithelial surface. The exocrine glands—including the sweat and oil glands, liver, and pancreas—are both external and internal; they will be discussed in conjunction with the organ systems to which their products are functionally related.

 Obtain slides of simple squamous, simple cuboidal, simple columnar, stratified squamous (nonkeratinizing), pseudostratified ciliated, and transitional epithelia. Examine each carefully, and compare your observations with the photomicrographs in Figure 6.3. Note any modifications for specific functions, such as cilia (motile cell projections that help to move substances along the cell surface), microvilli, which increase the surface area for absorption, and/or goblet cells, which secrete lubricating mucus.

CONNECTIVE TISSUE

Connective tissue is found in all parts of the body as discrete structures or as part of various body organs. It is the most abundant and widely distributed of the tissue types.

The connective tissues perform a variety of functions, but they primarily protect, support, and bind together other tissues of the body. For example, bones are composed of connective tissue (**osseous** tissue), and they protect and support other body tissues and organs. The ligaments and tendons (**dense connective tissue**) bind the bones together or bind skeletal muscles to bones.

Areolar connective tissue is a soft packaging material that cushions and protects body organs. Fat (**adipose**) tissue provides insulation for the body tissues and a source of stored food. Blood-forming (**hemopoietic**) tissue replenishes the body's supply of red blood cells. In addition, connective tissue also serves a vital function in the repair of all body tissues since many wounds are repaired by connective tissue in the form of scar tissue.

The characteristics of connective tissue include the following:

- With a few exceptions (cartilages, tendons, and ligaments), connective tissues are well vascularized.
- Connective tissues are composed of many types of cells.
- There is a great deal of nonliving material (matrix) between the cells of connective tissue.

The nonliving material between the cells—the **extracellular matrix**—deserves a bit more explanation. It is produced by the cells and then extruded. It may be liquid, semisolid, gellike, or very hard. The matrix is primarily responsible for the strength

(*Text continues on page 42.*)

(a) Simple squamous epithelium

Description: Single layer of flattened cells with disk-shaped central nuclei and sparse cytoplasm; the simplest of the epithelia.

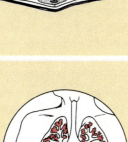

Location: Air sacs of lungs; kidney glomeruli; lining of heart, blood vessels, and lymphatic vessels; lining of ventral body cavity (serosae).

Function: Allows passage of materials by diffusion and filtration in sites where protection is not important; secretes lubricating substances in serosae.

Photomicrograph: Simple squamous epithelium forming walls of alveoli (air sacs) of the lung (280×).

Nucleus

Simple squamous epithelial cell

(b) Simple cuboidal epithelium

Description: Single layer of cubelike cells with large, spherical central nuclei; may have microvilli.

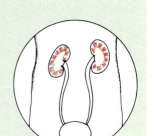

Location: Kidney tubules; ducts and secretory portions of small glands; ovary surface.

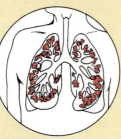

Function: Secretion and absorption.

Photomicrograph: Simple cuboidal epithelium in kidney tubules (260X)

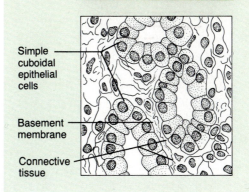

Simple cuboidal epithelial cells

Basement membrane

Connective tissue

Figure 6.3

Epithelial tissues.

(c) Simple columnar epithelium

Description: Single layer of tall cells with *oval* nuclei; some cells bear cilia; layer may contain mucus-secreting glands (goblet cells).

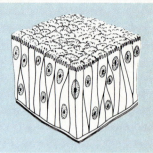

Location: Nonciliated type lines most of the digestive tract (stomach to anal canal), gallbladder and excretory ducts of some glands; ciliated variety lines small bronchi, uterine tubes, and some regions of the uterus.

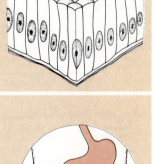

Function: Absorption; secretion of mucus, enzymes, and other substances; ciliated type propels mucus (or reproductive cells) by ciliary action.

Photomicrograph: Simple columnar epithelium of the stomach mucosa (280×).

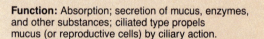

Connective tissue

Simple columnar epithelial cell

Basement membrane

(d) Pseudostratified epithelium

Description: Single layer of cells of differing heights, some not reaching the free surface; nuclei seen at different levels; may contain goblet cells and bear cilia.

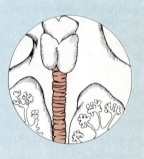

Location: Nonciliated type in ducts of large glands, parts of male urethra; ciliated variety lines the trachea, most of the upper respiratory tract.

Function: Secretion, particularly of mucus; propulsion of mucus by ciliary action.

Photomicrograph: Pseudostratified ciliated columnar epithelium lining the human trachea (430×).

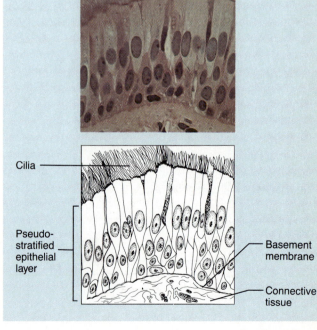

Cilia

Pseudo-stratified epithelial layer

Basement membrane

Connective tissue

Figure 6.3 *(continued)*

(e) Stratified squamous epithelium

Description: Thick membrane composed of several cell layers; basal cells are cuboidal or columnar and metabolically active; surface cells are flattened (squamous); in the keratinized type, the surface cells are full of keratin and dead; basal cells are active in mitosis and produce the cells of the more superficial layers.

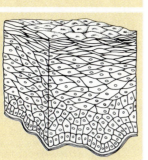

Location: Nonkeratinized type forms the moist linings of the esophagus, mouth, and vagina; keratinized variety forms the epidermis of the skin, a dry membrane.

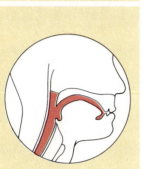

Function: Protects underlying tissues in areas subjected to abrasion.

Photomicrograph: Stratified squamous epithelium lining of the esophagus (173×).

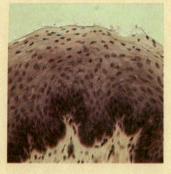

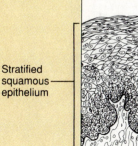

Stratified squamous epithelium — Nuclei — Basement membrane — Connective tissue

(f) Transitional epithelium

Description: Resembles both stratified squamous and stratified cuboidal; basal cells cuboidal or columnar; surfaces cells dome-shaped or squamous-like, depending on degree of organ stretch.

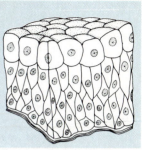

Location: Lines the ureters, bladder, and part of the urethra.

Function: Stretches readily and permits distension of urinary organ by contained urine.

Photomicrograph: Transitional epithelium lining of the bladder, relaxed state (170×); note the bulbous, or rounded, appearance of the cells at the surface; these cells flatten and become elongated when the bladder is filled with urine.

Basement membrane — Connective tissue — Transitional epithelium

Figure 6.3 (*continued*)

associated with connective tissue, but there is variation. At one extreme, hemopoietic and adipose tissues are composed mostly of cells. At the opposite extreme, bone and cartilage have very few cells and large amounts of matrix. When the matrix is firm, as in cartilage and bone, the connective tissue cells reside in cavities in the matrix called *lacunae*.

Fibers of various types and amounts are deposited in and form a part of the matrix material. They include **collagenic** (white) fibers, **elastic** (yellow) fibers, and **reticular** (fine collagenic) fibers.

There are four main types of connective tissue, all of which typically have large amounts of matrix—these are connective tissue proper (which includes areolar, adipose, and dense fibrous connective tissues), cartilage, bone, and blood. Figure 6.4 lists the general characteristics, location, and function of some of the connective tissues found in the body.

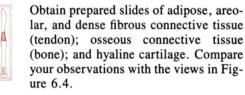

Obtain prepared slides of adipose, areolar, and dense fibrous connective tissue (tendon); osseous connective tissue (bone); and hyaline cartilage. Compare your observations with the views in Figure 6.4.

While examining the areolar connective tissue, notice how much "empty space" there appears to be and distinguish between the collagen fibers and the coiled elastic fibers. Also, try to locate a **mast cell,** which has large darkly staining granules in its cytoplasm. This cell type releases histamine that makes capillaries quite permeable during inflammatory reactions and allergies and thus is partially responsible for that "runny nose" of allergies.

In adipose tissue, locate a cell in which the nucleus can be seen pushed to one side by the large fat-filled vacuole (signet ring cell) and notice how little matrix there is in fat or adipose tissue.

Distinguish between the living cells and the matrix in the dense fibrous, bone, and hyaline cartilage preparations.

extracellular matrix
- fibers
- ground substance

3 fibers
collagenic (white)
elastic (yellow)
reticular (fine)

Connective Tissue Proper
(Supportive connective tissue (bone, cartilage)
→ fibroblasts
macrophage } hard to
mast cell } identify
fat cell
→ 1. loose c.t. (areolar, adipose, reticular)
2. dense c.t. (regular, irregular)

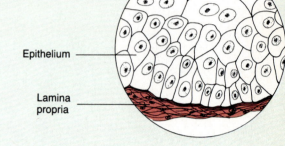

(a) Areolar connective tissue

Description: Gel-like matrix with all three fiber types; cells: fibroblasts, macrophages, mast cells, and some white blood cells.

Location: Widely distributed under epithelia of body, e.g., forms lamina propria of mucous membranes; packages organs; surrounds capillaries.

Epithelium

Lamina propria

Function: Wraps and cushions organs; its macrophages phagocytize bacteria; plays important role in inflammation; holds and conveys tissue fluid.

Photomicrograph: Areolar connective tissue, a soft packaging tissue of the body (170×).

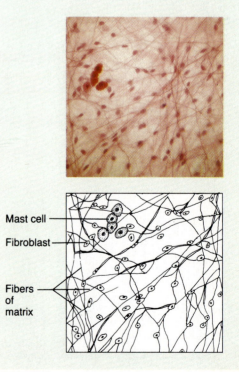

Mast cell

Fibroblast

Fibers of matrix

Figure 6.4

Connective tissues.

(b) Adipose tissue

Description: Matrix as in areolar, but very sparse; closely packed adipocytes, or fat cells, have nucleus pushed to the side by large fat droplet.

Location: Under skin; around kidneys and eyeballs; in bones and within abdomen; in breasts.

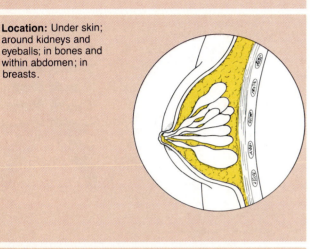

Function: Provides reserve food fuel; insulates against heat loss; supports and protects organs.

Photomicrograph: Adipose tissue from the subcutaneous layer under the skin (500×).

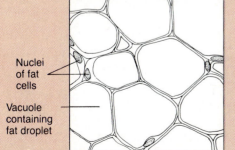

Nuclei of fat cells

Vacuole containing fat droplet

(c) Dense regular connective tissue

Description: Primarily parallel collagen fibers; a few elastin fibers; major cell type is the fibroblast.

Location: Tendons, most ligaments, aponeuroses.

Shoulder joint

Ligament

Tendon

Function: Attaches muscles to bones or to muscles; attaches bones to bones; withstands great tensile stress when pulling force is applied in one direction.

Photomicrograph: Dense regular connective tissue from a tendon (200×).

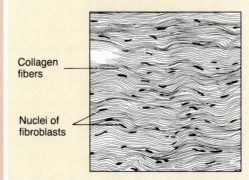

Collagen fibers

Nuclei of fibroblasts

Figure 6.4 (*continued*)

(d) Bone (osseous tissue)

Description: Hard, calcified matrix containing many collagen fibers; osteocytes lie in lacunae. Very well vascularized.

Location: Bones

Function: Bone supports and protects (by enclosing); provides levers for the muscles to act on; stores calcium and other minerals and fat; marrow inside bones is the site for blood cell formation (hematopoiesis).

Photomicrograph: Cross-sectional view of bone (100×).

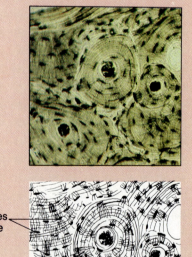

Osteocytes in lacunae

(e) Hyaline cartilage

Description: Amorphous but firm matrix; collagen fibers form an imperceptible network; chondroblasts produce the matrix and when mature (chondrocytes) lie in lacunae.

Location: Forms most of the embryonic skeleton; covers the ends of long bones in joint cavities; forms costal cartilages of the ribs; cartilages of the nose, trachea, and larynx.

Costal cartilages

Function: Supports and reinforces; has resilient cushioning properties; resists compressive stress.

Photomicrograph: Hyaline cartilage from the trachea (475×).

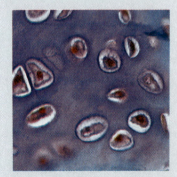

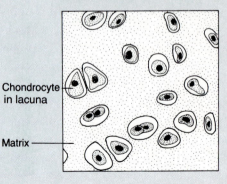

Chondrocyte in lacuna

Matrix

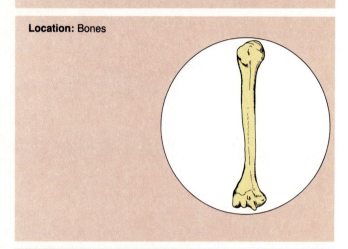

Figure 6.4 *(continued)*

MUSCLE TISSUE

Muscle tissue is highly specialized to contract (shorten) in order to produce movement of some body parts. As you might expect, muscle cells tend to be quite elongated, providing a long axis for contraction. The three basic types of muscle tissue are described briefly here; skeletal muscle is treated more completely in a later exercise.

Skeletal muscle, the "meat," or flesh, of the body, is attached to the skeleton. It is under voluntary control (consciously controlled), and its contraction moves the limbs and other external body parts. The cells of skeletal muscles are long, cylindrical, and multinucleate (more than one nucleus); they have obvious striations (stripes).

Cardiac muscle is found only in the heart. As it contracts, the heart acts as a pump, propelling the blood through the blood vessels. Structurally, cardiac muscle is quite similar to skeletal muscle (striations are also visible), but cardiac cells are branching cells that interdigitate (fit together) at tight junctions called **intercalated disks.** These structures allow the cardiac muscle to act as a unit. Cardiac muscle is under involuntary control, which means that we cannot voluntarily or consciously control the operation of the heart.

Smooth muscle, or visceral muscle, is found in the walls of hollow organs and blood vessels. Its contraction generally constricts or dilates the lumen (cavity) of the structure, thus propelling substances along predetermined pathways. Smooth muscle cells are quite different in appearance from those of skeletal or cardiac muscle. No striations are visible, and the uninucleate cells are spindle-shaped.

 Obtain and examine prepared slides of skeletal, cardiac, and smooth muscle. Note their similarities and dissimilarities in both your observations and in the illustrations in Figure 6.5.

Cartilage 3 types
① Hyaline most common lining of trachea
   chondrocytes in lacuna
   amorphous ground substance - not much fiber
   avascular - just cells

② Elastic - see elastic fibers

③ Fibrous cartilage
    lots of thick collagen fibers

Bone
  - ground substance solid
  - harversian system - concentric rings
  - osteocytes in lacuna
  - can alaculi

(a) Skeletal muscle

Description: Long, cylindrical, multinucleate cells; obvious striations.

Location: In skeletal muscles attached to bones or occasionally to skin.

Function: Voluntary movement; locomotion; manipulation of the environment; facial expression. Voluntary control.

Photomicrograph: Skeletal muscle (approx. 30x). Notice the obvious banding pattern and the fact that these large cells are multinucleate.

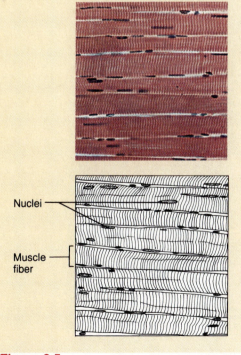

Nuclei

Muscle fiber

Figure 6.5

Muscle tissues.

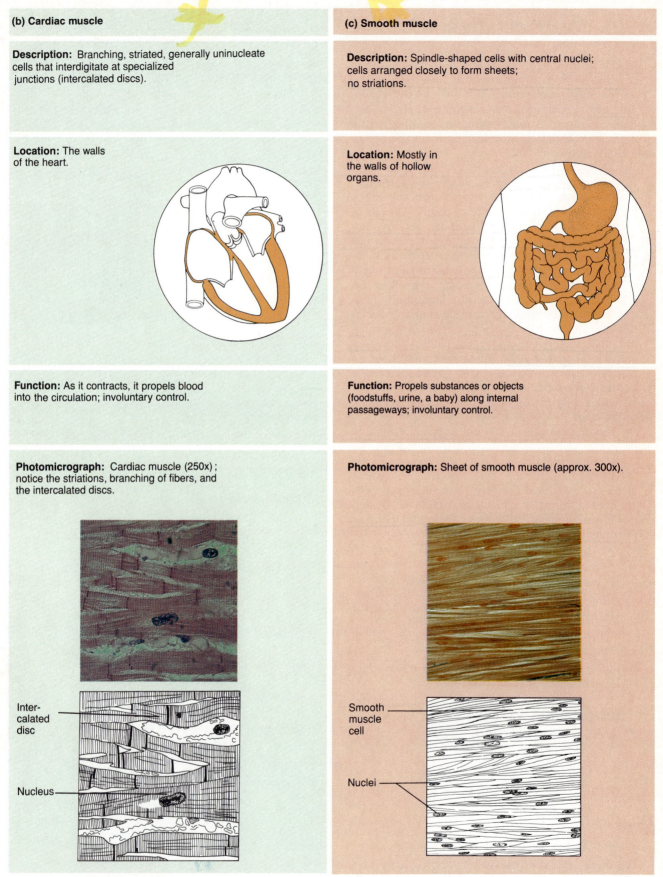

(b) Cardiac muscle

Description: Branching, striated, generally uninucleate cells that interdigitate at specialized junctions (intercalated discs).

Location: The walls of the heart.

Function: As it contracts, it propels blood into the circulation; involuntary control.

Photomicrograph: Cardiac muscle (250x); notice the striations, branching of fibers, and the intercalated discs.

Inter-calated disc

Nucleus

(c) Smooth muscle

Description: Spindle-shaped cells with central nuclei; cells arranged closely to form sheets; no striations.

Location: Mostly in the walls of hollow organs.

Function: Propels substances or objects (foodstuffs, urine, a baby) along internal passageways; involuntary control.

Photomicrograph: Sheet of smooth muscle (approx. 300x).

Smooth muscle cell

Nuclei

Figure 6.5 (continued)

Description: Neurons are branching cells; cell processes that may be quite long extend from the nucleus-containing cell body; also contributing to nervous tissue are nonirritable supporting cells (not illustrated).

Cell body

Neuron

Location: Brain, spinal cord, and nerves.

Function: Transmit electrical signals from sensory receptors and to effectors (muscles and glands) which control their activity.

Photomicrograph: Neuron (170×).

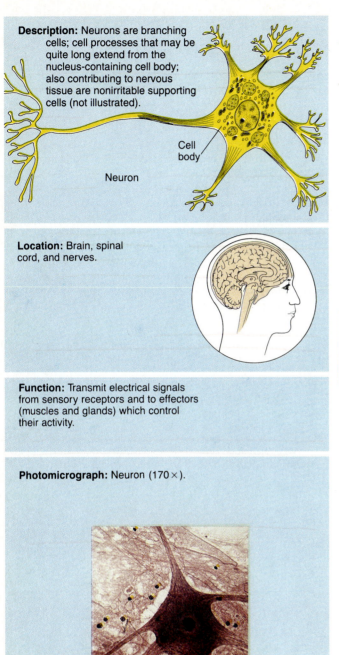

Nuclei of supporting cells

Cell body

Cell processes

Figure 6.6

Nervous tissue.

NERVOUS TISSUE

Nervous tissue is composed of two major cell populations. The neuroglia are special supporting cells that protect, support, and insulate the more delicate neurons. The **neurons** are highly specialized to receive stimuli (irritability) and to conduct waves of excitation, or impulses, to all parts of the body (conductivity). The structure of neurons is markedly different from that of all other body cells. They all have a nucleus-containing nerve cell body, and their cytoplasm is drawn out into long extensions (axons and dendrites) sometimes as long as 3 feet (about 1 m), which allows a single neuron to conduct an impulse over relatively long distances. More detail about the anatomy of the different classes of neurons and about the special supporting cells that also are a part of nervous tissue appears in Exercise 16.

 Obtain a prepared slide of a spinal cord smear. Locate a neuron and compare it to Figure 6.6. Keep the light dim—this will help you see the cellular extensions. The smaller cells surrounding the neurons are neuroglia cells.

Hemotoxylin + Eosin stains
cytoplasm red   nucleus purple

Simple Squamous E (Artery, vein, nerve)

Lumen

Simple Columnar

The Integumentary System

OBJECTIVES

1. To recount several important functions of the skin, or integumentary system.

2. To recognize and name from observation of an appropriate model, diagram, projected slide, or microscopic specimen the following skin structures: epidermis (and note relative positioning of the stratum basale and stratum corneum), dermis (papillary and reticular layers), hair follicles and hair, sebaceous glands, and sweat glands.

3. To name the four major layers of the epidermis and describe the characteristics of each.

4. To compare the properties of the epidermis to those of the dermis.

5. To describe the distribution and function of the skin derivatives—sebaceous glands, sweat glands, and hairs.

6. To differentiate between eccrine and apocrine sweat glands.

7. To enumerate the factors determining skin color.

8. To describe the function of melanin.

MATERIALS

Skin model (three-dimensional, if available)
Compound microscope
Prepared slide of human skin with hair follicles
Sheet of #20 bond paper ruled to mark off cm^2 areas
Scissors
Betadine swabs, or Lugol's iodine and cotton swabs
Adhesive tape

The **skin,** or **integument,** is often considered an organ system because of its extent and complexity; it is much more than an external body covering. Architecturally, the skin is a marvel; it is tough yet pliable. This characteristic enables it to withstand constant insult from outside agents.

The skin has many functions, most (but not all) concerned with protection. It insulates and cushions the underlying body tissues and protects the entire body from mechanical damage (bumps and cuts), chemical damage (acids, alkalis, and the like), thermal damage (heat), and bacterial invasion (by virtue of its acid mantle and continuous surface). The uppermost layer of the skin (the cornified layer) is hardened, preventing water loss from the body surface. The skin's abundant capillary network (under the control of the nervous system) plays an important role in regulating heat loss from the body surface.

The skin has other functions as well. For example, it acts as a miniexcretory system; urea, salts, and water are lost through the skin pores. The skin is also the site of vitamin D synthesis for the body. Finally, the cutaneous sense organs are located in the dermis.

BASIC STRUCTURE OF THE SKIN

Structurally, the skin consists of two kinds of tissue. The outer epidermis is made up of stratified squamous epithelium, and the underlying dermis is made up of dense irregular connective tissue. These layers are firmly cemented together. Friction, however, such as the rubbing of a poorly fitting shoe, may cause them to separate, resulting in a blister. Immediately under the dermis is the hypodermis or superficial fascia (primarily adipose tissue), which is not considered part of the skin. The main skin areas and structures are described below.

As you read, locate the following structures on Figure 7.1 and on a skin model.

Epidermis

The avascular epidermis consists of several layers of cells.

The uppermost layer is the **stratum corneum,** sometimes called the *horny layer* because it consists

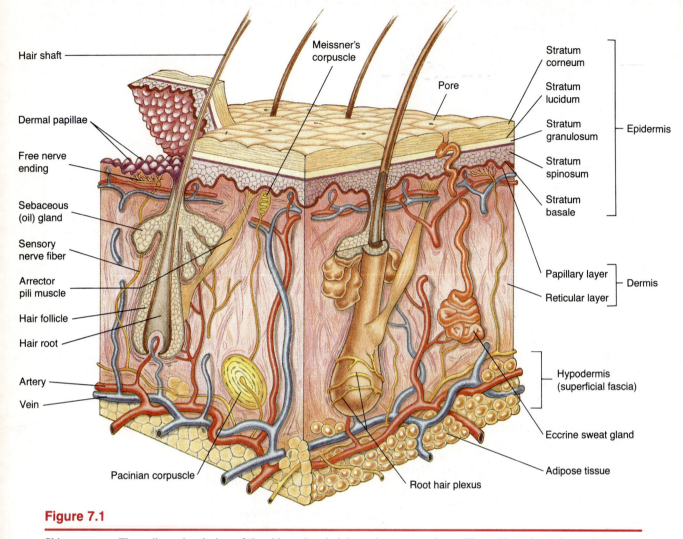

Hair shaft

Dermal papillae

Free nerve ending

Sebaceous (oil) gland

Sensory nerve fiber

Arrector pili muscle

Hair follicle

Hair root

Artery

Vein

Pacinian corpuscle

Meissner's corpuscle

Pore

Root hair plexus

Stratum corneum

Stratum lucidum

Stratum granulosum

Stratum spinosum

Stratum basale

Epidermis

Papillary layer

Reticular layer

Dermis

Hypodermis (superficial fascia)

Eccrine sweat gland

Adipose tissue

Figure 7.1

Skin structure. Three-dimensional view of the skin and underlying subcutaneous tissue. The epidermal and dermal layers have been pulled apart at the left front corner to reveal the dermal papillae.

of flattened keratinized cells. Keratin is a protein with waterproofing properties; thus this layer provides a natural "raincoat" for the body and prevents water loss from the deeper tissues. Keratinized cells are dead cells. They are constantly rubbing and flaking off and being replaced by the division of deeper cells.

Immediately below the stratum corneum is the **stratum lucidum,** the cells of which appear clear because of an accumulation of keratin fibrils.

The **stratum granulosum,** below the stratum lucidum, is the area in which the cells begin to die owing to their accumulation of granules of the keratin precursor keratohyalin and their increasing distance from the dermal blood supply.

The two deepest layers are the **stratum basale** immediately adjacent to the dermis and the more superficial **stratum spinosum.** These two strata thus contain the only epidermal cells that receive adequate nourishment (via diffusion of nutrients from the dermis). These cells (particularly those of the stratum basale) are constantly undergoing cell divi-

sion; millions of new cells are produced daily. As the daughter cells are pushed upward, away from the source of nutrition, they gradually die, and their soft protoplasm becomes increasingly keratinized.

Melanin, a brown pigment, is produced by special cells (melanocytes) found in the deeper layers of the epidermis. The skin tans because of an increase in melanin production when the skin is exposed to sunlight. The melanin provides a protective pigment "umbrella" over basal cell nuclei, thus shielding their genetic material (DNA) from the damaging effects of ultraviolet radiation. A concentration of melanin in one spot is commonly called a *freckle*.

Skin color is the result of two factors—the amount of melanin production and the degree of oxygenation of the blood. People with high levels of melanin production have brown-toned skin; in light-skinned people, who have less melanin, the dermal blood supply flushes through the rather transparent cell layers above, thus giving the skin a rosy glow. When the blood is poorly oxygenated or does not circulate well in the dermis, the skin takes on a bluish, or **cyanotic,** cast.

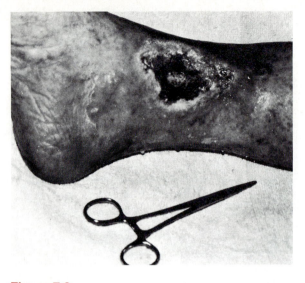

Figure 7.2

A decubitus ulcer on the ankle of a patient.

Skin color may be an important diagnostic tool. For example, flushed skin may indicate hypertension, fever, or embarrassment, whereas pale skin is typically seen in anemic individuals. When the blood is inadequately oxygenated, as during asphyxiation and serious lung disease, the skin becomes cyanotic as mentioned above. Jaundice, in which the tissues become yellowed, is almost always diagnostic for liver disease, whereas a bronzing of the skin hints that a person's adrenal cortex is hypoactive (*Addison's disease*). ■

Dermis

The connective tissue making up the dermis consists of two principal regions—the papillary and reticular areas—and like the epidermis, it varies in thickness. For example, the skin is particularly thick on the palms of the hands and soles of the feet and is quite thin on the eyelids.

The **papillary layer** is the upper dermal region. It is very uneven and has conelike projections from its superior surface, the **dermal papillae,** which attach it to the epidermis above. These projections are reflected in fingerprints, which are unique patterns of ridges that remain unchanged throughout life. Abundant capillary networks in the papillary layer furnish nutrients for the epidermal layers and allow heat to radiate to the skin surface. The touch receptors are also found here.

The **reticular layer** is the deepest skin layer. It contains many arteries and veins, sweat and sebaceous glands, and pressure receptors.

Both the papillary and reticular layers are heavily invested with collagenic and elastic fibers. The latter give skin its exceptional elasticity in youth. In old age, the number of elastic fibers decreases and the subcutaneous layers lose fat, which leads to wrinkling and inelasticity of the skin. Fibroblasts, adipose cells, various types of macrophages (which are important in the body's defense), and other cell types are found throughout the dermis.

The dermis has an abundant blood supply, which allows it to play a role in the regulation of body temperature. When body temperature is high, the arterioles dilate, and the capillary network of the dermis becomes engorged with the heated blood. Thus body heat is allowed to radiate from the skin surface. If the environment is cool and body heat must be conserved, the arterioles constrict so that blood bypasses the capillary networks.

Any restriction of the normal blood supply to the skin results in cell death and, if severe enough, skin ulcers (Figure 7.2). *Bedsores* (*decubitus ulcers*) occur in bedridden patients who are not turned regularly enough. The weight of the body exerts pressure on the skin, especially over bony projections, which leads to restriction of the blood supply and death of tissue. ■

The dermis is also richly provided with lymphatic vessels and a nerve supply. Many of the nerve endings bear highly specialized receptor organs that, when stimulated by environmental changes, transmit messages to the central nervous system for interpretation. Some of these receptors are shown in Figure 7.1.

APPENDAGES OF THE SKIN

The appendages of the skin—hair, nails, and cutaneous glands—are all derivatives of the epidermis, but they reside in the dermis. They originate from the stratum germinativum and grow downward into the deeper skin regions.

Cutaneous Glands

The cutaneous glands fall primarily into two categories: the sebaceous glands and the sudoriferous glands. The **sebaceous glands** are found nearly all over the skin, with the exception of the palms of the hands and the soles of the feet. Their ducts usually empty into a hair follicle, but some open directly onto the skin surface.

The product of the sebaceous glands, called **sebum,** is a mixture of oily substances and fragmented cells. The sebum is a lubricant that keeps the skin soft and moist (a natural skin cream) and keeps the hair from becoming brittle. The sebaceous glands become particularly active during puberty; thus, the skin tends to become oilier during this period of life. *Blackheads* are accumulations of dried sebum and bacteria; *acne* is due to active infection of the sebaceous glands.

Epithelial openings, called pores, are the outlets for the **sweat glands** (sudoriferous glands). These exocrine glands are widely distributed in the skin.

Sudoriferous glands are subcategorized on the basis of the composition of their secretions. The **eccrine glands,** which are distributed all over the body, produce clear perspiration, consisting primarily of water, salts (NaCl), and urea. The **apocrine glands,** found predominantly in the axillary and genital areas, secrete a milky protein-based substance (also containing water, salts, and urea) that is an ideal nutrient medium for microorganisms generally found on the skin.

The sweat glands, under the control of the nervous system, are an important part of the body's heat-regulating apparatus. They secrete perspiration when the external temperature or body temperature is high. When this water-based substance evaporates, it carries excess body heat with it. Thus perspiration acts as an emergency system of heat liberation when the capillary cooling system is not sufficient or is unable to maintain homeostasis.

Hair

Hairs are found over the entire body surface, with the exception of the palms of the hands, the soles of the feet, and the lips. A hair, enclosed in a hair **follicle,** is also an epithelial structure. The portion of the hair enclosed within the follicle is called the **root;** the portion projecting from the scalp surface is called the **shaft.** The hair is formed by mitosis of the well-nourished germinal epithelial cells at the basal end of the follicle (the hair **bulb**). As the daughter cells are pushed farther away from the growing region, they die and become keratinized; thus the bulk of the hair shaft, like the bulk of the epidermis, is dead material.

A hair consists of a central region (medulla) surrounded first by a protective cortex and then by the cuticle. Abrasion of the cuticle results in "split ends." Hair color is a manifestation of the amount of pigment (generally melanin) within the hair cortex.

If you look carefully at the structure of the hair follicle (see Figure 7.1), you will see that it generally is slanted. Small bands of smooth muscle cells—**arrector pili**—connect each side of the hair follicle to the papillary layer of the dermis. When these muscles contract (during cold or fright), the hair is pulled upright, dimpling the skin surface with "goose bumps." This phenomenon is especially dramatic in a scared cat, whose fur actually stands on end to increase its apparent size. The activity of the arrector pili muscles also exerts pressure on the sebaceous glands surrounding the follicle, causing a small amount of sebum to be released.

Nails

Nails, the hornlike derivatives of the epidermis, consist of a root (adhering to an epithelial nail bed) and a body. The germinal cells in the nail root divide to produce its growth; these cell divisions occur below the lunula, the crescent- or half-moon-shaped region at the proximal end of the nail. Like the hair shaft, the nail is composed of nonliving material. The nails

Hair follicles (24×)

Figure 7.3

Photomicrograph of skin.

are nearly colorless but appear pink because of the blood supply in the nail bed. When someone is cyanotic due to a lack of oxygen in the blood, the nails take on a blue cast.

EXAMINATION OF THE MICROSCOPIC STRUCTURE OF THE SKIN

 Obtain a prepared slide of human skin, and study it carefully under the microscope. Compare your tissue slide to the view shown in Figure 7.3, and identify as many of the structures diagrammed in Figure 7.1 as possible. Label Figure 7.3.

How is this stratified squamous epithelium different from that observed in Exercise 6? _____

_____

_____

How do these differences relate to the functions of these two similar epithelia? _____

_____

_____

_____

PLOTTING THE DISTRIBUTION OF SWEAT GLANDS

1. For this simple experiment you will need two squares of bond paper (each 1 cm × 1 cm), adhesive tape, and a betadine (iodine) swab *or* Lugol's iodine and a cotton-tipped swab. (The bond paper has been preruled in cm²—just cut along the lines to obtain the required squares.)

2. Paint the medial aspect of your left palm (avoid the crease lines) and a region of your left forearm with the iodine solution, and allow it to dry thoroughly. The painted area in each case should be slightly larger than the paper squares to be used.

3. Have your lab partner securely tape a square of bond paper over each iodine-painted area, and leave them in place for 1 hour.

4. After 1 hour, remove the paper squares, and count the number of blue-black dots on each square. The presence of a blue-black dot on the paper indicates an active sweat gland. (The iodine in the pore is dissolved in the sweat and reacts chemically with the starch in the bond paper to produce the blue-black color.) Thus "sweat maps" have been produced for the two skin areas.

5. Which skin area has the greater density of sweat

glands? _____

_____

Classification of Membranes

OBJECTIVES

1. To name and recognize by microscopic examination epidermal, mucous, and serous membranes.
2. To list the general functions of each membrane type and note its location in the body.
3. To compare the structure and function of the major membrane types.

MATERIALS

Compound microscope
Prepared slides of trachea (cross section) and small intestine (cross section)
Prepared slide of serous membrane (e.g., mesentery)
Longitudinally cut fresh beef joint (if available)

The body membranes, which cover surfaces, line body cavities, and form protective (and often lubricating) sheets around organs, fall into two major categories. These are the **epithelial membranes** (consisting of the cutaneous membrane (the skin) and the mucous and serous membranes) and the **synovial membranes** (composed entirely of connective tissue). The cutaneous membrane or skin has already been introduced in Exercise 7 in conjunction with a broader topic—the integumentary system. The other membranes are considered briefly in this exercise.

EPITHELIAL MEMBRANES

Mucous Membranes

The mucous membranes (mucosae) are composed of epithelial cells resting on a layer of loose connective tissue called the *lamina propria*. They line all body cavities that open to the body exterior—the respiratory, digestive, and urinary tracts. Although mucous membranes often secrete mucus, this is not a requirement. The mucous membranes of both the digestive and respiratory tracts secrete mucus, but that of the urinary tract does not.

 Examine a slide made from a cross section of the trachea and of the small intestine. Draw each in the appropriate circle, and fully identify each epithelial type. Remember to look for the epithelial cells at the free surface.

Goblet cells are columnar epithelial cells with a large mucus-containing vacuole (goblet) in the cytoplasm. Which of these mucous membranes contains goblet

cells? _____

_____

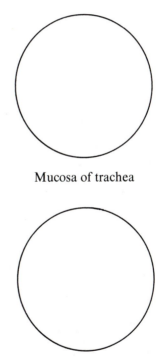

Mucosa of trachea

Mucosa of small intestine

How do the roles of these mucous membranes differ?

_____

_____

_____

How are they the same? _____

_____

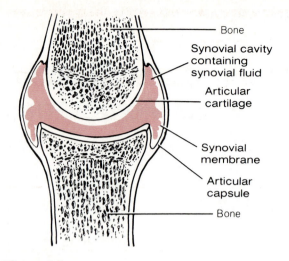

Figure 8.1

A typical synovial joint.

Would you say that mucous membranes have a high regenerative capacity? _____ Why or why not? _____

_____

_____

Serous Membranes

The serous membranes (serosae) are also epithelial membranes. They are composed of two layers—a layer of simple squamous epithelium on a scant layer of loose connective tissue. Serous membranes generally occur in twos. The parietal layer lines a body cavity, and the visceral layer covers the outside of the organs in that cavity. In contrast to the mucous membranes, which line open body cavities, the serous membranes line body cavities that are closed to the exterior (with the exception of the female peritoneal cavity, which the fallopian tubes enter, and the dorsal body cavity). The serosae secrete a thin fluid (serous fluid), which lubricates the organs and body walls and thus reduces friction as the organs slide across one another and against the body cavity walls. A serous membrane also lines the interior of blood vessels (endothelium) and the heart (endocardium). In capillaries, the entire wall is composed of serosa that serves as a selectively permeable membrane between the blood and the tissue fluid of the body.

 Examine a prepared slide of a serous membrane and diagram it in the circle provided here.

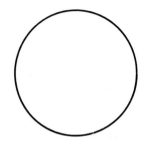

What kind of cells that you have seen many times before are these shaped like? _____

_____

What are the specific names of the serous membranes of the heart and heart cavity (respectively)? _____

_____

The lungs and thoracic cavity (respectively)? _____

_____

The abdominal viscera and visceral cavity (respectively)? _____

_____

SYNOVIAL MEMBRANES

Synovial membranes, unlike the mucous and serous membranes, are composed solely of connective tissue; they contain no epithelial cells. These membranes line the cavities surrounding the joints, providing a smooth surface and secreting a lubricating fluid. They also line smaller sacs of connective tissue (bursae) and tendon sheaths, both of which cushion structures moving against each other, as during muscle activity. Figure 8.1 illustrates the positioning of a synovial membrane in the joint cavity.

If a freshly sawed beef joint is available, examine the interior surface of the joint capsule to observe the smooth texture of the synovial membrane.

Bone Classification, Structure, and Relationships: An Overview

OBJECTIVES

1. To list at least three functions of the skeletal system.
2. To identify the four main kinds of bones.
3. To identify surface bone markings and function.
4. To identify the major anatomic areas on a sagitally cut long bone (or diagram of one).
5. To identify the major regions and structures of an osteon system in a histologic specimen of compact bone (or diagram of one).
6. To explain the role of the inorganic salts and organic matrix in providing flexibility and hardness to bone.

MATERIALS

Numbered disarticulated bones that demonstrate classic examples of the four bone classifications (long, short, flat, and irregular)

Long bone sawed longitudinally (beef bone from a slaughterhouse, if possible, or prepared laboratory specimen)

Compound microscope

Prepared slide of ground bone (cross section)

Long bone soaked in 10% nitric acid until flexible

Long bone baked at 250°F for more than 2 hours

The skeleton is constructed of two of the most supportive tissues found in the human body—cartilage and bone. In embryos, the skeleton is predominantly composed of hyaline cartilage, but in the adult, most of the cartilage is replaced by more rigid bone. Cartilage persists only in such isolated areas as the bridge of the nose, the larynx, the joints, and parts of the ribs.

Besides supporting and protecting the body as an internal framework, the skeleton also provides a system of levers with which the skeletal muscles work to move the body. In addition, the bones store such substances as lipids and calcium. Finally, the red marrow cavities of bones provide a site for hemopoiesis (blood cell formation).

The skeleton is made up of bones that are connected at joints, or articulations. The skeleton is subdivided into two divisions: the **axial skeleton** (those bones that lie around the body's center of gravity) and the **appendicular skeleton** (those of the limbs, or appendages) (Figure 9.1).

Before beginning your study of the skeleton, imagine for a moment that your bones have turned to putty. What if you were running when this metamorphosis took place? Now imagine your bones forming a continuous metal framework within your body, somewhat like a network of plumbing pipes. What problems could you envision with this arrangement? These images should help you understand how well the skeletal system provides support and protection, as well as facilitates movement.

BONE MARKINGS

Even a casual observation of the bones will reveal that bone surfaces are not featureless smooth areas but are scarred with an array of bumps, holes, and ridges. These bone markings reveal where muscles, tendons, and ligaments were attached and where blood vessels and nerves passed. Bone markings fall into two categories: projections, or processes which grow out from the bone, and depressions, or cavities, which are indentations in the bone.

The projections of bone are of several types:

Condyle: rounded convex projection that articulates with another bone

Crest: narrow ridge of bone serving as a site for muscle attachment

Epicondyle: raised area on a condyle

Head: extension carried on a narrow neck that takes part in forming a joint

Line: similar to a crest but less pronounced

Ramus: armlike bar, extending from the bone body, that takes part in joint formation

Spine: sharp, slender site of muscle attachment

Trochanter: very large, usually irregularly shaped site of muscle attachment

Tubercle: small rounded site of muscle attachment

Tuberosity: large rounded projection or roughened area that serves as a site of muscle attachment

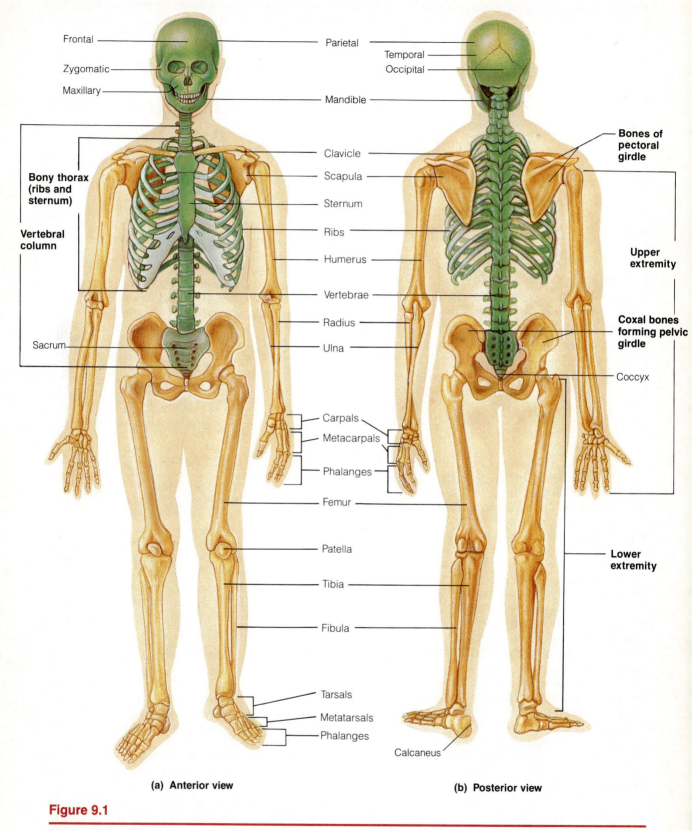

(a) Anterior view

(b) Posterior view

Figure 9.1

The human skeleton. The bones of the axial skeleton are shaded green; the bones of the appendicular skeleton are shaded gold.

There are also several types of depressions:

Fissure: narrow slitlike opening that often serves as a nerve passageway

Foramen: opening through a bone that serves as a passageway for blood vessels and nerves

Fossa: shallow depression that often forms a socket for another bone in a joint

Meatus: canallike passageway that often serves as a conduit for a nerve

Sinus: cavity within bone, often filled with air and lined with a mucous membrane

Sulcus: groove or furrow

CLASSIFICATION OF BONES

The 206 bones of the adult skeleton are composed of two basic kinds of osseous tissue that differ in their texture. **Compact** bone looks smooth and homogeneous; **spongy** (or cancellous) bone is composed of small spicules of bone and lots of open space. Bones may be classified further on the basis of their relative gross anatomy into four groups: long, short, flat, and irregular bones.

Long bones, such as the femur, humerus, and metatarsals (see Figure 9.1), are generally longer than wide, consisting of a shaft with heads at either end. Long bones are composed predominantly of compact bone. **Short bones** are generally cube-shaped, and they contain more spongy bone than compact bone. See the tarsals and carpals in Figure 9.1.

As the name implies, **flat bones,** like the sternum and bones of the skull (Figure 9.1), are generally thin and flattened, with two thin layers of compact bone sandwiching a layer of spongy bone between them. Bones that do not fall into one of the preceding categories are classified as **irregular bones.** The vertebrae are irregular bones (see Figure 9.1).

Some anatomists also recognize two other subcategories of bones. **Sesamoid bones** are small bones formed in tendons. The kneecaps (patellas) are sesamoid bones, but the other members of this group are found in different locations depending on the individual. **Wormian bones** are tiny bones between major cranial bones.

 Examine the isolated (disarticulated) bones (numbered) on display. See if you can find specific examples of the bone markings described above. Then classify each of the bones into one of the four anatomic groups by recording its number in the chart that follows. Verify your identifications with your instructor before you leave the laboratory.

Long	Short	Flat	Irregular

GROSS ANATOMY OF THE TYPICAL LONG BONE

 1. Obtain a fresh beef long bone that has been sawed along its longitudinal axis, or use a cleaned dry bone. With the help of Figure 9.2, identify the shaft, or **diaphysis.** Note its smooth surface, which is composed of compact bone. If you are using a fresh specimen, carefully pull away the **periosteum,** or fibrous membrane covering, to view the bone surface. Notice that many fibers of the periosteum penetrate into the bone. These fibers are called **Sharpey's fibers.** The periosteum is the source of the blood vessels and nerves that invade the bone.

2. Now inspect the **epiphysis,** the end of the long bone. Note that it is composed of a thin layer of compact bone filled with spongy bone.

3. Identify the **articular cartilage** covering the epiphyseal surface in place of the periosteum. Since it is composed of glassy hyaline cartilage, it provides a smooth surface to prevent friction at joint surfaces.

4. If the animal was still young and growing, you will be able to see a thin area of hyaline cartilage, the **epiphyseal plate,** which provides for longitudinal growth of the bone during youth. Once the long bone has stopped growing, these areas are replaced with bone and may appear as thin, barely discernible remnants—the **epiphyseal lines.**

5. In an adult animal, the interior or cavity of the shaft is essentially a storage region for adipose tissue, or **yellow marrow.** In the infant, this area is involved in forming blood cells, and so **red marrow** is found in the marrow cavities. In adult bone, the red marrow is confined to the interior of the epiphyses, where it occupies the spaces between the spicules of spongy bone.

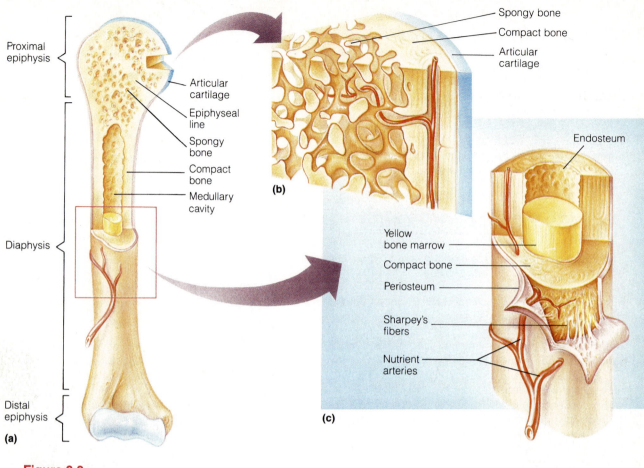

Figure 9.2

The structure of the long bone (humerus of the arm). (a) Anterior view with longitudinal section cut away at the proximal end. (b) Pie-shaped, three-dimensional view of spongy bone and compact bone of the epiphysis. (c) Cross section of shaft (diaphysis). Note that the external surface of the diaphysis is covered by a periosteum, but the articular surface of the epiphysis is covered with hyaline cartilage.

6.   If you are examining a fresh bone, look carefully to see if you can distinguish the delicate **endosteum** lining the shaft. In a living bone, osteoclasts (bone-destroying cells) are found on the inner surface of the endosteum, against the compact bone of the di-aphysis. As the bone grows in diameter on its exter-nal surface, it is constantly being broken down on its inner surface. Thus the thickness of the compact bone layer composing the shaft remains relatively constant.

Longitudinal bone growth at epiphyseal disks follows a predictable sequence and provides a reliable indicator of the age of children exhibiting normal growth. In cases in which problems of long-bone growth are suspected (for example, pituitary dwarfism), X-rays are taken to view the width of the growth plates. An abnormally thin epiphyseal plate indicates growth retardation. ■

MICROSCOPIC STRUCTURE OF COMPACT BONE

As you have seen, spongy bone has a spiky, open-work appearance, resulting from the arrangement of the spicules of bony material, or trabeculae, that compose it, while compact bone appears to be dense and homogeneous. Microscopic examination of com-pact bone, however, reveals that it is riddled with passageways carrying blood vessels, nerves, and lymphatic vessels that provide the living bone cells with needed substances and a way to eliminate wastes.

1.   Obtain a prepared slide of ground bone and examine it under low power. Using Figure 9.3 as a guide, focus on a **central (Haversian) canal.** The central canal runs parallel to the long axis of the bone and carries blood vessels, nerves, and lymph vessels through the bony matrix. Identify the **osteocytes** (mature bone cells) in **lacunae**

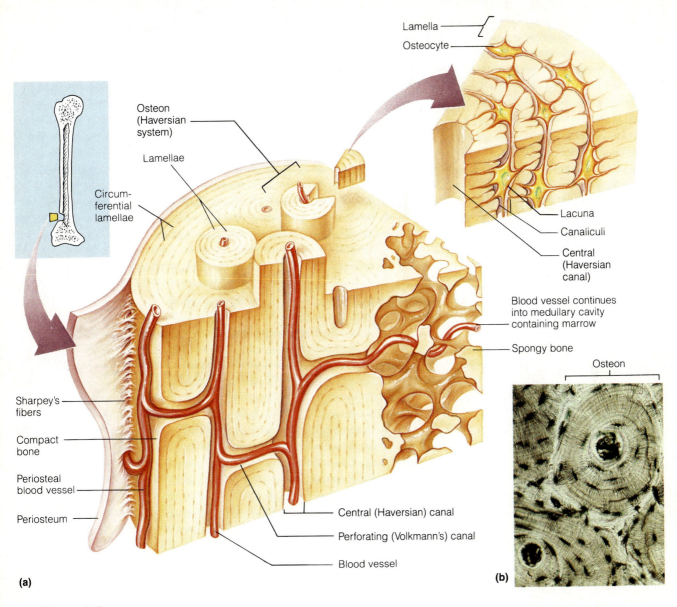

Figure 9.3

Microscopic structure of compact bone. (a) Diagrammatic view of a pie-shaped segment of compact bone, illustrating its structural units (osteons). The inset shows a more highly magnified view of a portion of one osteon. Note the position of osteocytes in lacunae (cavities in the matrix). (b) Photomicrograph of a cross-sectional view of one osteon.

(chambers), which are arranged in concentric circles (concentric **lamellae**) around the central canal. A central canal and all the concentric lamellae surrounding it are referred to as an **osteon**, or **Haversian system.** Also identify **canaliculi,** tiny canals radiating outward from a Haversian canal to the lacunae of the first lamella and then from lamella to lamella. The canaliculi form a dense transportation network through the hard bone matrix, connecting all the living cells of the osteon to the nutrient supply. You may need a higher-power magnification to see the fine canaliculi.

2.   Note the **perforating (Volkmann's) canals** in Figure 9.3. These canals run into the compact bone

and marrow cavity from the periosteum at right angles to the shaft. With the Haversian canals, the perforating canals complete the communication pathway between the bone interior and its external surface.

CHEMICAL COMPOSITION OF BONE

Bone is one of the hardest materials in the body. Although relatively light, bone has a remarkable ability to resist tension and shear forces that continually act on it. An engineer would tell you that a cylinder (like a long bone) is one of the strongest structures for its mass.

Thus nature has given us an extremely strong, exceptionally simple (almost crude), and flexible supporting system without sacrificing mobility.

The hardness of bone is due to the inorganic calcium phosphate salts deposited in its ground substance. Its flexibility comes from the organic constituents of the matrix, particularly the collagenic fibers.

Obtain a bone sample that has been soaked in nitric acid and one that has been baked. Heating removes the organic part of bone, while acid dissolves out the minerals. Do the treated bones retain the structure of untreated specimens? _____

Gently apply pressure to each bone sample. What happens to the heated bone? _____

_____

_____

The bone treated with acid? _____

_____

What does the nitric acid appear to remove from the bone? _____

_____

What does baking appear to do to the bone?

_____

_____

_____

In rickets, the bones are not properly calcified. Which of the demonstration specimens would more closely resemble the bones of a child with rickets? _____

_____

_____

The Axial Skeleton

<table>
<tr><td>

OBJECTIVES

1. To name the three bone groups composing the axial skeleton (skull, bony thorax, and vertebral column).

2. To identify the bones composing the axial skeleton, either by examining the isolated bones or by pointing them out on an articulated skeleton, or skull, and to name the important bone markings on each.

3. To distinguish by examination the different types of vertebrae.

4. To discuss the importance of the intervertebral fibrous discs and spinal curvatures.

5. To differentiate among lordosis, kyphosis, and scoliosis.

</td><td>

MATERIALS

Intact skull and Beauchene skull
X-rays of individuals with scoliosis, lordosis, and kyphosis (if available)
Articulated skeleton, articulated spinal column
Isolated cervical, thoracic, and lumbar vertebrae, sacrum, and coccyx

</td></tr>
</table>

The axial skeleton (the green portion of Figure 9.1) can be divided into three parts: the skull, the vertebral column, and the bony thorax.

THE SKULL

The skull is composed of two sets of bones: the **cranium,** or cranial vault, encloses and protects the fragile brain tissue; the **facial bones** present the eyes in an anterior position and form the base for the facial muscles, which make it possible for us to present our feelings to the world. All but one of the bones of the skull are joined by interlocking joints called *sutures;* the mandible, or jawbone, is attached to the rest of the skull by a freely movable joint.

The bones of the skull, shown in Figures 10.1 through 10.4, are described below. As you read through this material, identify each bone on an intact (and/or Beauchene) skull. Note that important bone markings are listed beneath the bones on which they appear and that a color–coding dot before each bone name indicates its color in the figures.

The Cranium

The cranium is composed of eight large flat bones. *With the exception of two paired bones (the parietals and the temporals), all are single bones.* Sometimes the six ossicles of the middle ear are also considered part of the cranium. Because the ossicles are functionally

part of the hearing apparatus, their consideration is deferred to Exercise 20, Special Senses: Hearing and Equilibrium.

○ **FRONTAL** See Figures 10.1, 10.2, and 10.4. Anterior portion of cranium; forms the forehead, superior part of the orbit, and floor of anterior cranial fossa.

Supraorbital foramen: opening above each orbit allowing blood vessels and nerves to pass.
Glabella: smooth area between the eyes.

○ **PARIETAL** See Figures 10.1 and 10.2. Posterolateral to the frontal bone, forming sides of cranium.

Sagittal suture: midline articulation point of the two parietal bones.
Coronal suture: point of articulation of parietals with frontal bone.

○ **TEMPORAL** See Figures 10.1 through 10.4. Inferior to parietal on lateral skull. The temporals can be divided into two major parts: the **squamous portion** adjoins the parietals; the **petrous portion** forms the lateral inferior aspect of the skull.

Squamous suture: point of articulation of temporal with parietal bone.
External auditory meatus: canal leading to eardrum and middle ear.

61

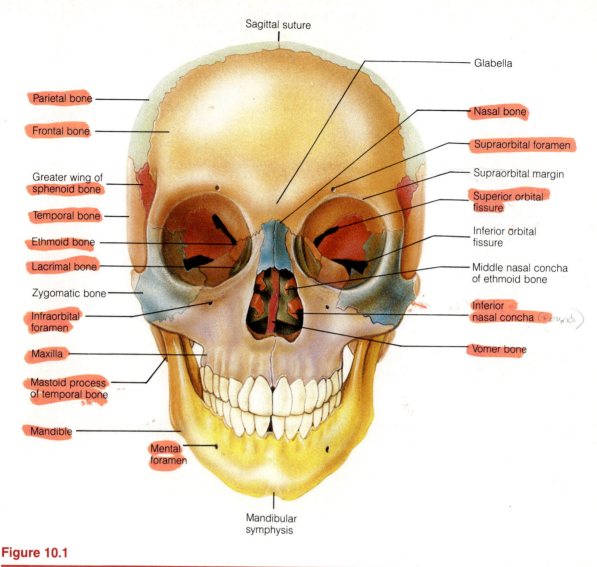

Sagittal suture

Glabella

Parietal bone

Frontal bone

Nasal bone

Supraorbital foramen

Supraorbital margin

Greater wing of sphenoid bone

Superior orbital fissure

Temporal bone

Inferior orbital fissure

Ethmoid bone

Lacrimal bone

Middle nasal concha of ethmoid bone

Zygomatic bone

Infraorbital foramen

Inferior nasal concha (Founda)

Maxilla

Vomer bone

Mastoid process of temporal bone

Mandible

Mental foramen

Mandibular symphysis

Figure 10.1

Anatomy of the anterior aspect of the skull.

Styloid process: needlelike projection inferior to external auditory meatus; attachment point for muscles and ligaments.
Zygomatic process: bridgelike projection joining the zygomatic bone (cheekbone) anteriorly.
Mastoid process: rough projection inferior and posterior to external auditory meatus; attachment site for muscles.

The mastoid process is full of air cavities and is so close to the middle ear, a trouble spot for infections, that it often becomes infected too, a condition referred to as *mastoiditis*. Should this troublesome condition persist, the mastoid process may be surgically removed. Because this area is separated from the brain by only a very thin layer of bone, an ear infection that has spread to the mastoid process can inflame the brain coverings or the meninges. This latter condition is known as *meningitis*. ■

Mandibular fossa: rounded depression anterior to external auditory meatus; forms the socket for the mandibular condyle, the point where the mandible (lower jaw) joins the cranium.
Jugular foramen: opening medial to styloid process through which blood vessels and cranial nerves IX, X, and XI pass.
Carotid canal: opening, medial to the styloid process, through which the internal carotid artery passes to reach the brain.
Stylomastoid foramen: tiny opening between the mastoid and styloid processes through which the seventh cranial nerve leaves the cranium.
Internal acoustic meatus: opening on posterior aspect of temporal bone allowing passage of two cranial nerves VII and VIII (see Figure 10.4).

OCCIPITAL   See Figures 10.2, 10.3, 10.4. Most posterior bone of cranium—forms floor and back wall. Joins the sphenoid bone anteriorly via its basioccipital region.

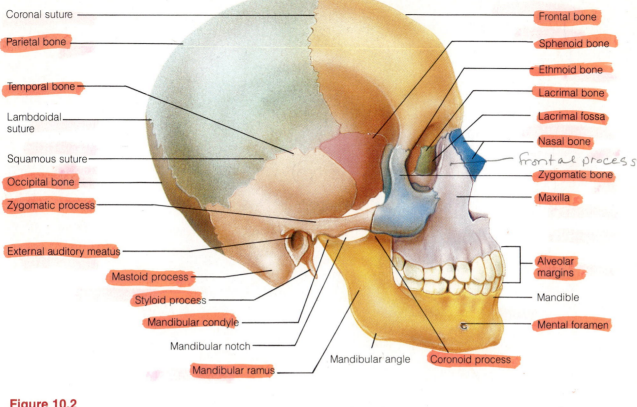

Coronal suture

Parietal bone

Temporal bone

Lambdoidal suture

Squamous suture

Occipital bone

Zygomatic process

External auditory meatus

Mastoid process

Styloid process

Mandibular condyle

Mandibular notch

Mandibular ramus

Frontal bone

Sphenoid bone

Ethmoid bone

Lacrimal bone

Lacrimal fossa

Nasal bone

frontal process

Zygomatic bone

Maxilla

Alveolar margins

Mandible

Mental foramen

Mandibular angle

Coronoid process

Figure 10.2

External anatomy of the right lateral aspect of the skull.

Lambdoidal suture: site of articulation of occipital bone and parietal bones.

Foramen magnum: large opening in base of occipital, which allows the spinal cord to join with the brain.

Occipital condyles: rounded projections lateral to the foramen magnum, which articulate with the first cervical vertebra (atlas).

External occipital crest and protuberance: midline prominences posterior to the foramen magnum.

Superior and inferior nuchal lines: inconspicuous ridges that serve as sites of muscle attachment.

● SPHENOID See Figures 10.1 through 10.4. Bat-shaped bone forming the anterior plateau of the middle cranial cavity across the width of the skull.

Greater wings: portions of the sphenoid seen exteriorly anterior to the temporal and forming a portion of the orbits of the eyes.

Superior orbital fissures: jagged openings in orbits providing passage for cranial nerves III, IV, V, and VI to enter the orbit where they serve the eye.

The sphenoid bone can be seen in its entire width if the top of the cranium (calverium) is removed (see Figure 10.4).

Sella turcica (Turk's saddle): small depression in sphenoid midline in which the pituitary gland rests in the living person.

Lesser wings: bat-shaped portion of sphenoid anterior to sella turcica.

Optic foramina: openings in the base of the lesser wings through which the optic nerves enter the orbits to serve the eyes.

Foramen rotundum: opening lateral to sella turcica providing passage for a branch of the fifth cranial nerve.

Foramen ovale: opening posterior to the sella turcica providing passage for a branch of the fifth cranial nerve.

Foramen lacerum: a jagged opening between the temporal bone and the sphenoid providing passage for a number of small nerves.

● ETHMOID See Figures 10.1, 10.2, and 10.4. Irregularly shaped bone anterior to the sphenoid. Forms the roof of the nasal cavity, upper nasal septum, and part of the medial orbit walls.

Crista galli (cock's comb): vertical projection providing a point of attachment for the dura mater (outermost membrane covering of the brain).

Cribriform plates: bony plates lateral to the crista galli through which olfactory fibers pass to the brain from the nasal mucosa.

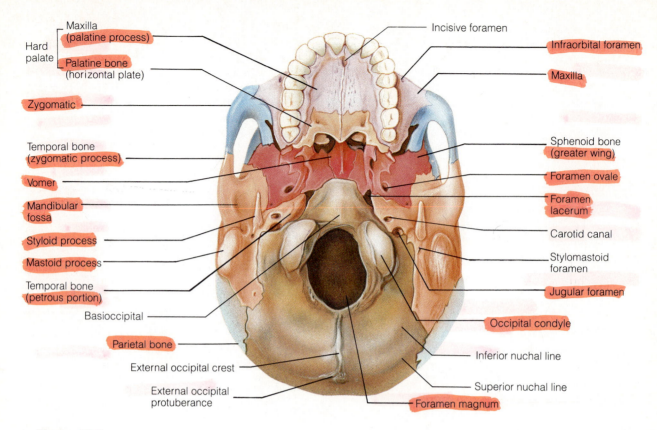

Figure 10.3

Anatomy of the inferior superficial view of the skull; mandible removed.

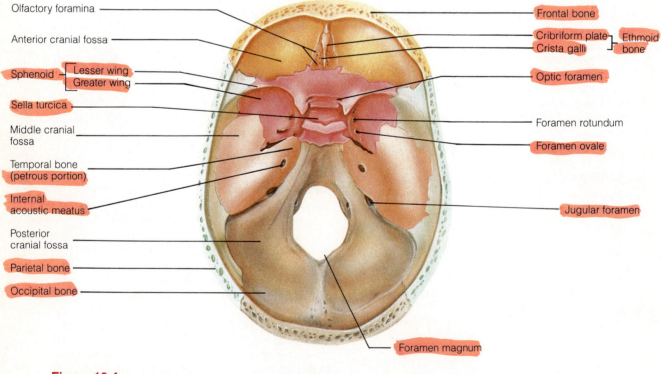

Figure 10.4

Anatomy of the superior view of the floor of the cranial cavity; calvaria has been removed.

Superior and middle conchae (turbinates): thin plates of bone extending laterally from the lateral masses of the ethmoid into the nasal cavity. The shelf-like conchae provide for a more efficient air flow through the nasal cavity and greatly increase the surface area of the mucosa that covers them, thus increasing the ability of the mucosa to warm and humidify the incoming air.

Facial Bones

There are 14 bones composing the face, and 12 of these are paired. *Only the mandible and vomer are single bones.* An additional bone, the hyoid bone, although not a facial bone, is considered here because of its location. Refer to Figures 10.1 through 10.4 to find the structures described below.

○ **MANDIBLE**   See Figures 10.1 and 10.2. The lower jaw bone, which articulates with the temporal bones, providing the only freely movable joints of the skull.

Body: horizontal portion; forms the chin.
Rami: extend superiorly from the body on either side.
Mandibular condyle: point of articulation of the mandible with the mandibular fossa of the temporal bone.
Coronoid process: jutting anterior portion of the ramus; site of muscle attachment.
Angle: point at which ramus meets the body.
Mental foramina: prominent openings on the body that transmit blood vessels and nerves.
Alveoli: sockets on superior margin of mandible in which the teeth lie.
Mandibular symphysis: anterior median depression indicating point of mandibular fusion.

○ **MAXILLAE**   See Figures 10.1, 10.2, and 10.3. Two bones fused in a median suture, which form the upper jawbone and part of the orbits. All facial bones, with the exception of the mandible, join the maxillae. Thus they represent the main, or keystone, bones of the face.

Alveoli: sockets on the inferior margin in which teeth lie.
Palatine processes: form the anterior parts of the hard palate.
Infraorbital foramen: opening under the orbit carrying the infraorbital nerves to the nasal region.

○ **PALATINE**   See Figure 10.3. Paired bones posterior to the palatine processes; form posterior hard palate.

○ **ZYGOMATIC**   See Figures 10.1, 10.2, and 10.3. Lateral to the maxilla; forms the portion of the face commonly called the cheekbone, and forms part of the lateral orbit.

○ **LACRIMAL**   See Figures 10.1 and 10.2. Fingernail-sized bones forming a part of the medial orbit walls between the maxilla and the ethmoid. Each lacrimal bone is pierced by an opening, the **lacrimal fossa,** which serves as a passageway for tears (*lacrima* means "tear").

○ **NASAL**   See Figures 10.1 and 10.2. Small rectangular bones forming the bridge of the nose.

○ **VOMER**   See Figure 10.1. Irregularly shaped bone in median plane of nasal cavity; forms the posterior and inferior nasal septum.

○ **INFERIOR CONCHAE** (turbinates)   See Figure 10.1. Thin curved bones protruding from the lateral walls of the nasal cavity; serve the same purpose as the turbinate portions of the ethmoid bone (described earlier).

● **HYOID BONE**   See Figure 10.5 on page 66. Not really considered a skull bone. Located in the throat above the larynx; serves as a point of attachment for many tongue and mouth muscles. Does not articulate with any other bone, and is thus unique. Horseshoe-shaped with a body and two pairs of horns, or **cornua.**

Palpate the following areas on yourself:

- Mastoid process of the temporal bone
- Temporomandibular joints (Open and close your jaws to locate these.)
- Hyoid bone (Place a thumb and finger high behind the lateral edge of the mandible and squeeze medially.)

Four skull bones (maxillary, sphenoid, ethmoid, and frontal) contain sinuses (air cavities lined with mucosa), which lead into the nasal passages (see Figure 10.6 on page 66). These paranasal sinuses lighten the facial bones and may act as resonance chambers for speech. The maxillary sinus is the largest of the sinuses found in the skull.

THE VERTEBRAL COLUMN

The vertebral column, extending from the skull to the pelvis, forms the major axial support of the body. Additionally, it surrounds and protects the delicate spinal cord while allowing the spinal nerves to issue from the cord via openings between adjacent vertebrae. To some people the term *vertebral column* might suggest a rather rigid supporting rod, but this is far from the truth. The vertebral column consists of 24 single bones (**vertebrae**) and two composite, or fused,

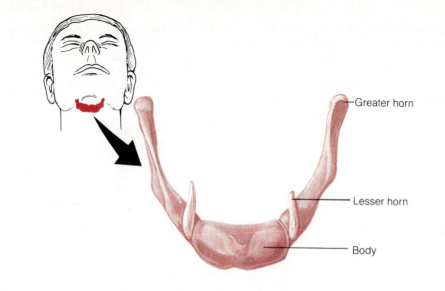

Greater horn

Lesser horn

Body

Figure 10.5

Anatomic location and structure of the hyoid bone. The hyoid bone is suspended in the midanterior neck by ligaments attached to the lesser cornua and the styloid processes of the temporal bones.

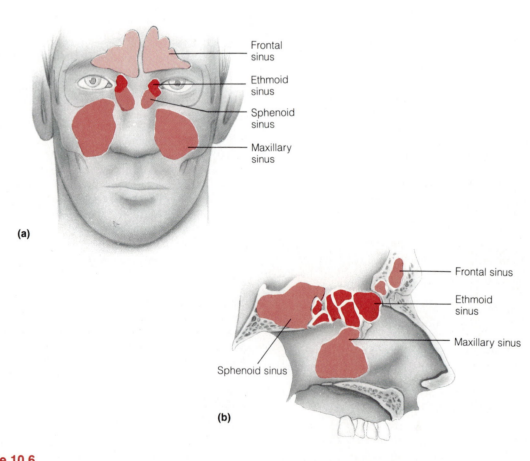

Frontal sinus

Ethmoid sinus

Sphenoid sinus

Maxillary sinus

(a)

Frontal sinus

Ethmoid sinus

Maxillary sinus

Sphenoid sinus

(b)

Figure 10.6

Paranasal sinuses: (a) anterior "see-through" view; (b) as seen in a sagittal section of the head.

bones (the sacrum and coccyx) that are connected in such a way as to provide a flexible curved structure (Figure 10.7).

The vertebrae are separated by pads of fibrocartilage, **intervertebral discs,** that cushion the vertebrae and absorb shocks. Each disc is composed of two major regions, a central gelatinous *nucleus pulpolsus* that behaves like a fluid, and an outer ring of encircling collagen fibers called the *annulus fibrosis,* which stabilizes the disc and contains the pulpolsus.

As a person ages, the water content of the discs decreases (as it does in other tissues throughout the body), and the discs become thinner and less compressible. This situation, along with other degenerative changes such as weakening of the ligaments and tendons of the vertebral column, predisposes older people to *ruptured discs*. A ruptured disc is a situation in which the nucleus pulpolsus herniates through the annulus portion and compresses adjacent nerves. ■

The presence of the discs and the S-shaped or springlike construction of the vertebral column prevent shock to the head in walking and running and allow for flexibility in the movement of the body trunk. The thoracic and sacral curvatures of the spine are referred to as primary curvatures, since they are present at birth. Later the secondary curvatures are formed. The cervical curvature appears when the baby begins to raise its head, and the lumbar curvature develops when the baby begins to walk.

1. Note the normal curvature of the vertebral column in your laboratory specimen, and then examine Figure 10.8, which depicts three abnormal spinal curvatures—scoliosis, kyphosis, and lordosis. These abnormalities may result from disease or poor posture. Also examine X-rays, if they are available, showing these same conditions in a living patient.

2. Using an articulated spinal column (or an articulated skeleton), examine the freedom of movement between two lumbar vertebrae separated by an intervertebral disc.

When the disc is properly positioned, are the spinal cord or peripheral nerves impaired in any way?

_____

Remove the disc and put the two vertebrae back together. What happens to the nerve?

_____

What would happen to the spinal nerves in areas of

malpositioned or "slipped" discs? _____

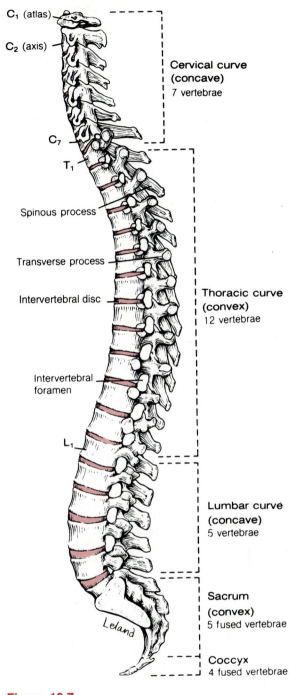

Figure 10.7

The vertebral column. Notice the curvatures in the lateral view. (The terms *convex* and *concave* refer to the curvature of the posterior aspect of the vertebral column.)

_____

_____

_____

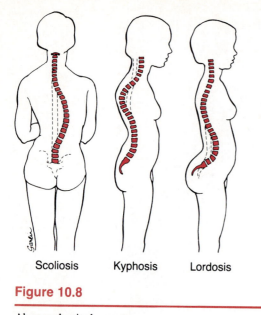

Scoliosis   Kyphosis   Lordosis

Figure 10.8

Abnormal spinal curvatures.

Structure of a Typical Vertebra

Although they differ in size and specific features, all vertebrae have some structures in common (Figure 10.9).

Body (or centrum): rounded central portion of the vertebra, which faces anteriorly in the human vertebral column.

Vertebral arch: composed of pedicles, laminae, and a spinous process, it represents the junction of all posterior extensions from the vertebral body.

Vertebral foramen: conduit for the spinal cord.

Transverse processes: two lateral projections from the vertebral body.

Spinous process: single medial and posterior projection.

Superior and inferior articular processes: paired projections lateral to the vertebral foramen that allow articulation with adjacent vertebrae. The superior articular processes typically face toward the spinous process, whereas the inferior articular processes face away from the spinous process.

Intervertebral foramina: notches on the inferior surfaces of the pedicles that create openings for spinal nerves to leave the spinal cord between adjacent vertebrae.

Figures 10.10 and 10.11 show how specific vertebrae differ; refer to them as you read the following sections.

Cervical Vertebrae

The seven cervical vertebrae (referred to as C_1 to C_7) form the neck portion of the vertebral column. The first two cervical vertebrae (atlas and axis) are highly modified to perform special functions (see

Figure 10.10). The **atlas** (C_1 in a widespread system of reference) lacks a body, and its lateral processes contain large concave depressions on their superior surfaces that receive the occipital condyles of the skull. This joint enables you to nod yes. The **axis** (C_2) acts as a pivot for the rotation of the atlas (and skull) above. It bears a large vertical process, the **odontoid process,** or **dens,** which serves as the pivot point. The articulation between C_1 and C_2 allows you to rotate your head from side to side.

The "typical" cervical vertebrae (C_3 through C_7) are distinguished from the thoracic and lumbar vertebrae by several features (see Figure 10.11(a)). They are the smallest, lightest vertebrae; the vertebral foramen is triangular; and the spinous process is short and usually bifurcated, or divided into two branches. The spinous process of C_7 is not branched, however, and is substantially longer than that of the other cervical vertebrae. Transverse processes of the cervical vertebrae are wide, and they contain foramina through which the vertebral arteries pass superiorly on their way to the occipital region of the brain. Any time you see these foramina in a vertebra, you should know immediately that it is a cervical vertebra.

Thoracic Vertebrae

The 12 thoracic vertebrae (referred to as T_1 through T_{12}) may be recognized by the following structural characteristics. As you can see in Figure 10.11(b), they have a larger body than the cervical vertebrae. The body is somewhat heart-shaped, with two articulating surfaces, or demifacets, on each side (one superior, the other inferior) close to the origin of the vertebral arch. The vertebral foramen is oval or round, and the spinous process is long, with a sharp downward hook. The closer the thoracic vertebra is to the lumbar region, the less sharp and shorter is the spinous process. Articular facets on the transverse processes articulate with the tubercles of the ribs.

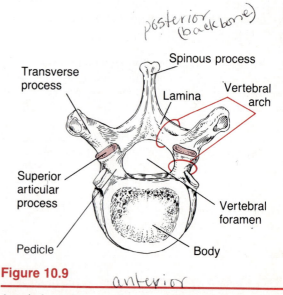

posterior (backbone)

Transverse process

Spinous process

Lamina

Vertebral arch

Superior articular process

Vertebral foramen

Pedicle

Body

Figure 10.9

anterior

A typical vertebra, superior view (inferior articulating surfaces not shown).

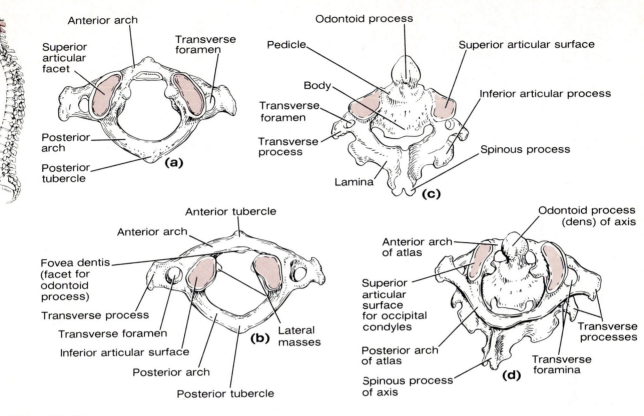

Figure 10.10

Cervical vertebrae C1 and C2. (a) Superior view of the atlas (C1); (b) inferior view of the atlas (C1); (c) superior view of the axis (C2); (d) superior lateral view of the articulated atlas and axis.

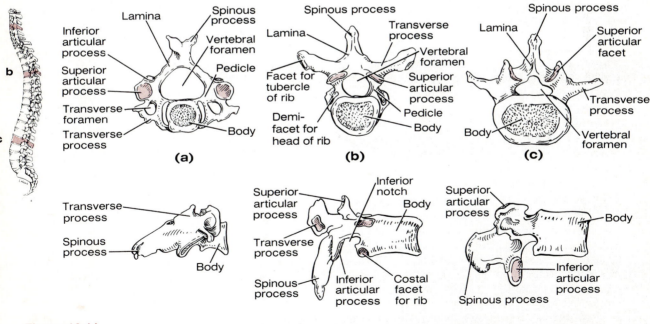

Figure 10.11

Comparison of typical cervical, thoracic, and lumbar vertebrae (superior view above, lateral view below): (a) cervical vertebra; (b) thoracic vertebra; (c) lumbar vertebra.

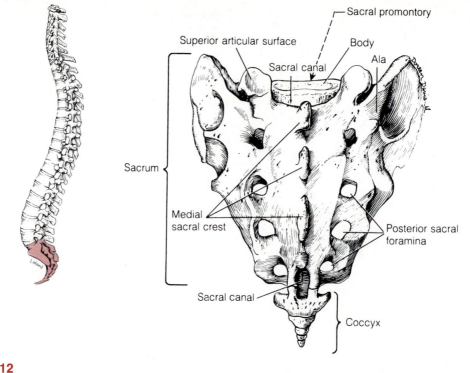

Figure 10.12

Sacrum and coccyx, posterior view.

Lumbar Vertebrae

The five lumbar vertebrae (L_1 through L_5) have massive blocklike bodies and short, thick, hatchet-shaped spinous processes extending backward horizontally (see Figure 10.11(c)). The superior articular facets are directed posteromedially; the inferior ones are directed anterolaterally. Since most of the stress of the vertebral column occurs in the lumbar region, these are the sturdiest of the vertebrae.

The spinal cord ends at the superior edge of L_2, but the outer covering of the cord, filled with cerebrospinal fluid, extends an appreciable distance beyond. Thus a lumbar puncture (for examination of the cerebrospinal fluid) or the administration of "saddle block" anesthesia for childbirth is normally done between L_3 and L_4 or L_4 and L_5, where there is no chance of injuring the delicate spinal cord. ■

The Sacrum

The sacrum (Figure 10.12) is a composite bone formed from the fusion of five vertebrae. Superiorly it articulates with L_5, and inferiorly it connects with the coccyx. The **median sacral crest** is a remnant of the spinous processes of the fused vertebrae. The winglike **alae,** formed from the fusion of the transverse processes, articulate laterally with the hip bones. The sacrum is slightly concave anteriorly and forms the posterior border of the pelvis. Four ridges cross the anterior part of the sacrum, and **sacral foramina** are located at either end of these ridges. These foramina allow for the passage of blood vessels and nerves.

The vertebral canal continues inside the sacrum as the **sacral canal.** The **sacral promontory** (anterior border of the body of S_1) is an important anatomic landmark for obstetricians.

The Coccyx

The coccyx (see Figure 10.12) is formed from the fusion of three to five small irregularly shaped vertebrae. It is literally the human tailbone, a vestige of the tail that other vertebrates have. The coccyx is attached to the sacrum by ligaments.

Obtain examples of each type of vertebra and examine them carefully, comparing them to Figures 10.10, 10.11, and 10.12.

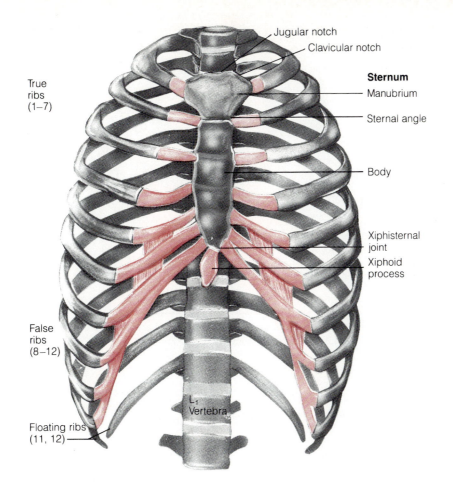

Jugular notch
Clavicular notch

True
ribs
(1–7)

Sternum
Manubrium

Sternal angle

Body

Xiphisternal
joint
Xiphoid
process

False
ribs
(8–12)

L₁
Vertebra

Floating ribs
(11, 12)

Figure 10.13

Skeleton of the bony thorax, anterior view.

THE BONY THORAX

The bony thorax is composed of the sternum, ribs, and thoracic vertebrae (Figure 10.13). It is referred to as the **thoracic cage** because of its appearance and because it forms a protective cone-shaped enclosure around the organs of the thoracic cavity (heart and lungs, for example).

The Sternum

The sternum (breastbone), a typical flat bone, is a result of the fusion of the **manubrium, body (gladiolus),** and **xiphoid process.** It is attached to the first seven pairs of ribs. The sternum has two important bony landmarks—the **jugular notch** and the **sternal angle.** The jugular notch (concave upper border of the manubrium) can be palpated easily; generally it is at the level of the third thoracic vertebra. The sternal angle is a result of the manubrium and body meeting at a slight angle to each other, so that a transverse ridge is formed at the level of the second ribs, providing a handy reference point for counting ribs.

Because of its accessibility, the sternum is a favored site for obtaining samples of blood-forming (hemopoietic) tissue for the diagnosis of suspected blood diseases. A needle is inserted into the marrow of the sternum and the sample withdrawn (sternal puncture). ■

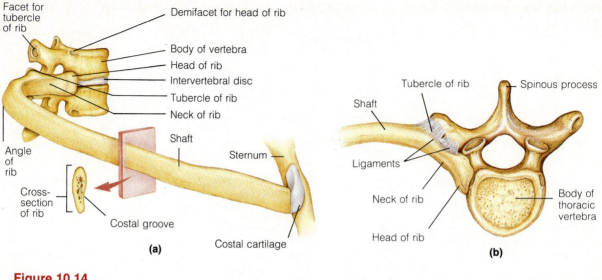

Facet for
tubercle
of rib

Demifacet for head of rib

Body of vertebra

Head of rib

Intervertebral disc

Tubercle of rib

Neck of rib

Angle
of
rib

Shaft

Sternum

Cross-
section
of rib

Costal groove

Costal cartilage

(a)

Tubercle of rib

Spinous process

Shaft

Ligaments

Neck of rib

Head of rib

Body of
thoracic
vertebra

(b)

Figure 10.14

Structure of a "typical" true rib and its articulations: (a) vertebral and sternal articulations of a typical true rib; (b) superior view of the articulation between a rib and a thoracic vertebra.

The Ribs

The 12 pairs of ribs form the walls of the thoracic cage (see Figures 10.13 and 10.14). All of the ribs articulate posteriorly with the vertebral column via their heads and tubercles and then curve downward and toward the anterior body surface. The *"true ribs,"* the first seven pairs, attach directly to the sternum by their "own" costal cartilages. *"False ribs,"* the next five pairs, have indirect cartilage attachments to the sternum or no sternal attachment at all, as in the case of the last two pairs (which are also called floating ribs).

First take a deep breath to expand your chest. Notice how your ribs seem to move outward and how your sternum rises. Then examine an articulated skeleton to observe the relationship between the ribs and the vertebrae.

The Appendicular Skeleton

OBJECTIVES

1. To identify on an articulated skeleton the bones of the shoulder and pelvic girdles and their attached limbs.

2. To arrange unmarked, disarticulated bones in their proper relative position to form the entire skeleton.

3. To differentiate between a male and a female pelvis.

4. To discuss the common features of the human appendicular girdles (pectoral and pelvic), and to note how their structure relates to their specialized functions.

5. To identify specific bone markings in the appendicular skeleton.

MATERIALS

Articulated skeletons
Disarticulated skeletons (complete)
Articulated pelves (male and female for comparative study)

The appendicular skeleton (the gold-colored portion of Figure 9.1) is composed of the 126 bones of the appendages and the pectoral and pelvic girdles, which attach the limbs to the axial skeleton.

 Carefully examine each of the bones described and identify the characteristic bone markings of each. The markings aid in determining whether a bone is the right or left member of its pair. *This is a very important instruction because, before completing this laboratory exercise, you will be constructing your own skeleton.*

BONES OF THE SHOULDER GIRDLE AND UPPER EXTREMITY

The Shoulder Girdle

The **paired shoulder** or **pectoral girdles** (Figure 11.1) each consist of two bones—the anterior clavicle and the posterior scapula—and function to attach the arms to the axial skeleton. In addition, the bones of the shoulder girdles serve as attachment points for many trunk and neck muscles.

The **clavicle,** or collarbone, is a slender doubly curved bone, rounded on the medial end, which attaches to the sternal manubrium, and flattened on the lateral end, where it articulates with the scapula to form a part of the shoulder joint. The clavicle serves as an anterior brace, or strut, to hold the arm away from the top of the thorax and provides attachment points for many thoracic and shoulder muscles.

The **scapulae,** or shoulder blades, are generally triangular and are commonly called the "wings" of humans. Each scapula has a flattened body and two important processes—the **acromion** (the enlarged end of the spine of the scapula) and the beaklike **coracoid process.** The scapular notch at the base of the coracoid process allows nerves to pass. The acromion connects with the clavicle; the coracoid process points anteriorly over the tip of the shoulder joint and serves as a point of attachment for some of the muscles of the upper limb. The scapula has no direct attachment to the axial skeleton but is loosely held in place by trunk muscles. The scapula has three angles: superior, inferior, and lateral. The scapula also has three named borders: the superior; the medial, or vertebral; and the lateral, or axillary. Several shallow depressions (fossae) appear on both sides of the scapula and are named according to location. The **glenoid cavity,** a shallow socket that receives the head of the arm bone, is located in the lateral angle.

The shoulder girdle is exceptionally light and allows the upper limb a degree of mobility not

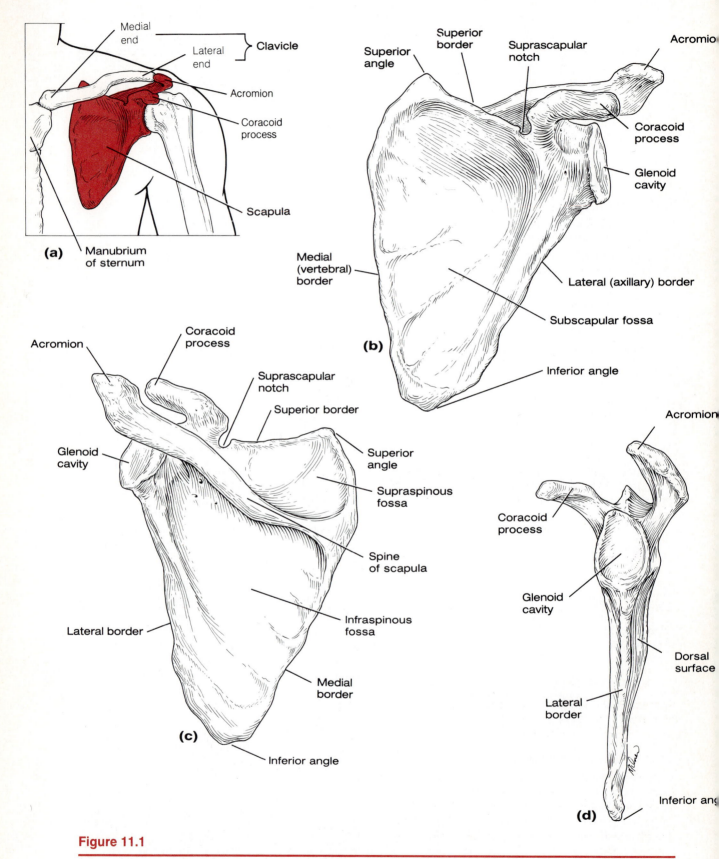

(a)

Medial end
Lateral end
Clavicle
Acromion
Coracoid process
Scapula
Manubrium of sternum

(b)

Superior angle
Superior border
Suprascapular notch
Acromion
Coracoid process
Glenoid cavity
Medial (vertebral) border
Lateral (axillary) border
Subscapular fossa
Inferior angle

(c)

Acromion
Coracoid process
Suprascapular notch
Superior border
Superior angle
Supraspinous fossa
Glenoid cavity
Spine of scapula
Lateral border
Infraspinous fossa
Medial border
Inferior angle

(d)

Acromion
Coracoid process
Glenoid cavity
Dorsal surface
Lateral border
Inferior angle

Figure 11.1

Bones of the shoulder girdle: (a) left shoulder girdle articulated to show the relationship of the girdle to the bones of the thorax and upper arm; (b) left scapula, anterior view; (c) left scapula, posterior view; (d) left scapula, lateral view.

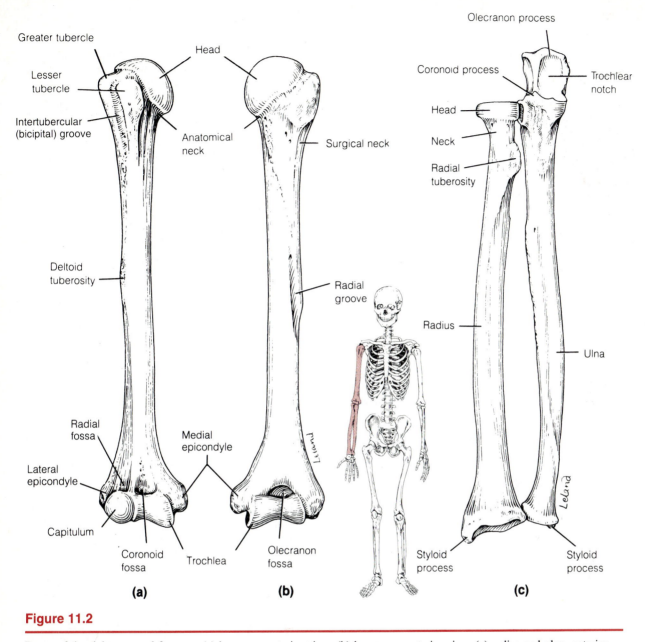

Figure 11.2

Bones of the right arm and forearm: (a) humerus, anterior view; (b) humerus, posterior view; (c) radius and ulna, anterior view.

observed anywhere else in the body. This is due to the following factors:

- There is but a single point on each shoulder girdle for attachment to the axial skeleton—the sternoclavicular joints.
- The relative looseness of the scapular attachment allows it to slide back and forth against the thorax with muscular activity.
- The glenoid cavity is shallow, and reinforcement of the shoulder joint is relatively poor.

However, this exceptional flexibility exacts a price: the shoulder girdle is very susceptible to dislocation.

The Arm

The arm (Figure 11.2) consists of a single bone—the **humerus,** a typical long bone. At its proximal end is the rounded head, which fits into the shallow glenoid fossa of the scapula. The head is separated from the shaft by the anatomic neck and the more constricted surgical neck, which is a common site of fracture. Opposite the head are two prominences, the **greater** and **lesser tubercles** (from lateral to medial aspect), separated by a groove (the **intertubercular** or **bicipital groove**) that guides the tendon of the biceps muscle to its point of attachment (the rim of the glenoid fossa). In the midpoint of the shaft is a roughened area, the **deltoid tuberosity,** where the large fleshy shoulder muscle, the deltoid, attaches. Just inferior to the deltoid tuberosity is the radial groove, which indicates the pathway of the radial nerve.

At the distal end of the humerus are two condyles—the medial **trochlea** (looking rather like a spool), which articulates with the ulna and the lateral **capitulum,** which articulates with the radius of the forearm. This condyle pair is flanked medially by the **medial epicondyle** and laterally by the **lateral epicondyle.**

The medial epicondyle is commonly referred to as the "funny bone." The large ulnar nerve runs in a groove beneath the medial epicondyle; and when this region is sharply bumped, we are quite likely to experience a temporary, but excruciatingly painful, tingling sensation. This event is called "hitting the funny bone," a strange expression, because it is certainly *not* "funny"!

Above the trochlea on the anterior surface is a depression, the **coronoid fossa;** on the posterior surface is the **olecranon fossa.** These two depressions allow the corresponding processes of the ulna to move freely when the elbow is flexed and extended. A small radial fossa, lateral to the coronoid fossa, receives the head of the radius when the elbow is flexed.

The Forearm

Two bones, the radius and the ulna, compose the skeleton of the forearm, or antibrachium (see Figure 11.2c). When the body is in the anatomic position, the **radius** is in the lateral position in the forearm. Proximally, the disk-shaped head of the radius articulates with the capitulum of the humerus. Just below the head, on the medial aspect of the shaft, is a prominence called the **radial tuberosity,** the point of attachment for the tendon of the biceps muscle of the arm.

The **ulna** is the medial bone of the forearm. Its proximal end bears the anterior **coronoid process** and the posterior **olecranon process,** which are separated by the **trochlear notch;** together they grip the trochlea of the humerus in a plierslike joint. The smaller distal end of the ulna bears a small medial **styloid process,** which serves as a point of attachment for the ligaments of the wrist.

The Wrist

The wrist (Figure 11.3) is referred to anatomically as the **carpus,** and the eight bones composing it are the **carpals.** The carpals are arranged in two irregular rows of four bones each. In the proximal row (lateral to medial) are the scaphoid, lunate, triangular, and pisiform bones; the scaphoid and lunate articulate with the distal end of the radius. In the distal row are the trapezium, trapezoid, capitate, and hamate. The carpals are bound closely together by ligaments, which restrict movements between them.

The Hand

The hand, or manus (see Figure 11.3), consists of two groups of bones: the **metacarpals** (bones of the palm) and the **phalanges** (bones of the fingers). The metacarpals are numbered 1 to 5 from the thumb side of the hand toward the little finger. When the fist is clenched, the heads of the metacarpals become prominent as the knuckles. Each hand contains 14 phalanges. There are 3 phalanges in each finger except the thumb, which has only proximal and distal phalanges.

BONES OF THE PELVIC GIRDLE AND LOWER EXTREMITY

The Pelvic Girdle

The pelvic girdle (Figure 11.4) is formed by two **coxal,** or os coxa, **bones** (hip bones). The two coxal bones, together with the sacrum and coccyx, form the bony pelvis. In contrast to the bones of the shoulder girdle, those of the pelvic girdle are heavy and massive, and they are attached securely to the axial skeleton. The sockets for the heads of the femurs (thighbones) are deep and heavily reinforced by ligaments to ensure a stable, strong limb attachment. The ability to bear weight is more important here than extreme mobility and flexibility; the combined weight of the upper body rests on the pelvis (specifically, where it meets the L_5 vertebra).

Each coxal bone results from the fusion of three bones—the ilium, ischium, and pubis—which are distinguishable in the young child. The **ilium,** which connects posteriorly with the sacrum (at the **sacroiliac joint**), is a large flaring bone forming the major portion of a coxal bone; when you put your hands on your hips, you are palpating the ilia. The upper margin of the ilium, the **iliac crest,** is roughened; it terminates in the **anterior superior spine** and the **posterior superior spine.** Two inferior spines are located below these. The shallow **iliac fossa** marks its internal surface, and a shallow ridge, the **arcuate line,** outlines the pelvic inlet.

The **ischium** is the "sit-down" bone, forming the most inferior and posterior portion of the os coxa. The most outstanding marking on the ischium is the **ischial tuberosity,** which receives the weight of the body when sitting. The **ischial spine,** superior to the ischial tuberosity, is an important anatomic landmark of the pelvic cavity. Two other anatomic features are the **lesser** and **greater sciatic notches,** which allow passage of nerves and blood vessels to and from the leg. The sciatic nerve passes through the latter.

The **pubis** is the most anterior portion of the os coxa. The fusion of the **rami** of the pubic bone anteriorly and the ischium posteriorly forms a bar of bone enclosing the **obturator foramen,** through which blood vessels and nerves run from the pelvic cavity into the leg. The pubic bones of each hipbone fuse

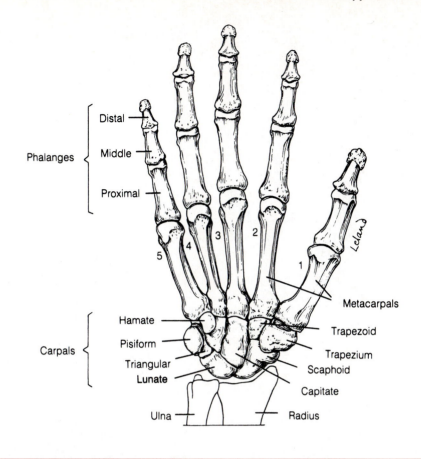

Phalanges { Distal / Middle / Proximal

5   4   3   2   1   *Leland*

Carpals { Hamate / Pisiform / Triangular / **Lunate**

Metacarpals
Trapezoid
Trapezium
Scaphoid
Capitate

Ulna   Radius

Figure 11.3

Bones of the right hand and wrist, anterior view.

anteriorly at the ridgelike **pubic crest** to form a cartilaginous joint called the **pubic symphysis.**

The ilium, ischium, and pubis fuse at the deep hemispherical socket called the **acetabulum** (literally, "vinegar cup"), which receives the head of the thighbone.

 Before continuing with the bones of the lower limbs, take the time to examine an articulated pelvis. Note how each os coxa articulates with the sacrum posteriorly and how the two coxal bones join at the pubic symphysis. The sacroiliac joint, because of the pressure it must bear, is often a site of lower back problems.

COMPARISON OF THE MALE AND FEMALE PELVES   Although bones of males are usually larger and heavier and have more prominent bone markings, the bones of the male and female skeletons are very similar. The outstanding exception to this generalization is pelvic structure (Figure 11.5).

Most importantly, a female pelvis reflects modifications for childbearing. Speaking in generalities, the female pelvis is wider, more shallow, lighter, and rounder than that of the male. Not only must her pelvis support the increasing size of a fetus, but it

must also be large enough to allow the infant's head (its largest dimension) to descend through the birth canal at birth.

To describe pelvic sex differences, a few more terms must be introduced. Anatomically, the pelvis can be described in terms of a **false pelvis** and a **true pelvis.** The false pelvis is that portion superior to the arcuate line; it is bounded by the alae of the ilia laterally and the lumbar vertebrae posteriorly. Although the false pelvis supports the abdominal viscera, it does not restrict childbirth in any way. The true pelvis is the region inferior to the arcuate line that is almost entirely surrounded by bone. Its posterior boundary is formed by the sacrum; the ilia, ischia, and pubic bones define its limits laterally and anteriorly.

The dimensions of the true pelvis, particularly its inlet and outlet, are critical if delivery of a baby is to be nonproblematic; and they are carefully measured by the obstetrician. The **pelvic inlet,** or **pelvic brim,** is the opening delineated by the arcuate lines and is the superiormost margin of the true pelvis. Its widest dimension is from left to right, that is, along the frontal plane. The **pelvic outlet** is the inferior margin of the true pelvis. It is bounded anteriorly by the pelvic arch, laterally by the ischia, and posteriorly by the sacrum and coccyx. Since both the coc-

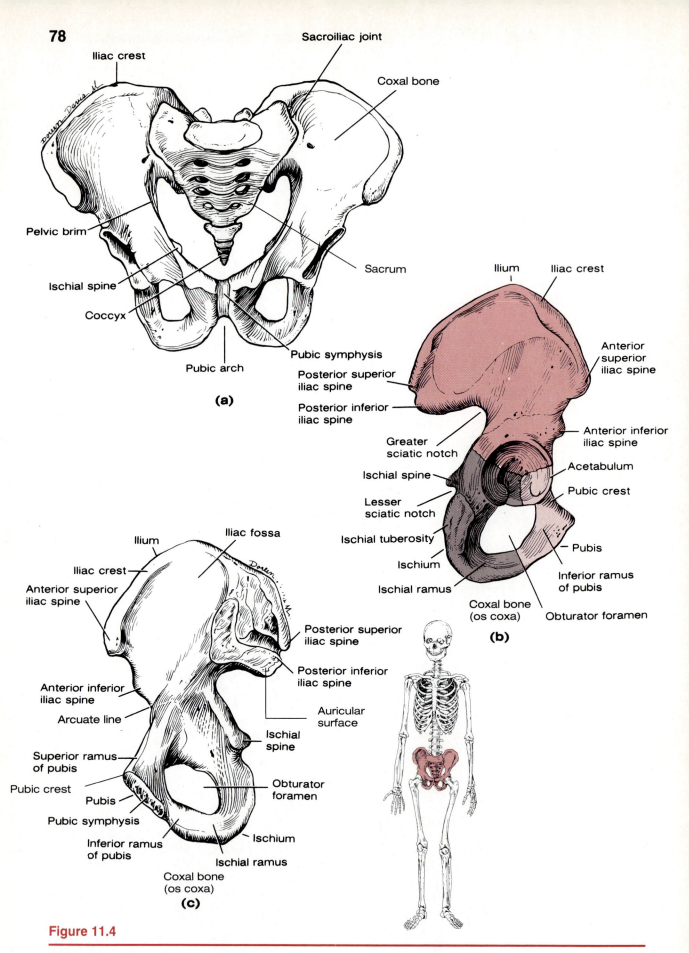

Iliac crest

Sacroiliac joint

Coxal bone

Pelvic brim

Sacrum

Ischial spine

Coccyx

Pubic symphysis

Pubic arch

(a)

Ilium **Iliac crest**

Anterior superior iliac spine

Posterior superior iliac spine

Posterior inferior iliac spine

Anterior inferior iliac spine

Greater sciatic notch

Acetabulum

Pubic crest

Ischial spine

Lesser sciatic notch

Ischial tuberosity

Pubis

Ischium

Inferior ramus of pubis

Ischial ramus

Coxal bone (os coxa)

Obturator foramen

(b)

Ilium **Iliac fossa**

Iliac crest

Anterior superior iliac spine

Posterior superior iliac spine

Anterior inferior iliac spine

Posterior inferior iliac spine

Arcuate line

Auricular surface

Superior ramus of pubis

Ischial spine

Pubic crest

Obturator foramen

Pubis

Pubic symphysis

Inferior ramus of pubis

Ischium

Ischial ramus

Coxal bone (os coxa)

(c)

Figure 11.4

Bones of the pelvic girdle: (a) articulated pelvis, showing the two os coxae and the sacrum; (b) right os coxa, lateral view, showing the point of fusion of the ilium, ischium, and the pubic bones; (c) right os coxa, medial view.

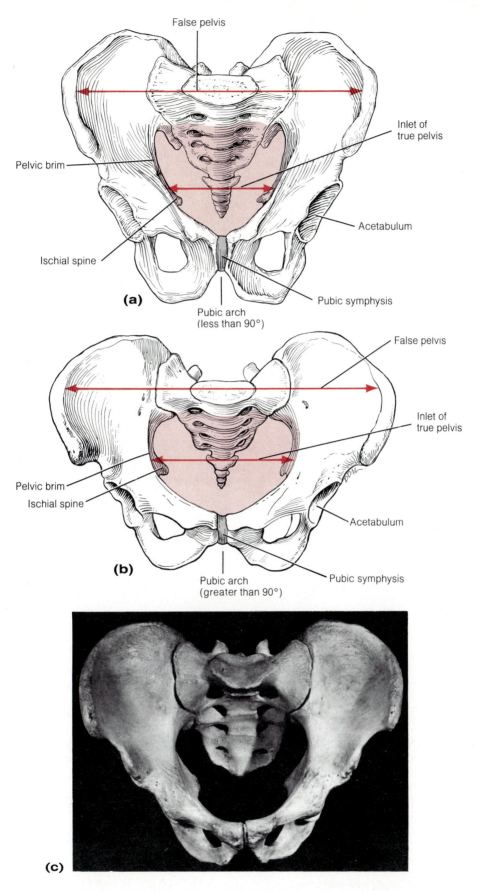

False pelvis

Inlet of
true pelvis

Pelvic brim

Acetabulum

Ischial spine

Pubic symphysis

(a)

Pubic arch
(less than 90°)

False pelvis

Inlet of
true pelvis

Pelvic brim

Ischial spine

Acetabulum

(b)

Pubic arch
(greater than 90°)

Pubic symphysis

(c)

Figure 11.5

Comparison of male and female pelves, anterior views: (a) male; (b) female; (c) photograph of male pelvis.

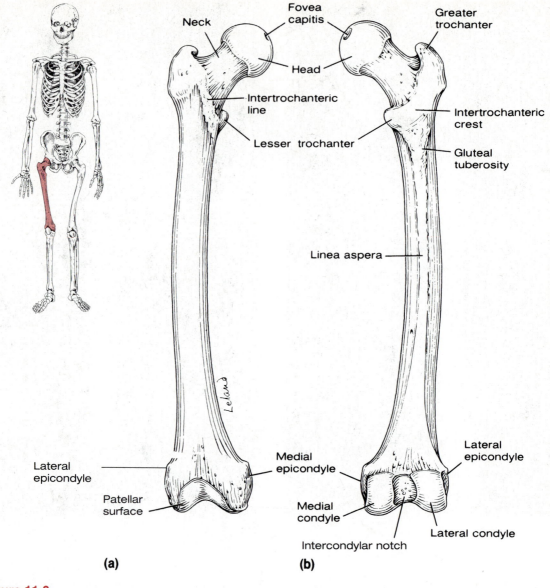

Figure 11.6

Bones of the right thigh: (a) femur, anterior view; (b) posterior view.

cyx and the ischial spines protrude into the outlet opening, a sharply angled coccyx or large sharp spines can dramatically narrow the outlet. The largest dimension of the outlet is the anterior–posterior diameter.

The major differences between the male and female pelves are summarized below (see Figure 11.5).

 Examine male and female pelves for the following differences:

- The female inlet is larger and more circular.

- The female pelvis as a whole is shallower, and the bones are lighter and thinner.
- The female sacrum is broader and less curved, and the pubic arch is more rounded.
- The female acetabula face more anteriorly and are farther apart, and the ilia flare more laterally.
- The female ischial spines are shorter, thus enlarging the pelvic outlet.

The Thigh

The **femur,** or thighbone (Figure 11.6), is the sole bone of the thigh and is the heaviest, strongest bone in the body. Its proximal end bears a ball-like head, a neck, and **greater** and **lesser trochanters** (separated posteriorly by the **intertrochanteric crest** and anteriorly by the **intertrochanteric line**). The head of the femur articulates with the acetabulum of the hipbone

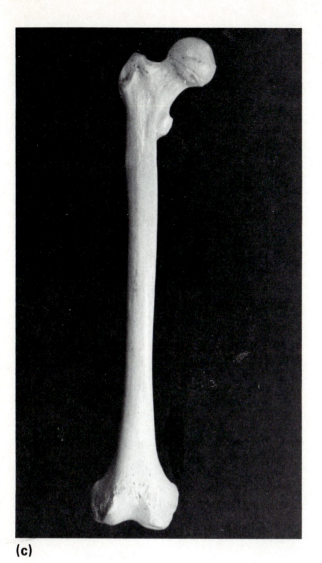

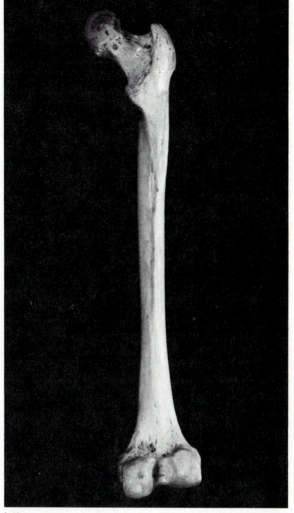

(c)   **(d)**

Figure 11.6 (*continued*)

Photos of bones of the right thigh: (c) femur, anterior view; (d) posterior view.

in a deep secure socket that is heavily reinforced with ligaments. However, the neck of the femur is a common fracture site, especially in the elderly.

The femur inclines medially as it runs downward to the lower leg bones; this brings the knees in the line with the body's center of gravity or maximum weight. The medial course of the femur is even more noticeable in females because of the wider female pelvis.

Distally the femur terminates in the **lateral** and **medial epicondyles.** The more inferior lateral and medial condyles, which are separated by the **intercondylar fossa,** articulate with the tibia below.

The trochanters and trochanteric crest, as well as the **gluteal tuberosity** and the **linea aspera** located on the shaft, serve as sites for muscle attachment.

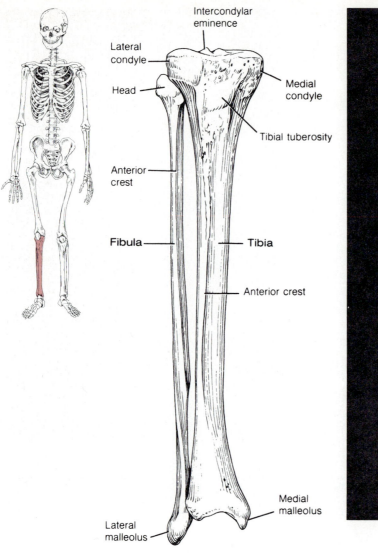

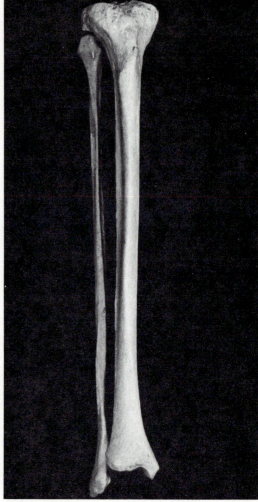

Figure 11.7

Bones of the right leg: tibia and fibula, anterior view.

The Leg

Two bones, the tibia and the fibula, form the skeleton of the leg (see Figure 11.7). The **tibia,** or shinbone, is the larger and more medial of the two leg bones. At the proximal end, the **medial** and **lateral condyles** (separated by the **intercondylar eminence**) receive the distal end of the femur to form the knee joint. The **tibial tuberosity,** a roughened protrusion on the anterior tibial surface (just below the condyles), serves as the site for attachment of the patellar (kneecap) ligament. Small facets on its superior and inferior lateral surface articulate with the fibula. Distally, a process called the **medial malleolus** forms the inner bulge of the ankle, and the smaller distal end articulates with the talus bone of the foot. The anterior surface of the tibia is a sharpened ridge (anterior crest) that is relatively unprotected by muscles; thus it is easily felt beneath the skin.

The **fibula,** which lies parallel to the tibia, takes no part in forming the knee joint. Its proximal head articulates with the lateral condyle of the tibia. The fibula is thin and sticklike with a sharp anterior crest. It terminates distally in the **lateral malleolus,** which forms the outer part of the ankle.

The Ankle

The ankle, or **tarsus** (Figure 11.8), is composed of seven tarsal bones—the **calcaneus, talus,** navicular, cuboid, and lateral, medial, and intermediate cuneiforms. Body weight is concentrated on the two largest tarsals, the calcaneus (heel bone), and the talus, which lies between the tibia and the calcaneus.

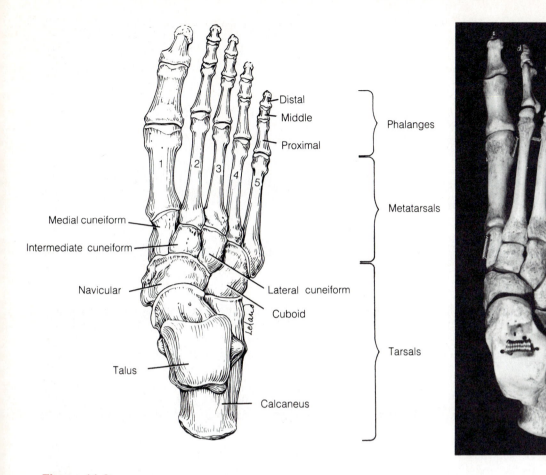

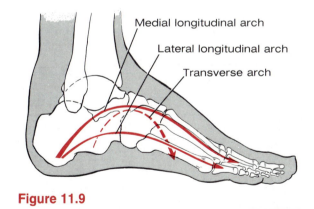

Figure 11.8

Bones of the right ankle and foot, superior view.

The Foot

The bones of the foot include the five **metatarsals,** which form the instep, and 14 **phalanges,** which form the toes (see Figure 11.8). Like the fingers of the hand, each toe has three phalanges except the large toe, which has two.

The bones in the foot are arranged to produce three strong arches—two longitudinal arches (medial and lateral) and one transverse arch (Figure 11.9). Ligaments, binding the foot bones together, and tendons of the foot muscles hold the bones firmly in the arched position but still allow a certain degree of give. Weakened arches are referred to as fallen arches or flat feet.

1. When you have finished examining the disarticulated bones of the appendicular skeleton, arrange them on the laboratory bench in their proper relative positions to form an entire skeleton. Careful observations of the bone markings should help you distinguish between right and left members of bone pairs.

Figure 11.9

Arches of the foot.

2. When you believe that you have correctly accomplished this task, ask the instructor to check your arrangement to ensure that it is correct. If it is not, go to the articulated skeleton and check your bone arrangements. Also review the descriptions of the bone markings as necessary to correct your bone arrangement.

The Fetal Skeleton

OBJECTIVES

1. To define *fontanel* and discuss its function and fate in the fetus.
2. To demonstrate important differences between the fetal and adult skeletons.

MATERIALS

Isolated fetal skull
Fetal skeleton
Adult skeleton

The human fetus about to be born has 275 bones, many more than the 206 bones found in the adult skeleton. This is because many of the bones described as single bones in the adult skeleton (for example, the os coxa, sternum, and sacrum) have not yet fully ossified and fused in the fetus.

 1. Obtain a fetal skull and study it carefully. Make observations as needed to answer the following questions. Does it have the same bones as the adult skull? How does the size of the fetal face relate to the cranium? How does this compare to what is seen in the adult?

2. Indentations between the bones of the fetal skull, called **fontanels,** are fibrous membranes. These areas will become bony (ossify) as the fetus ages, completing the process by the age of 20 to 22 months. The fontanels allow the fetal skull to be compressed slightly during birth and also allow for brain growth during late fetal life. Locate the following fontanels on the fetal skull with the aid of Figure 12.1: anterior (or frontal) fontanel, mastoid fontanel, sphenoidal fontanel, and posterior (or occipital) fontanel.

3. Note that some of the cranial bones have conical protrusions. These are growth centers. Notice also that the frontal bone is still bipartite, and the temporal bone is incompletely ossified, little more than a ring of bone in the fetus.

4. Obtain a fetal skeleton and examine it carefully, noting differences between it and an adult skeleton. Pay particular attention to the vertebrae, sternum, frontal bone of the cranium, patellae (kneecaps), os coxa, carpals and tarsals, and rib cage.

5. Check the questions in the review section before completing this study to ensure that all of the necessary observations have been made.

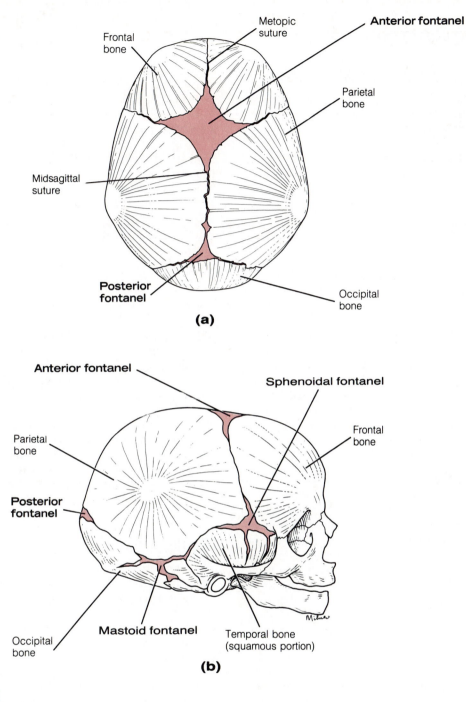

(a)

(b)

Figure 12.1

The fetal skull: (a) superior view; (b) lateral view.

Articulations and Body Movements

OBJECTIVES

1. To name the three major categories of joints, and to compare the structure and mobility of the categories.
2. To identify the types of movement seen in synovial joints.
3. To identify some of the causes of joint problems.
4. To define *origin* and *insertion*.
5. To demonstrate or identify the various body movements.

MATERIALS

Articulated skeleton, skull
Diarthrotic beef joint (fresh)
Anatomical chart of joint types (if available)
X-rays of normal and arthritic joints (if available)
Disposable plastic gloves

With one exception (the hyoid bone), every bone in the body is connected to or forms a joint with at least one other bone. Joints, or articulations, perform two functions for the body: holding the skeletal bones together and allowing the rigid skeletal system some flexibility so that gross body movements can occur.

TYPES OF JOINTS

Joints may be classified structurally or functionally. The *structural classification* is based on whether there is connective tissue fibers, cartilage, or a joint cavity binding the bones together. Structurally there are *fibrous, cartilaginous,* and *synovial joints. The functional classification* focuses on the amount of movement allowed by the joint. On this basis, there are *synarthroses* (immovable joints), *amphiarthroses* (slightly movable joints), and *diarthroses* (freely movable joints). Freely movable joints predominate in the limbs, where movement is important. The immovable and slightly movable joints are largely restricted to the axial skeleton, where firm bony attachments and protection of enclosed organs are a priority. Since the structural categories are more clearcut, we will use the structural classification here, indicating functional properties as appropriate.

Fibrous Joints

The **fibrous joints** are synarthroses that allow essentially no movement (Figure 13.1).

The two major types of fibrous joints are sutures and syndesmoses. In **sutures** the irregular edges of the bones interlock and are united by fibrous connective tissue, as in most joints of the skull. In **syn-**

desmoses the articulating bones are connected by a cord or sheet of fibrous tissue, and the bones do not interlock. The joint at the distal end of the tibia and fibula is an example of a syndesmosis.

 Examine a human skull again. Note that adjacent bone surfaces do not actually touch, but are separated by fibrous connective tissue. Also examine a skeleton and anatomic chart of joint types for examples of syndesmoses.

Cartilaginous Joints

The **cartilaginous joints** (see Figure 13.1) are amphiarthroses that allow a slight degree of movement. In **symphyses** (*symphyses* means "a growth together") the bones are connected by a broad, flat disk of fibrocartilage. The intervertebral joints and the pubic symphysis of the pelvis are symphyses. In **synchondroses** the bony portions are united by hyaline cartilage. Synchondroses are usually temporary joints, eventually replaced by bone. The best examples of synchondroses are the epiphyseal plates seen in the long bones of growing children (Figure 13.2) and the articulation of the first rib and the sternum.

 Identify the amphiarthroses on a human skeleton and on an anatomical chart of joint types.

Synovial Joints

Synovial joints (see Figure 13.1), are characterized by the greatest degree of freedom or flexibility, and

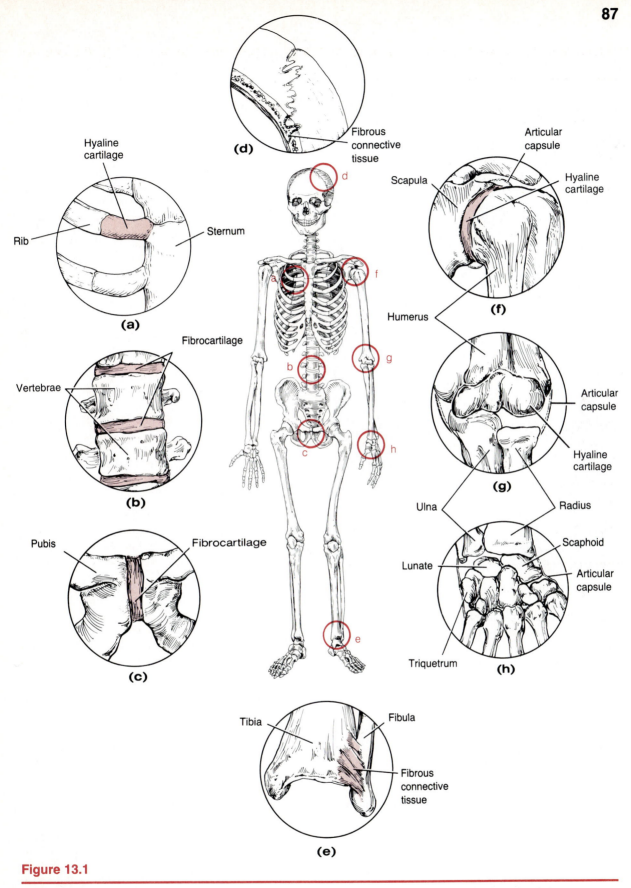

(d) Fibrous connective tissue

Hyaline cartilage

Rib

Sternum

(a)

Fibrocartilage

Vertebrae

(b)

Pubis

Fibrocartilage

(c)

Articular capsule

Scapula

Hyaline cartilage

Humerus

(f)

Articular capsule

Hyaline cartilage

(g)

Ulna

Radius

Lunate

Scaphoid

Articular capsule

Triquetrum

(h)

Tibia

Fibula

Fibrous connective tissue

(e)

Figure 13.1

Types of joints. Joints shown to the left of the skeleton are cartilaginous joints; joints above and below the skeleton are fibrous joints; joints to the right of the skeleton are synovial joints. (a) Synchondrosis (joint between the costal cartilage of the rib 1 and the sternum); (b) symphysis (intervertebral discs of fibrocartilage connecting the vertebrae); (c) symphysis (cartilaginous pubic symphysis connecting the pubic bones anteriorly); (d) suture (fibrous connective tissue connecting the interlocking skull bones); (e) syndesmosis (fibrous connective tissue connecting the distal ends of the tibia and fibula); (f) synovial joint (multiaxial shoulder joint); (g) synovial joint (uniaxial elbow joint); (h) synovial joints (biaxial intercarpal joints of the hand).

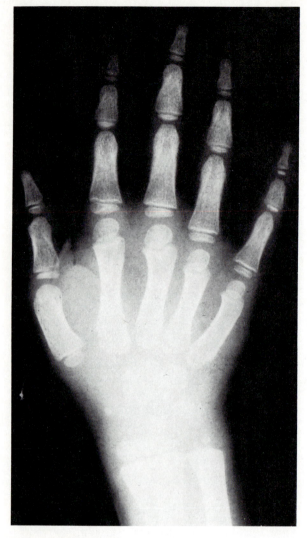

Figure 13.2

X ray of the hand of a child. Notice the cartilaginous epiphyseal plates, examples of temporary synchondroses.

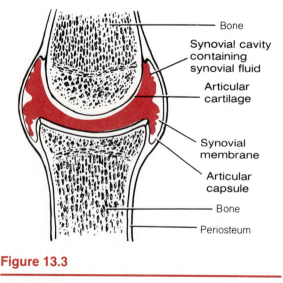

Bone

Synovial cavity containing synovial fluid

Articular cartilage

Synovial membrane

Articular capsule

Bone

Periosteum

Figure 13.3

Major structural components of synovial joints.

all are diarthrotic. However, this flexibility varies; some can move in only one plane, and others can move in several directions (multiaxial movement). Most joints in the body are synovial joints.

All synovial joints are characterized by the following structural characteristics (Figure 13.3):

- The joint surfaces are enclosed by an *articular capsule* of fibrous connective tissue.
- The interior of this capsule is lined with a smooth connective tissue membrane, called *synovial membrane,* which produces a lubricating fluid to reduce friction.
- Articulating surfaces of the bones forming the joint are covered with hyaline (articular) cartilage.
- The fibrous capsule may or may not be reinforced with ligaments and may or may not contain bursae (fluid-filled sacs that reduce friction where tendons cross bone).
- Fibrocartilage pads may or may not be present within the capsule.

1. Examine a beef joint to identify the general structural features of synovial joints. If the joint is fresh and you will be handling it, don plastic gloves before beginning your observations.

2. Compare and contrast the structure of the two synovial joints (hip and elbow) shown in Figure 13.4 relative to their stability, their subclass, and the degree of movement allowed.

Because there are so many types of synovial joints, they have been divided into the following subcategories on the basis of movement allowed:

- Gliding: articulating surfaces are flat or slightly curved, allowing sliding movements in one or two planes. Examples are the intercarpal and intertarsal joints and the vertebrocostal joints.
- Hinge: the rounded process of one bone fits into the concave surface of another to allow movement in one plane (uniaxial), usually flexion and extension. Examples are the knee and elbow.
- Pivot: the rounded or conical surface of one bone articulates with a shallow depression or foramen in another bone to allow uniaxial rotation, as in the joint between the atlas and axis (C_1 and C_2).
- Condyloid: the oval condyle of one bone fits into an ellipsoidal depression in another bone, allowing biaxial (two-way) movement. The wrist joint and the metacarpal–phalangeal joints (knuckles) are examples.
- Saddle: articulating surfaces are saddle-shaped; the articulating surface of one bone is convex, and the reciprocal surface is concave. Saddle joints, which are biaxial, include the joint between the thumb metacarpal and the trapezium of the wrist.
- Ball and socket: the ball-shaped head of one bone fits into a cuplike depression of another. These are multiaxial joints, allowing movement in all

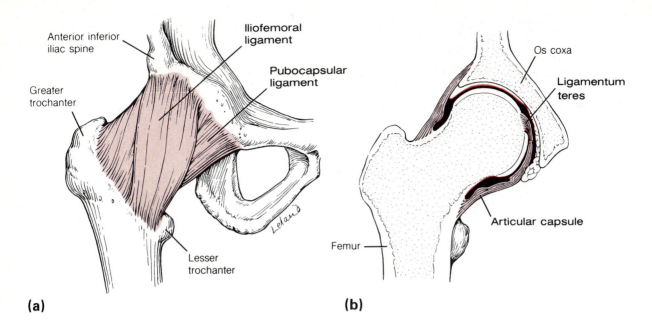

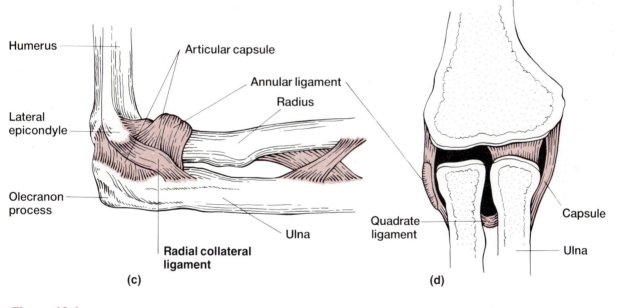

Figure 13.4

Comparative anatomy of the hip and elbow joints: (a) ligamentous reinforcements of right hip joint, anterior view; (b) right hip joint, frontal section view; (c) ligaments of the elbow joint, lateral view; (d) frontal section through an elbow joint.

directions and pivotal rotation. Examples are the shoulder and hip joints.

 Examine the articulated skeleton, anatomic charts, and yourself to identify the subcategories of diarthrotic joints. Make sure you understand the terms *uniaxial, biaxial,* and *multiaxial.*

JOINT DISORDERS

Most of us don't think about our joints until something goes wrong with them. Joint pains and malfunctions may be caused by a variety of things. For example, a hard blow to the knee can cause bursitis, or "water on the knee," due to damage to or inflammation of the bursa or synovial membrane. Slippage of a fibrocartilage pad or the tearing of a

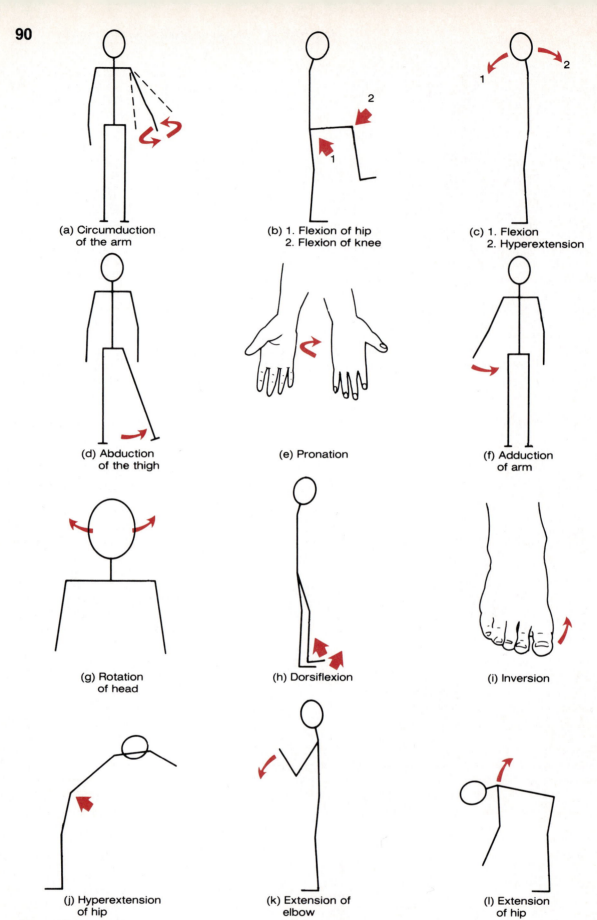

(a) Circumduction
of the arm

(b) 1. Flexion of hip
2. Flexion of knee

(c) 1. Flexion
2. Hyperextension

(d) Abduction
of the thigh

(e) Pronation

(f) Adduction
of arm

(g) Rotation
of head

(h) Dorsiflexion

(i) Inversion

(j) Hyperextension
of hip

(k) Extension of
elbow

(l) Extension
of hip

Figure 13.5

Movements occurring at synovial joints of the body.

ligament may result in a painful condition that persists over a long period, since these poorly vascularized structures heal so slowly.

Sprains and dislocations are other types of joint problems. In a sprain, the ligaments (and possibly tendons) reinforcing a joint are damaged by excessive stretching or are torn away from the bony attachment. Since both ligaments and tendons are cords of dense connective tissue with a poor blood supply, sprains heal slowly and are quite painful.

Dislocations occur when bones are forced out of their normal position in the joint cavity. They are normally accompanied by torn or stressed ligaments and considerable inflammation. The process of returning the bone to its proper position, called reduction, should be done only by a physician. Attempts by the untrained person to "snap the bone back into its socket" are often more harmful than helpful.

Advancing years also take their toll on joints. Weight-bearing joints in particular eventually begin to degenerate. Adhesions (fibrous bands) may form between the surfaces where bones join, and extraneous bone tissue (spurs) may grow along the joint edges. Such degenerative changes lead to the complaint so often heard from the elderly: "My joints are getting so stiff. . . ."

● If possible compare an X-ray of an arthritic joint to one of a normal joint. ■

BODY MOVEMENTS

Every muscle of the body is attached to bone (or other connective tissue structures) at two points—the **origin** (the stationary, immovable, or less movable attachment) and the **insertion** (the movable attachment). Body movement occurs when muscles contract across diarthrotic synovial joints. When the muscle contracts and its fibers shorten, the insertion moves toward the origin. The type of movement depends on the construction of the joint (uniaxial, biaxial, or multiaxial) and on the placement of the muscle relative to the joint. The most common types of body movements are described below and illustrated in Figure 13.5.

Attempt to demonstrate each movement as you read through the following material:

Flexion: a movement, generally in the sagittal plane, that decreases the angle of the joint and lessens the distance between the two bones. Flexion is typical of hinge joints (bending the knee or elbow), but is also common at ball-and-socket joints (bending forward at the hip).

Extension: a movement that increases the angle of a joint and the distance between two bones or parts of the body (straightening the knee or elbow). Extension is the opposite of flexion. If extension is greater than 180 degrees (bending the trunk backward), it is termed *hyperextension.*

Abduction: movement of a limb away from the midline or median plane of the body, generally on the frontal plane, or the fanning movement of fingers or toes when they are spread apart.

Adduction: movement of a limb toward the midline of the body. Adduction is the opposite of abduction.

Rotation: movement of a bone around its longitudinal axis without lateral or medial displacement. Rotation, a common movement of ball-and-socket joints, also describes the movement of the atlas around the odontoid process of the axis.

Circumduction: a combination of flexion, extension, abduction, and adduction commonly observed in ball-and-socket joints like the shoulder. The proximal end of the limb remains stationary, and the distal end moves in a circle. The limb as a whole outlines a cone.

Pronation: movement of the palm of the hand from an anterior or upward-facing position to a posterior or downward-facing position. This action moves the distal end of the radius across the ulna.

Supination: movement of the palm from a posterior position to an anterior position (the anatomic position). Supination is the opposite of pronation. During supination, the radius and ulna are parallel.

The last four terms refer to movements of the foot:

Inversion: a movement that results in the medial turning of the sole of the foot.

Eversion: a movement that results in the lateral turning of the sole of the foot; the opposite of inversion.

Dorsiflexion: a movement of the ankle joint in a dorsal direction (standing on one's heels).

Plantarflexion: a movement of the ankle joint in which the foot is flexed downward (standing on one's toes or pointing the toes).

EXERCISE 14

Microscopic Anatomy, Organization, and Classification of Skeletal Muscle

OBJECTIVES

1. To describe the structure of skeletal muscle from gross to microscopic levels.

2. To define and explain the role of the following:

actin	perimysium
myosin	aponeurosis
fiber	tendon
myofibril	endomysium
myofilament	epimysium

3. To describe the structure of a myoneural junction and to explain its role in muscle function.

4. To define: *prime mover* (agonist), *antagonist, synergist,* and *fixator.*

5. To cite criteria used in naming skeletal muscles.

MATERIALS

Three-dimensional model of skeletal muscle cells (if available)

Forceps

Dissecting needles

Microscope slides and coverslips

0.9% saline solution in dropper bottles

Chicken breast or thigh muscle (freshly obtained from the meat market)

Compound microscope

Histological slides of skeletal muscle (longitudinal and cross-sectional) and skeletal muscle showing myoneural junctions

Three-dimensional model of skeletal muscle showing myoneural junction (if available)

The bulk of the body's muscle is called skeletal muscle because it is attached to the skeleton (or underlying connective tissue). Skeletal muscle influences body contours and shape, allows you to smile and frown, provides a means of locomotion, and enables you to manipulate the environment. The balance of the body's muscle—smooth and cardiac muscle—is the major component of the walls of hollow organs and the heart, where it is involved with the transport of materials within the body.

Each of the three muscle types has a structure and function uniquely suited to its task in the body. However, because the term *muscular system* applies specifically to skeletal muscle, the primary objective of this unit is to investigate the structure and function of skeletal muscle.

Skeletal muscle is also known as *voluntary muscle* (because it is under our conscious control) and as *striated muscle* (because it appears to be striped). As you might guess from both of these alternative names, skeletal muscle has some very special characteristics. Thus an investigation of skeletal muscle should begin at the cellular level.

THE CELLS OF SKELETAL MUSCLE

Skeletal muscle is composed of relatively large, long cylindrical cells ranging from 10 to 100 μm in diameter and up to 4 cm in length. However, the cells of large, hard-working muscles like the antigravity muscles of the hip are extremely coarse, ranging up to 30 cm in length, and can be seen with the naked eye.

Skeletal muscle cells (Figure 14.1(a)) are multinucleate: Multiple oval nuclei can be seen just beneath the plasma membrane (called the sarcolemma in these cells). The nuclei are pushed peripherally by the longitudinally arranged **myofibrils,** which nearly fill the sarcoplasm (Figure 14.1(b)). Alternating light (I) and dark (A) bands along the length of the perfectly aligned myofibrils give the muscle fiber as a whole its striped appearance.

Electron microscope studies have revealed that the myofibrils are made up of even smaller threadlike structures called **myofilaments** (Figure 14.1(b) and (d)). The myofilaments are composed of two varieties of contractile proteins—**actin** and **myosin**—which

Figure 14.1

Structure of skeletal muscle cells: (a) muscle fibers, longitudinal and transverse views, and photomicrograph; (b) a portion of a skeletal muscle cell, where one myofibril has been extended and disrupted to indicate its myofilament composition; (c) one sarcomere of the myofibril; (d) banding pattern in the sarcomere. (From H. E. Huxley, "The Contraction of Muscle." © November 1958 by Scientific American, Inc. All rights reserved.)

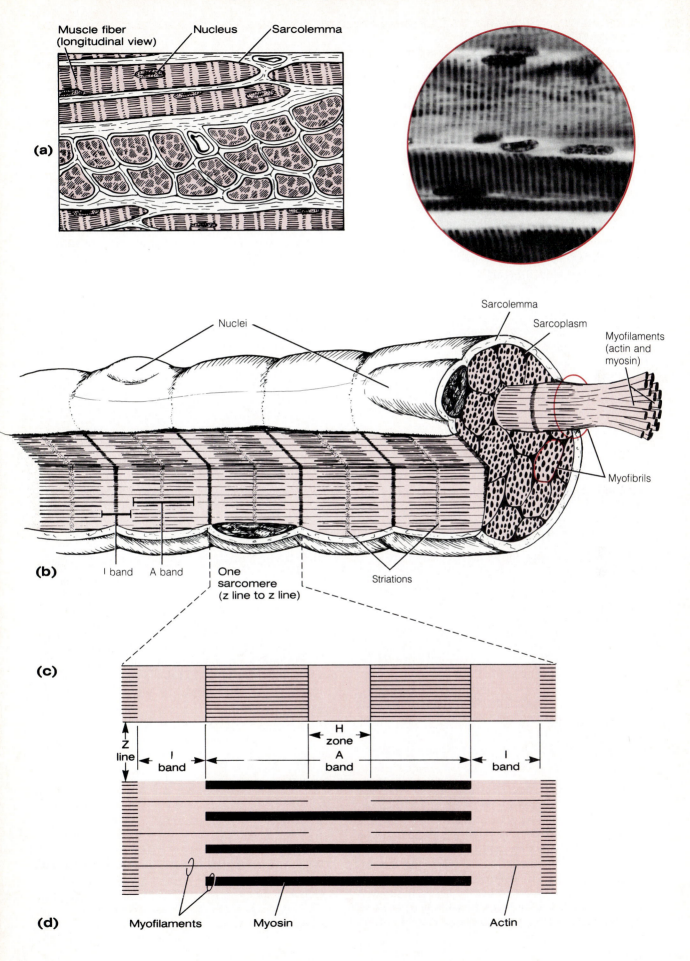

(a) Muscle fiber (longitudinal view), Nucleus, Sarcolemma

(b) Nuclei, Sarcolemma, Sarcoplasm, Myofilaments (actin and myosin), Myofibrils, I band, A band, One sarcomere (z line to z line), Striations

(c) Z line, I band, H zone, A band, I band

(d) Myofilaments, Myosin, Actin

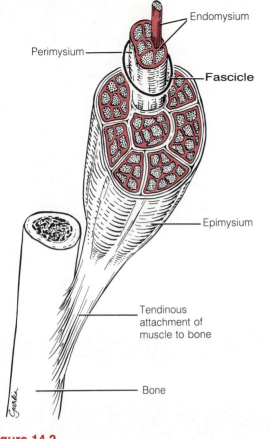

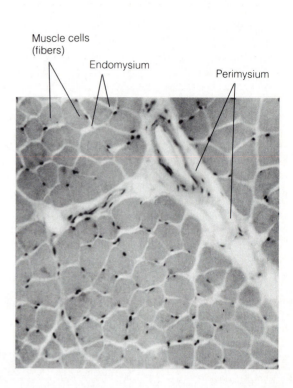

Figure 14.2

Connective tissue coverings of skeletal muscle (64 ×).

slide past each other during muscle activity to bring about shortening or contraction of the muscle cells. It is the highly specific arrangement of the myofilaments within the myofibrils that is responsible for the banding pattern in skeletal muscle. The actual contractile units of muscle, called **sarcomeres,** extend from the middle of one I band (its Z line) to the middle of the next along the length of the myofibrils. (See Figure 14.1(c) and (d).)

1.   Look at the three-dimensional model of skeletal muscle cells, noting the relative shape and size of the cells. Identify the nuclei, myofibrils, and light and dark bands.

2.   Obtain forceps, two dissecting needles, slide and coverslip, and a dropper bottle of saline solution. With forceps, remove a very small piece of muscle from the chicken breast or thigh. Place the tissue on a clean microscope slide, and add a drop of the saline solution. Pull the muscle fibers apart with the dissecting needles (tease them) until you have a fluffy-looking mass of tissue. Cover the teased tissue with a coverslip, and observe under the high-power lens of a microscope. Look for the banding pattern. Reg-

ulate the light carefully to obtain the highest possible contrast.

3.   Compare your observations with what can be seen with professionally prepared muscle tissue. Obtain a slide of skeletal muscle (longitudinal section), and view it under high power. From your observations, draw a small section of a muscle fiber in the space provided here. Label the nuclei, cell membrane, and A and I bands.

What structural details become apparent with the

prepared slide? _____

_____

ORGANIZATION OF SKELETAL MUSCLE CELLS INTO MUSCLES

Muscle fibers are soft and surprisingly fragile. Thus thousands of muscle fibers are bundled together with connective tissue to form the organs we refer to as skeletal muscles (Figure 14.2). Each muscle fiber is enclosed in a delicate, connective tissue sheath called an **endomysium.** Several sheathed muscle fibers are wrapped by a collagenic membrane called a **perimysium,** forming a bundle of fibers called a **fascicle,** or **fasciculus.** A large number of fascicles are bound together by a substantially coarser "overcoat" of connective tissue wrappings called an **epimysium,** or **deep fascia,** which sheathes the entire muscle. These epimysia blend into the strong cordlike **tendons** or sheetlike **aponeuroses,** which attach muscles to each other or indirectly to bones.

The tendons perform several functions, two of the most important being to provide durability and to conserve space. Because tendons are tough collagenic connective tissue, they can span rough bony prominences that would destroy the more delicate muscle tissues. And because of their relatively small size, more tendons than fleshy muscles can pass over a joint.

In addition to supporting and binding the muscle fibers, and providing strength to the muscle as a whole, the connective tissue wrappings provide a route for the entry and exit of nerves and blood vessels that serve the muscle fibers. The larger, more powerful muscles have relatively more connective tissue than muscles involved in fine or delicate movements.

As we age, the amount of muscle fiber decreases, and the amount of connective tissue increases; thus the skeletal muscles gradually become more sinewy, or "stringier."

 Obtain a slide showing a cross section of skeletal muscle tissue. Using Figure 14.2 as a reference, identify the muscle fibers, endomysium, perimysium and epimysium (if visible).

THE NEUROMUSCULAR JUNCTION

Voluntary muscle cells are always stimulated by nerve impulses via motor neurons. The junction between a nerve fiber (axon) and a muscle cell is called a **neuromuscular,** or **myoneural, junction** (Figure 14.3).

Each motor axon breaks up into many branches before it terminates, and each of these branches participates in forming a neuromuscular junction with a single muscle cell. Thus a single neuron may stimulate many muscle fibers. Together, a neuron and all the muscle cells it stimulates make up the functional

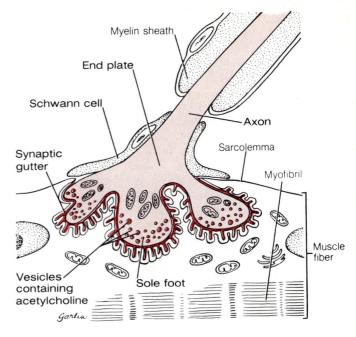

Figure 14.3

The neuromuscular junction.

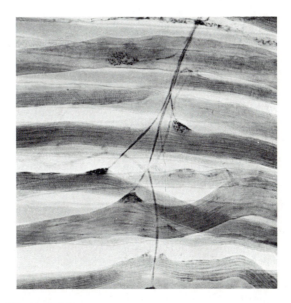

Figure 14.4

Photomicrograph of a portion of a motor unit (50×).

structure called the **motor unit.** Part of a motor unit is shown in Figure 14.4.

Each motor axon terminal, called an **end plate,** has numerous projections called **sole feet.** The neuron and muscle fiber membranes, close as they are, do not actually touch. They are separated by a small gap of 300 to 500 Å, called the **synaptic cleft** or **synaptic gutter** (see Figure 14.3).

Within the sole foot are many mitochondria and vesicles containing a neurotransmitter called acetylcholine. When a nerve impulse reaches the end plate, a few of these vesicles liberate their contents into the synaptic cleft. The acetylcholine rapidly diffuses across the junction and combines with the receptors on the sarcolemma. If sufficient acetylcholine has been released, a transient change in the permeability of the sarcolemma briefly allows more sodium ions to diffuse into the muscle fiber, resulting in the depolarization of the sarcolemma and subsequent contraction of the muscle fiber.

 1.   If possible, examine a three-dimensional model of skeletal muscle cells that illustrates the myoneural junction. Identify the structures described above.

2.   Obtain a slide of skeletal muscle stained to show the end plates of a motor unit. Examine the slide under high power to identify the axonal fibers extending leashlike to the muscle cells. Follow one of the axonal fibers to its terminus to identify the oval-shaped motor end plate. Compare your observations to Figure 14.4. Sketch a small section in the space provided, labeling the motor axon, its terminal branches, motor end plates, and muscle fibers.

CLASSIFICATION OF SKELETAL MUSCLES

Naming Skeletal Muscles

Remembering the names of the skeletal muscles is a monumental task, but certain clues help. Muscles are named on the basis of the following criteria:

- Direction of muscle fibers: Some muscles are named in reference to some imaginary line, usually the midline of the body or the longitudinal axis of a limb bone. A muscle with fibers running parallel to that imaginary line will have the term *rectus* (straight) in its name. For example, the rectus abdominis is the straight muscle of the abdomen. Likewise, the terms *transverse* and *oblique* indicate that the muscle fibers run at right angles and obliquely (respectively) to the imaginary line.
- Relative size of the muscle: Terms such as *maximus* (largest), *minimus* (smallest), *longus* (long), and *brevis* (short) are often used in the names of

muscles—as in gluteus maximus and gluteus minimus.
- Location of the muscle: Some muscles are named according to the bone with which they are associated. For example, the temporalis muscle overlies the temporal bone.
- Number of origins: When the term *biceps, triceps,* or *quadriceps* forms part of a muscle name, you can generally assume that the muscle has two, three, or four origins (respectively). For example, the biceps muscle of the arm has two heads, or origins.
- Location of the muscle's origin and insertion: For example, the sternocleidomastoid muscle has its origin on the sternum (*sterno*) and clavicle (*cleido*), and inserts on the mastoid process of the temporal bone.
- Shape of the muscle: For example, the deltoid muscle is roughly triangular (*deltoid* means "triangle"), and the trapezius muscle resembles a trapezoid.
- Action of the muscle: For example, all the adductor muscles of the anterior thigh bring about its adduction, and all the extensor muscles of the wrist extend the wrist.

Types of Muscles

Most often, body movements are not a result of the contraction of a single muscle but instead reflect the coordinated action of several muscles acting together. Muscles that are primarily responsible for producing a particular movement are called **prime movers,** or **agonists.**

Muscles that oppose or reverse a movement are called **antagonists.** When a prime mover is active, the fibers of the antagonist are stretched and in the relaxed state. The antagonist can also regulate the prime mover by providing some resistance, to prevent overshoot or to stop its action.

It should be noted that antagonists can be prime movers in their own right; for example, the biceps muscle of the arm (a prime mover of elbow flexion) is antagonized by the triceps (a prime mover of elbow extension).

Synergists contribute substantially to the action of agonists by reducing undesirable or unnecessary movement. Contraction of a muscle crossing two or more joints would cause movement at all joints spanned if the synergists were not there to stabilize them. For example, you can make a fist without bending your wrist only because synergist muscles stabilize the wrist joint and allow the prime mover to exert its force at the finger joints.

Fixators, or fixation muscles, are specialized synergists. They immobilize the origin of a prime mover so that all the tension is exerted at the insertion. Muscles that help maintain posture are fixators, so too are muscles of the back that stabilize or "fix" the scapula during arm movements.

Gross Anatomy of the Muscular System

OBJECTIVES

1. To name and locate the major muscles of the human body (on a torso model, laboratory chart, or diagram) and state the action of each.

2. To explain how muscle actions are related to their location.

3. To name muscle origins and insertions as required by the instructor.

4. To identify antagonists of the major prime movers.

5. To name and locate muscles on a dissection animal.

6. To recognize similarities and differences between human and cat musculature.

MATERIALS

Disposable gloves or protective skin cream
Preserved and injected cat (one for every two to four students)
Dissecting trays and instruments
Name tag and large plastic bag
Paper towels
Embalming fluid (formalin)
Human torso model or large anatomic chart showing human musculature

IDENTIFICATION OF HUMAN MUSCLES

Muscles of the Head and Neck

The muscles of the head serve many specific functions. For instance, the muscles of facial expression differ from most skeletal muscles because they insert into the skin (or other muscles) rather than into bone. As a result, they move the facial skin, allowing a wide range of emotions to be shown on the face. Other muscles of the head are the muscles of mastication, which manipulate the mandible during chewing, and the six muscles of the orbit, which aim the eye. (The orbital muscles are studied in conjunction with the anatomy of the eye in Exercise 19.) Neck muscles are primarily concerned with the movement of the head and shoulder girdle. Figures 15.1 and 15.2 are summary figures illustrating the superficial musculature of the body as a whole. The head and neck muscles are discussed in Tables 15.1 and 15.2 and shown in Figures 15.3 and 15.4.

Carefully read the description of each muscle and attempt to visualize what happens when the muscle contracts. Once you have read through the tables and have identified the head and neck muscles in Figures 15.3 and 15.4, use a torso model or an anatomic chart to again identify as many of these muscles as possible. Then carry out the following palpations on yourself:

● To demonstrate how the temporalis works, clench your teeth. The masseter can also be palpated at this time at the angle of the jaw.

Muscles of the Trunk

The trunk musculature includes muscles that move the vertebral column; anterior thorax muscles that act to move ribs, head, and arms; and muscles of the abdominal wall that play a role in the movement of the vertebral column but more importantly form the "natural girdle," or the major portion of the abdominal body wall.

 The trunk muscles are described in Tables 15.3 and 15.4 and shown in Figures 15.5 and 15.6. As before, identify the muscles in the figure as you read the tabular descriptions and then identify them on the torso or laboratory chart. When you have completed this study, work with a partner to demonstrate the operation of the following muscles. One of you can demonstrate the movement (the following steps are addressed to this partner); the other can supply the necessary resistance and palpate the muscle being tested.

1. Start by fully abducting the arm and extending the elbow. Now try to adduct the arm against resistance. You are exercising the latissimus dorsi.

2. To observe the deltoid, attempt to abduct your shoulder against resistance. Now attempt to elevate your shoulder against resistance; you are contracting the upper portion of the trapezius.

3. The pectoralis major comes into play when you press your hands together at chest level with your elbows widely abducted.

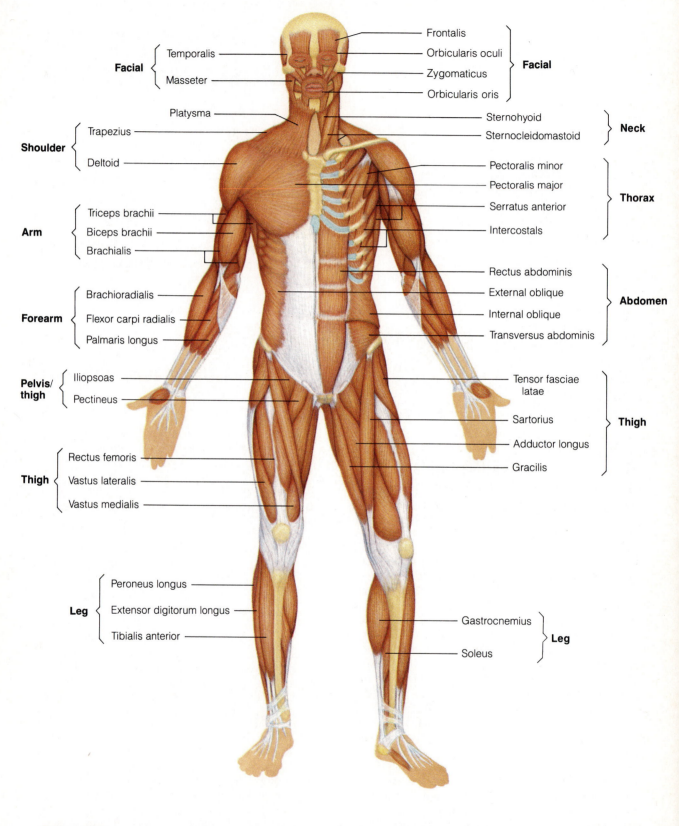

Facial
- Temporalis
- Masseter

Shoulder
- Trapezius
- Deltoid

Arm
- Triceps brachii
- Biceps brachii
- Brachialis

Forearm
- Brachioradialis
- Flexor carpi radialis
- Palmaris longus

Pelvis/ thigh
- Iliopsoas
- Pectineus

Thigh
- Rectus femoris
- Vastus lateralis
- Vastus medialis

Leg
- Peroneus longus
- Extensor digitorum longus
- Tibialis anterior

Platysma

- Frontalis
- Orbicularis oculi
- Zygomaticus
- Orbicularis oris

Facial

- Sternohyoid
- Sternocleidomastoid

Neck

- Pectoralis minor
- Pectoralis major
- Serratus anterior
- Intercostals

Thorax

- Rectus abdominis
- External oblique
- Internal oblique
- Transversus abdominis

Abdomen

- Tensor fasciae latae
- Sartorius
- Adductor longus
- Gracilis

Thigh

- Gastrocnemius
- Soleus

Leg

Figure 15.1

Anterior view of superficial muscles of the body. The abdominal surface has been partially dissected on the right side to show somewhat deeper muscles.

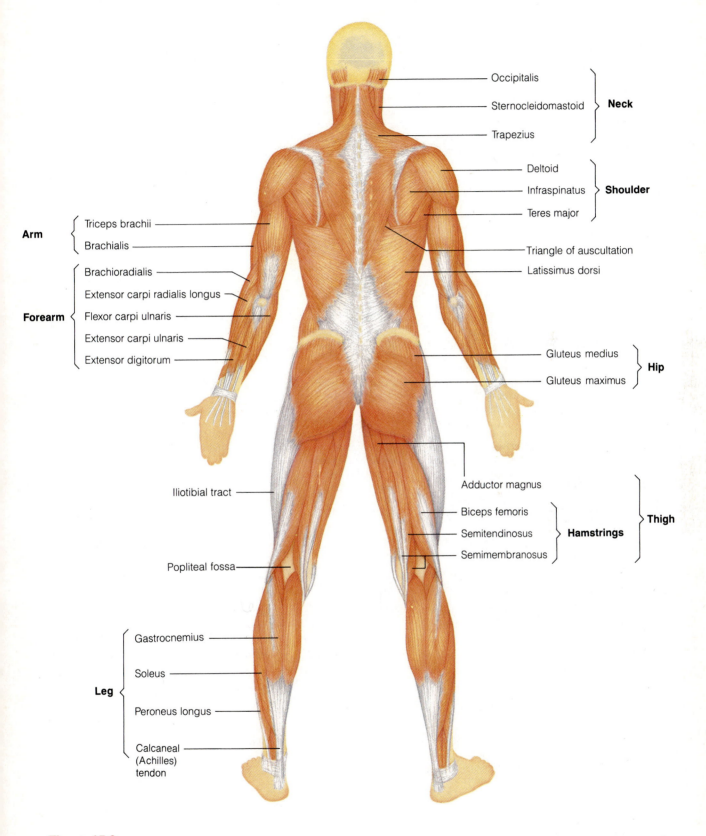

Figure 15.2

Posterior view of superficial muscles of the body, diagrammatic view.

TABLE 15.1   Major Muscles of Human Head (see Figure 15.3)

Muscle	Comments	Origin	Insertion	Action
Figure 15.3(a)				
Facial Expression				
Epicranius: frontalis and occipitalis	Bipartite muscle consisting of frontalis and occipitalis, which covers dome of skull	From the galea aponeurotica, a cranial aponeurosis (frontalis); occipital bone (occipitalis)	Skin of eyebrows and root of nose (frontalis) and cranial aponeurosis (occipitalis)	With aponeurosis fixed, frontalis raises eyebrows; occipitalis fixes aponeurosis and pulls scalp posteriorly
Orbicularis oculi	Sphincter muscle of eyelids	Frontal and maxillary bones and ligaments around orbit	Encircles orbit and inserts in tissue of eyelid	Various parts can be activated individually; closes eyes, produces blinking, squinting, and draws eyebrows downward
Orbicularis oris	Multilayered sphincter muscle of lips with fibers that run in many different directions	Arises indirectly from maxilla and mandible; fibers blended with fibers of other oral muscles	Encircles mouth; inserts into muscle and skin at angles of mouth	Closes mouth; purses and protrudes lips (kissing muscle)
Corrugator supercilii	Small muscle; activity associated with that of orbicularis oculi	Arch of frontal bone above nasal bone	Skin of eyebrow	Draws eyebrows medially; wrinkles skin of forehead vertically
Levator labii Superioris	Thin muscle between orbicularis oris and inferior eye margin	By three heads from the zygomatic, infraorbital margin of maxilla, root of nose	Skin and muscle of upper lip and border of nostril	Raises and furrows upper lip; flares nostril (as in disgust)
Zygomaticus— major and minor	Extends diagonally from corner of mouth to cheekbone	Zygomatic bone	Skin and muscle at corner of mouth	Raises lateral corners of mouth upward (smiling muscle)
Depressor labii inferioris	Small muscle from lower lip to jawbone	Body of mandible lateral to its midline	Skin and muscle of lower lip	Draws lower lip downward
Depressor anguli oris	Small muscle lateral to depressor labii inferioris	Body of mandible below incisors	Skin and muscle at angle of mouth below insertion of zygomaticus	Zygomaticus antagonist; draws corners of mouth downward and laterally
Mentalis	One of muscle pair forming V-shaped muscle mass on chin	Incisor fossa of mandible	Skin of chin	Protrudes lower lip; wrinkles chin
Buccinator	Principal muscle of cheek; runs horizontally, deep to the masseter	Molar region of maxilla and mandible	Orbicularis oris	Draws corner of mouth laterally; compresses cheek (as in whistling); holds food between teeth during chewing
Mastication				
Masseter	Extends across jawbone; can be palpated on forcible closure of jaws	Zygomatic process and arch	Angle and ramus of mandible	Closes jaw and elevates mandible

(continued on p. 102)

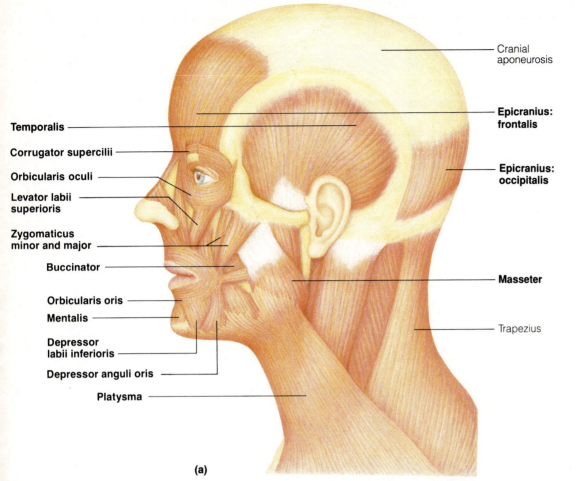

Temporalis

Corrugator supercilii

Orbicularis oculi

Levator labii
superioris

Zygomaticus
minor and major

Buccinator

Orbicularis oris

Mentalis

Depressor
labii inferioris

Depressor anguli oris

Platysma

Cranial
aponeurosis

**Epicranius:
frontalis**

**Epicranius:
occipitalis**

Masseter

Trapezius

(a)

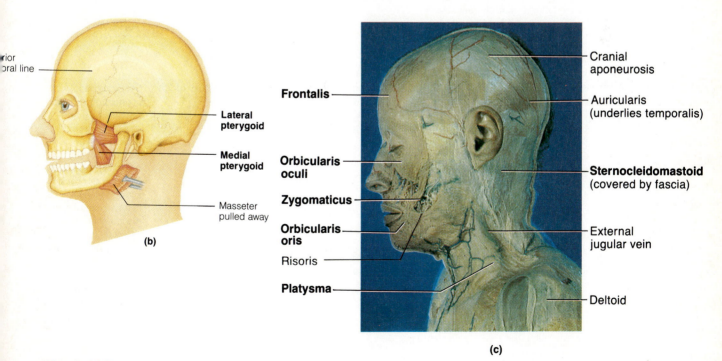

rior
oral line

Lateral
pterygoid

Medial
pterygoid

Masseter
pulled away

(b)

Frontalis

**Orbicularis
oculi**

Zygomaticus

**Orbicularis
oris**

Risoris

Platysma

Cranial
aponeurosis

Auricularis
(underlies temporalis)

Sternocleidomastoid
(covered by fascia)

External
jugular vein

Deltoid

(c)

Figure 15.3

Muscles of the scalp, face, and neck: (a) lateral view; (b) lateral view of the deep chewing muscles and the medial and lateral pterygoid muscles; (c) photo of superficial structures of head and neck, lateral view.

TABLE 15.1   *(continued)*

Muscle	Comments	Origin	Insertion	Action
Temporalis	Fan-shaped muscle over temporal bone	Temporal fossa	Coronoid process of mandible	Closes jaw and elevates mandible
Buccinator	(See muscles of facial expression.)			

Figure 15.3(b)

Muscle	Comments	Origin	Insertion	Action
Pterygoid—medial	Runs along internal (medial) surface of mandible (thus largely concealed by that bone)	Sphenoid, palatine, and maxillary bones	Medial surface of mandibular ramus and angle	Synergist of temporalis and masseter; closes and elevates mandible; in conjunction with lateral pterygoid, aids in grinding movements of jaw
Pterygoid—lateral	Superior to medial pterygoid	Greater wing of sphenoid bone	Mandibular condyle	Protracts jaw (moves it forward); in conjunction with medial pterygoid, aids in grinding movements of jaw

TABLE 15.2   Anterolateral Muscles of Human Neck (see Figure 15.4)

Muscle	Comments	Origin	Insertion	Action
Superficial				
Platysma	Unpaired muscle: thin, sheetlike superficial neck muscle, not strictly a head muscle but plays role in facial expression (see Fig. 15.3)	Fascia of chest (over pectoral muscles and deltoid)	Lower margin of mandible, skin, and muscle at corner of mouth	Depresses mandible; pulls lower lip back and down; i.e., produces downward sag of the mouth
Sternocleidomastoid	Two-headed muscle located deep to platysma on anterolateral surface of neck; fleshy parts on either side indicate the limits of anterior and posterior triangles of neck	Manubrium of sternum and medial portion of clavicle	Mastoid process of temporal bone	Simultaneous contraction of both muscles of pair causes flexion of neck forward, generally against resistance (as when lying on the back); acting independently, rotate head toward shoulder on opposite side
Scalenes—anterior, medial, and posterior	Located more on lateral than anterior neck; deep to platysma (see Fig. 15.4b)	Transverse processes of C_1 to C_5	Anterolaterally on first two ribs	Flex and slightly rotate neck; elevate first two ribs (aid in inspiration)

(continued on p. 104)

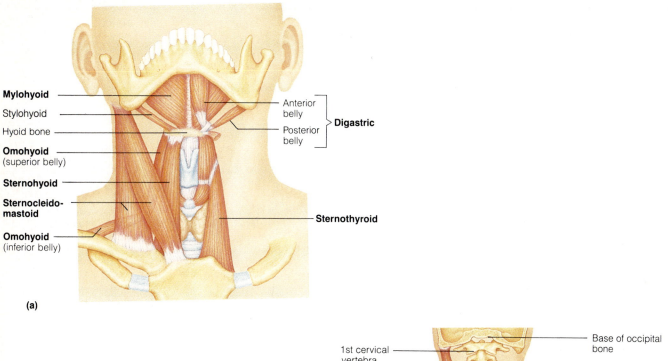

Mylohyoid
Stylohyoid
Hyoid bone
Omohyoid
(superior belly)
Sternohyoid
**Sternocleido-
mastoid**
Omohyoid
(inferior belly)

Anterior
belly
Posterior
belly
} **Digastric**

Sternothyroid

(a)

1st cervical
vertebra

Base of occipital
bone

**Sternocleido-
mastoid**

**Middle
scalene**
**Anterior
scalene**
**Posterior
scalene**

(b)

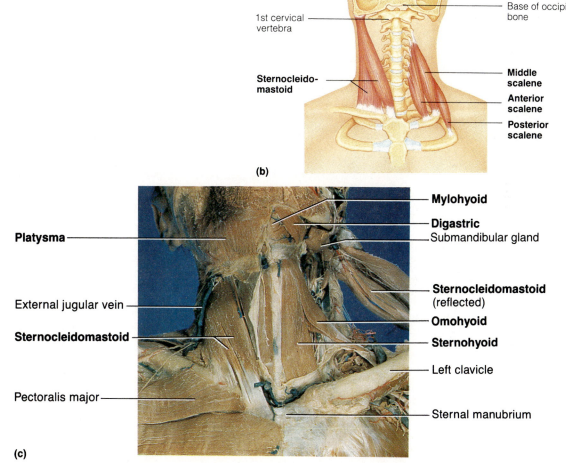

Mylohyoid
Digastric
Submandibular gland

Platysma

Sternocleidomastoid
(reflected)
Omohyoid
Sternohyoid

External jugular vein

Sternocleidomastoid

Left clavicle

Pectoralis major

Sternal manubrium

(c)

Figure 15.4

Muscles of neck and throat: (a) anterior view of the suprahyoid and infrahyoid muscles; (b) muscles of the anterolateral neck. The superficial platysma muscle and the deeper neck muscles have been removed to show the origins and insertions of the sternocleidomastoid and scalene muscles clearly; (c) photo of the anterior and lateral regions of the neck. The fascia has been partially removed (left side of photo) to expose the sternocleidomastoid muscle. On the right side of the photo, the sternocleidomastoid muscle is reflected to expose the sternohyoid and omohyoid muscles.

TABLE 15.2   *(continued)*

Muscle	Comments	Origin	Insertion	Action

Figure 15.4(a)

Deep

Muscle	Comments	Origin	Insertion	Action
Digastric	Consists of two bellies united by an intermediate tendon; assumes a V-shaped configuration under chin; attached by a connective tissue loop to hyoid bone	Lower margin of mandible (anterior belly) and mastoid process (posterior belly)	Hyoid bone	Acting in concert, elevate hyoid bone and may depress mandible
Mylohyoid	Just deep to digastric; forms floor of mouth	Medial surface of mandible	Hyoid bone	Elevate hyoid bone and base of tongue
Sternohyoid	Runs most medially along neck; straplike	Posterior surface of manubrium	Lower margin of body of hyoid bone	Acting with sternothyroid and omohyoid (all below hyoid bone), depresses larynx and hyoid bone if mandible is fixed; may also flex skull
Sternothyroid	Lateral to sternohyoid; straplike	Manubrium and medial end of clavicle	Thyroid cartilage of larynx	(See sternohyoid, above)
Omohyoid	Straplike with two bellies; lateral to sternohyoid	Superior surface of scapula	Hyoid bone	(See sternohyoid, above)

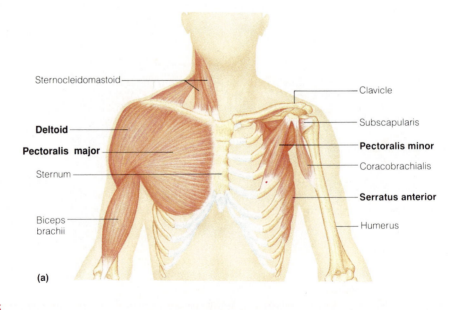

(a)

Figure 15.5

Anterior muscles of the thorax, shoulder, and abdominal wall. (a) Anterior thorax. The superficial pectoralis major and deltoid muscles that effect arm movements are illustrated on the left. These muscles have been removed on the right side of the figure to illustrate (1) the pectoralis minor and serratus anterior muscles and (2) the subscapularis muscle.

TABLE 15.3 Anterior Muscles of Human Thorax, Shoulder, and Abdominal Wall (see Figure 15.5)

Muscle	Comments	Origin	Insertion	Action
Figure 15.5(a)				
Pectoralis major	Large fan-shaped muscle covering upper portion of chest	Clavicle, sternum, and cartilage of first six ribs and aponeurosis of external oblique muscle	Fibers converge to insert by short tendon into greater tubercle of humerus	Flexes adducts, medially rotates arm; with arm fixed, pulls chest upward (thus also acts in forced inspiration)
Serratus anterior	Deep and superficial portions; beneath and inferior to pectoral muscles on lateral rib cage	Lateral aspect of first to eighth (or ninth) ribs	Vertebral border of anterior surface of scapula	Moves scapula forward toward chest wall; moves inferior angle of scapula laterally and upward
Deltoid	Fleshy triangular muscle forming shoulder muscle mass	Lateral third of clavicle; acromion process and spine of scapula	Deltoid tuberosity of humerus	Acting as a whole, abducts arm; when only specific fibers are active, can aid in flexion, extension, and rotation of humerus
Pectoralis minor	Flat, thin muscle directly beneath and obscured by pectoralis major	Upper border of third, fourth, and fifth ribs, near their costal cartilages	Corocoid process of scapula	With ribs fixed, draws scapula forward and downward; with scapula fixed, draws rib cage upward
Intercostals—external	11 pairs lie between ribs; fibers run obliquely toward sternum	Inferior border of rib above (not shown in figure)	Superior border of rib below	Pulls ribs toward one another to elevate rib cage; aids in inspiration
Intercostals—internal	11 pairs lie between ribs; fibers run at right angles to those of external intercostals	Superior border of rib below	Inferior border of rib above (not shown in figure)	Draws ribs together to depress rib cage; aids in expiration; antagonistic to external intercostals
Figure 15.5(b)				
Rectus abdominis	Medial superficial muscle, extends from pubis to rib cage; ensheathed by fascia of oblique muscles; segmented	Pubic crest	Xiphoid process and costal cartilages of fifth through seventh ribs	Flexes vertebral column; increases abdominal pressure; depresses ribs
External oblique	Most superficial lateral muscles; fibers run downward and medially; ensheathed by an aponeurosis	Anterior surface of last eight ribs	Linea alba* and iliac crest	As for Rectus abdominis; also aids muscles of back in trunk rotation and lateral flexion
Internal oblique	Fibers run at right angles to those of external oblique, which it underlies	Lumbodorsal fascia, iliac crest, and inguinal ligament	Linea alba, pubic crest, and costal cartilages of last three ribs	As for external oblique
Transversus abdominis	Innermost muscle of abdominal wall; fibers run horizontally	Inguinal ligament, iliac crest, and cartilages of last five or six ribs	Linea alba and pubic crest	Compresses abdominal contents

*The linea alba ("white line") is a narrow, tendinous sheath that runs along the middle of the abdomen from the sternum to the pubic symphysis. It is formed by the fusion of the aponeurosis of the external oblique and transversus muscles.

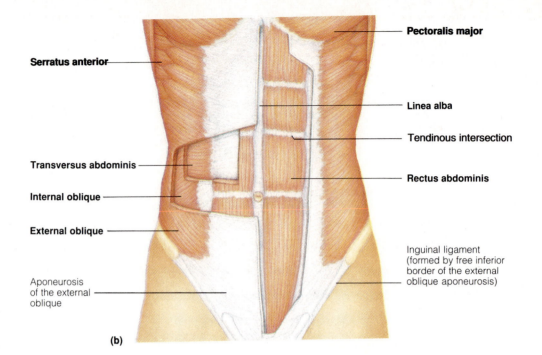

Pectoralis major

Serratus anterior

Linea alba

Tendinous intersection

Transversus abdominis

Rectus abdominis

Internal oblique

External oblique

Inguinal ligament
(formed by free inferior
border of the external
oblique aponeurosis)

Aponeurosis
of the external
oblique

(b)

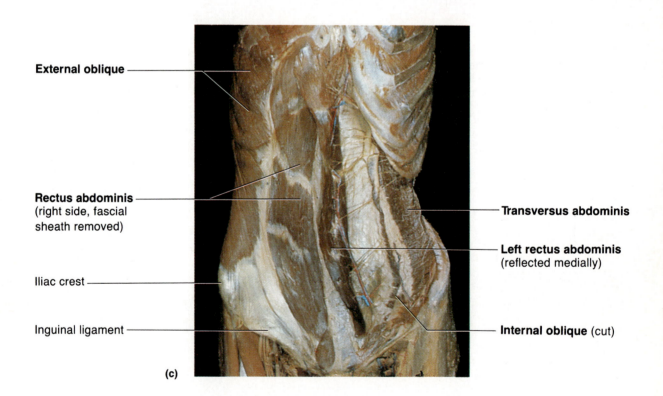

External oblique

Rectus abdominis
(right side, fascial
sheath removed)

Transversus abdominis

Left rectus abdominis
(reflected medially)

Iliac crest

Inguinal ligament

Internal oblique (cut)

(c)

Figure 15.5 *(continued)*

Anterior muscles of the thorax, shoulder, and abdominal wall. (b) Anterior view of the muscles forming the anterolateral abdominal wall. The superficial muscles have been partially cut away on the left side of the diagram to reveal the deeper internal oblique and transversus abdominis muscles. (c) Photo of hte anterolateral abdominal wall.

TABLE 15.4 Posterior Muscles of Human Trunk (see Figure 15.6)

Muscle	Comments	Origin	Insertion	Action
Figure 15.6(a)				
Trapezius	Most superficial muscle of posterior neck and trunk; very broad origin and insertion	Occipital bone ligamentum nuchae and spines of C_7 and all thoracic vertebrae	Acromion and spinous process of scapula; lateral third of clavicle	Extends head; adducts scapula and stabilizes it; upper fibers elevate scapula; lower fibers depress it
Latissimus dorsi	Broad flat muscle of lower back (lumbar region); extensive superficial origins	Indirect attachment to spinous processes of lower six thoracic vertebrae, lumbar vertebrae, lower 3 to 4 ribs, and iliac crest	Floor of intertubercular groove of humerus	Extends, adducts, and medially rotates arm; depresses scapula; brings arm down in power stroke
Infraspinatus	Partially covered by deltoid and trapezius	Infraspinous fossa of scapula	Greater tubercle of humerus	Lateral rotation of humerus; helps hold head of humerus in glenoid cavity
Teres minor	Small muscle inferior to infraspinatus	Lateral margin of scapula	Greater tuberosity of humerus	As for infraspinatus
Teres major	Located inferiorly to teres minor	Posterior surface at inferior angle of scapula	Crest of lesser tubercle of humerus	Extends, medially rotates, and adducts humerus
Supraspinatus	Obscured by trapezius and deltoid	Supraspinous fossa of scapula	Greater tubercle of humerus	Assists abduction of humerus
Levator scapulae	Located at back and side of neck, deep to trapezius	Transverse processes of C_1 through C_4	Superior vertebral border of scapula	Raises and adducts scapula; with fixed scapula, flexes neck to the same side
Rhomboids—major and minor	Beneath trapezius and posterior to levator scapulae; run from vertebral column to scapula	Spinous processes of C_7 and T_1 through T_5	Vertebral border of scapula	Pull scapula medially (retraction) and elevate it
Figure 15.6(b)				
Semispinalis	Deep composite muscle of the back—thoracis, cervicis, and capitis portions	Transverse processes of C_7-T_{12}	Occipital bone and spinous process of cervical vetebrae and T_1-T_4	Acting together, extend head and vertebral column; acting independently causes rotation toward the opposite side
Erector spinae	Tripartite muscle composed of iliocostalis (lateral), longissimus, and spinalis (medial); superficial to semispinalis muscles; extends from pelvis to head	Sacrum, iliac crest, transverse processes of lumbar, thoracic and cervical vertebrae, and/or ribs 3-12 depending on specific part	Ribs and transverse processes of vertebrae about six segments above origin. Longissimus so inserts into mastoid process	All act to extend and abduct vertebral column; fibers of the longissimus also extend head
Splenius (Not shown in Fig. 15.6(b))	Superficial muscle (capitis and cervicis parts) just deep to levator scapulae and superficial to erector spinae	Ligamentum nuchae and spinous processes of C_7-T_6	Mastoid process, occipital bone, and transverse processes of C_2-C_4	As a group, extend or hyperextend head; with one side is active, head is rotated and bent toward same side
Quadratus lumborum	Forms greater portion of posterior abdominal wall (See also Fig. 15.9.)	Iliac crest and iliolumbar fascia	Inferior border twelfth rib	Each flexes vertebral column laterally; acting jointly, extend lumbar vertebral column and fix ribs

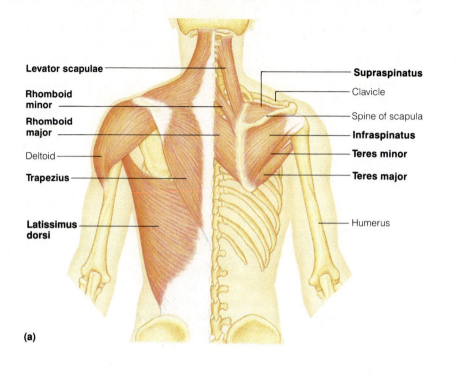

(a)

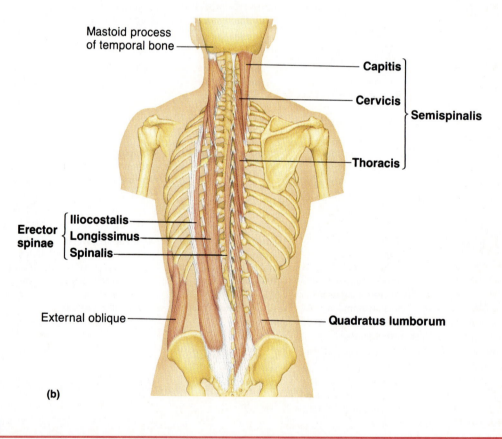

(b)

Figure 15.6

Muscles of the neck, shoulder, and thorax, posterior view. (a) The extensive superficial muscles of hte back are shown on the left. The superficial muscles are removed on the right to reveal (1) the deeper levator scapulae and rhomboid muscles acting on the scapula and (2) the supraspinatus, teres major, and teres minor muscles. (b) The erector spinae and semispinalis muscles which respectively form the intermediate and deep muscle layers of the back associated with the vertebral column.

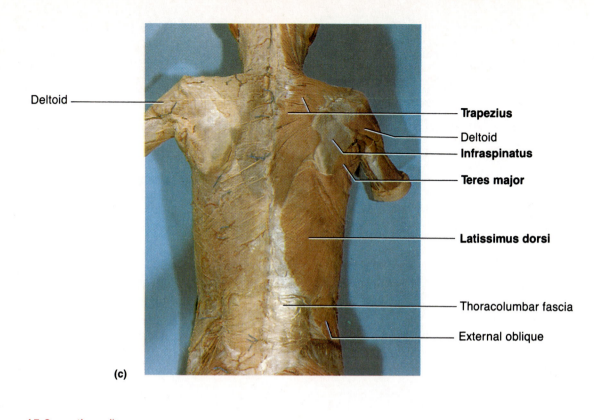

Deltoid ———————————

Trapezius
Deltoid
Infraspinatus
Teres major

Latissimus dorsi

Thoracolumbar fascia
External oblique

(c)

Figure 15.6 *continued)*

Muscles of the neck, shoulder, and thorax, posterior view. (c) Photo of superficial muscles of the back. The fascia has been partially removed on the right side.

Muscles of the Upper Extremity

The muscles that act on the upper extremity fall into three groups: those that move the arm, those causing movement at the elbow, and those effecting movements of the wrist and hand.

The muscles that cross the shoulder joint to insert on the humerus and cause movement of the arm are primarily trunk muscles (subscapularis, supraspinatus and infraspinatus, deltoid, and so on) that originate on the axial skeleton or shoulder girdle. These muscles are included with the trunk muscles.

The second group of muscles, which cross the elbow joint to bring about movement of the forearm, consists of the muscles forming the musculature of the humerus. These muscles arise primarily from the humerus and insert in forearm bones. They are responsible for flexion, extension, pronation, and supination. The origins, insertions, and actions of these muscles are summarized in Table 15.5 and the muscles are shown in Figure 15.7.

The third group composes the musculature of the forearm. For the most part, these muscles insert on the digits, producing movements at the wrist and fingers. In general, muscles acting on the wrist and hand can more easily be identified if their insertion tendons are located first. These muscles are described in Table 15.6 and illustrated in Figure 15.8.

First study the tables and figures, then see if you can identify these muscles on a torso model or anatomic chart.

Complete this portion of the exercise with palpation demonstrations as outlined next.

● To observe the biceps brachii, attempt to flex your forearm (hand supinated) against resistance. The tendons can also be felt.

● If you acutely flex your elbow and then try to extend it against resistance, you can demonstrate the action of your triceps brachii.

TABLE 15.5 Muscles of Human Humerus that Act on the Forearm (see Figure 15.7)

Muscle	Comments	Origin	Insertion	Action
Triceps brachii	Sole, large fleshy muscle of posterior humerus; three-headed origin	Long head: inferior margin of glenoid fossa; lateral head: posterior humerus; medial head: distal radial groove	Olecranon process of ulna	Powerful forearm extensor; antagonist of brachialis and biceps brachii
Biceps brachii	Most familiar muscle of anterior humerus because it bulges when forearm is flexed	Short head: coracoid process; tendon of long head runs in intertubercular groove over top of shoulder joint	Radial tuberosity	Flexion (powerful) of elbow and supination of hand; flexion of the shoulder; "it turns the corkscrew and pulls the cork"
Brachioradialis	Superficial muscle of lateral forearm	Lateral ridge at distal end of humerus	Base of styloid process of radius	Forearm flexor (weak)
Brachialis	Immediately deep to biceps brachii	Distal portion of anterior humerus	Coronoid process of ulna	A major flexor of forearm

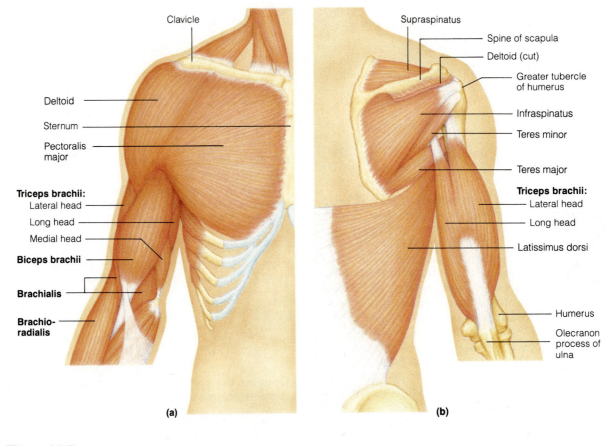

Figure 15.7

Muscles causing movements of the forearm: (a) superficial muscles of the anterior thorax, shoulder, and arm, anterior view; (b) Posterior aspect of the arm showing the lateral and long heads of the triceps brachii muscle.

TABLE 15.6 Muscles of Human Forearm that Act on Hand and Fingers (see Figure 15.8)

Muscle	Comments	Origin	Insertion	Action
Anterior Compartment (Figure 15.8(a,b,c)):				
Flexor carpi radialis	Superficial; runs diagonally across forearm	Medial epicondyle of humerus	Base of second and third metacarpal	Powerful flexor of wrist; abducts hand
Palmaris longus	Small fleshy muscle with a long tendon; medial to flexor carpi radialis	Medial epicondyle of humerus	Palmar aponeurosis	Flexes wrist (weak)
Flexor carpi ulnaris	Superficial; medial to palmaris longus	Medial epicondyle of humerus and olecranon process of ulna	Base of fifth metacarpal	Powerful flexor of wrist; adducts hand
Flexor digitorum superficialis	Deeper muscle; overlain by muscles named above; visible at distal end of forearm	Medial epicondyle of humerus, medial surface of ulna, and anterior border of radius	Middle phalanges of second through fifth fingers	Flexes wrist and middle phalanges of second through fifth fingers
Flexor pollicis longus	Deep muscle of anterior forearm; distal to and paralleling lower margin of flexor digitorum superficialis	Anterior surface of radius, and interosseous membrane	Distal phalanx of thumb	Flexes thumb (*pollix* is Latin for "thumb"); weak flexor of wrist
Flexor digitorum profundus	Deep muscle; overlain entirely by flexor digitorum superficialis	Anteromedial surface of ulna and interosseous membrane	Distal phalanges of second through fifth fingers	Sole muscle that flexes distal phalanges; assists in wrist flexion
Pronator teres	Seen in a superficial view between proximal margins of brachioradialis and flexor carpi ulnaris	Medial epicondyle of humerus and coronoid process of ulna	Midshaft of radius	Flexes forearm and acts synergistically with pronator quadratus to pronate hand
Pronator quadratus	Deep muscle of distal forearm	Distal portion of ulna	Anterior surface of radius, distal end	Pronates hand
Posterior Compartment (Figure 15.8(d,e)):				
Extensor carpi radialis longus	Superficial; parallels brachioradialis on lateral forearm	Lateral supracondylar ridge of humerus	Base of second metacarpal	Extends and abducts wrist (extends forearm)
Extensor carpi radialis brevis	Posterior to extensor carpi radialis longus	Lateral epicondyle of humerus	Base of third metacarpal	Extends and abducts wrist; steadies wrist during finger flexion
Extensor carpi ulnaris	Superficial; medial posterior forearm	Lateral epicondyle of humerus	Base of fifth metacarpal	Extends and adducts wrist

(continued)

TABLE 15.6   *(continued)*

Muscle	Comments	Origin	Insertion	Action
Extensor digitorum	Superficial; between extensor carpi ulnaris and extensor carpi radialis brevis	Lateral epicondyle of humerus	By four tendons into distal phalanges of second through fifth fingers	Extends fingers and wrist; adducts fingers
Extensor pollicis longus and brevis	Deep muscle overlain by extensor carpi ulnaris	Dorsal shaft of ulna, radius, and interosseous membrane	Distal phalanx (longus) and proximal phalanx (brevis) of thumb	Extends thumb
Abductor pollicis longus	Deep muscle; lateral and parallel to extensor pollicis longus	Posterior surface of radius and ulna; interosseous membrane	First metacarpal	Abducts and extends thumb
Supinator	Deep muscle at posterior aspect of elbow	Lateral epicondyle of humerus	Proximal end of radius	Supinates hand; antagonistic to pronator muscles

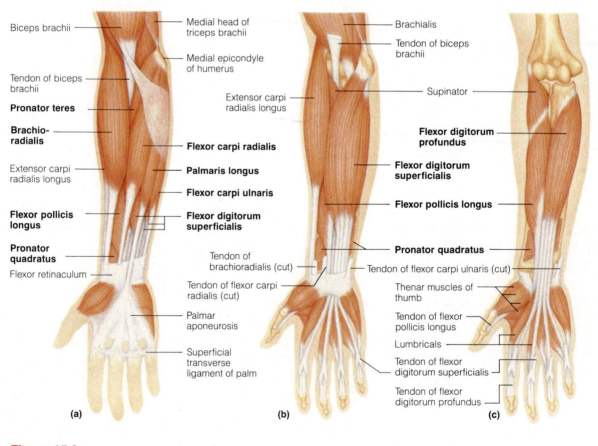

(a)

Biceps brachii

Tendon of biceps brachii

Pronator teres

Brachio-radialis

Extensor carpi radialis longus

Flexor pollicis longus

Pronator quadratus

Flexor retinaculum

Medial head of triceps brachii

Medial epicondyle of humerus

Flexor carpi radialis

Palmaris longus

Flexor carpi ulnaris

Flexor digitorum superficialis

Tendon of brachioradialis (cut)

Tendon of flexor carpi radialis (cut)

Palmar aponeurosis

Superficial transverse ligament of palm

(b)

Brachialis

Tendon of biceps brachii

Extensor carpi radialis longus

Tendon of flexor carpi radialis (cut)

(c)

Supinator

Flexor digitorum profundus

Flexor digitorum superficialis

Flexor pollicis longus

Pronator quadratus

Tendon of flexor carpi ulnaris (cut)

Thenar muscles of thumb

Tendon of flexor pollicis longus

Lumbricals

Tendon of flexor digitorum superficialis

Tendon of flexor digitorum profundus

Figure 15.8

Muscles of the forearm and wrist. (a) Superficial anterior view of muscles of right forearm and hand. (b) The brachioradialis, flexors carpi radialis and ulnaris, and palmaris longus muscles have been removed to reveal the position of the somewhat deeper flexor digitorum superficialis. (c) Deep muscles of the anterior compartment. Superficial muscles have been removed.

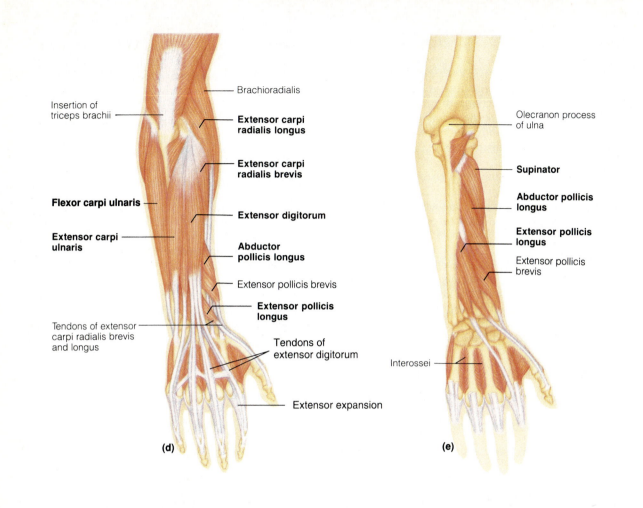

(d)

Insertion of
triceps brachii

Flexor carpi ulnaris

**Extensor carpi
ulnaris**

Tendons of extensor
carpi radialis brevis
and longus

Brachioradialis

**Extensor carpi
radialis longus**

**Extensor carpi
radialis brevis**

Extensor digitorum

**Abductor
pollicis longus**

Extensor pollicis brevis

**Extensor pollicis
longus**

Tendons of
extensor digitorum

Extensor expansion

(e)

Olecranon process
of ulna

Supinator

**Abductor pollicis
longus**

**Extensor pollicis
longus**

Extensor pollicis
brevis

Interossei

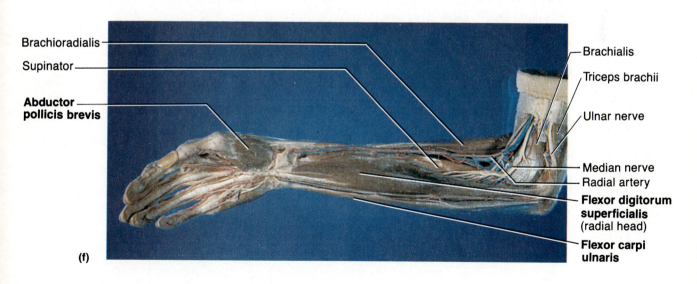

(f)

Brachioradialis

Supinator

**Abductor
pollicis brevis**

Brachialis

Triceps brachii

Ulnar nerve

Median nerve

Radial artery

**Flexor digitorum
superficialis**
(radial head)

**Flexor carpi
ulnaris**

Figure 15.8 *(continued)*

(d) Superficial muscles, posterior view. (e) Deep posterior muscles; superficial muscles have been removed. The interossei, the deepest layer of intrinsic hand muscles, are also illustrated. Note: The thenar muscles of the thumb and the lumbricals that help move the fingers illustrated here are not described in Table 15.6. (f) Photo of muscles of the anteromedial aspect of the forearm. The pronator teres and flexor carpi radialis muscles have been removed.

Muscles of the Lower Extremity

Muscles that act on the lower extremity cause movement at the hip, knee, and foot joints. Since the human pelvic girdle is composed of heavy fused bones that allow little movement, no special group of muscles is necessary to stabilize it. This is unlike the shoulder girdle, where many muscles (mainly trunk muscles) are necessary to stabilize the scapulae.

Muscles acting on the thigh (femur) cause various movements at the multiaxial hip joint (flexion, extension, rotation, abduction, and adduction). These include the iliopsoas, the adductor group, and other muscles summarized in Tables 15.7 and 15.8 and illustrated in Figures 15.9 and 15.10.

Muscles acting on the leg form the major musculature of the thigh. (Anatomically the term *leg* refers only to that portion between the knee and the ankle.) The thigh muscles cross the knee to allow its flexion and extension. They include the hamstrings and the quadriceps and, along with the muscles acting on the

thigh, are described in Tables 15.7 and 15.8 and illustrated in Figures 15.9 and 15.10. Since some of these muscles also have attachments on the pelvic girdle, they can cause movement at the hip joint.

 The muscles originating on the leg and acting on the foot and toes are described in Table 15.9 and shown in Figures 15.11 and 15.12. Identify the muscles as instructed previously. Complete this exercise by performing the following palpation demonstrations.

- You can demonstrate the quadriceps femoris by trying to extend the knee against resistance. Note how the patellar tendon reacts. The biceps femoris comes into play when you flex your knee against resistance.
- Now stand on your toes. Have your partner palpate the lateral and medial heads of the gastrocnemius and follow it to its insertion in the Achilles tendon.

TABLE 15.7 Muscles Acting on Human Thigh and Leg, Anterior Aspect (see Figure 15.9)

Muscle	Comments	Origin	Insertion	Action
Iliopsoas—iliacus and psoas major	Two closely related muscles; fibers pass under inguinal ligament to insert into femur via a common tendon	Iliac fossa (iliacus); transverse processes, bodies, and discs of T_{12} and lumbar vertebrae (psoas major)	Lesser trochanter of femur	Flex trunk on thigh; major flexor of hip (or thigh on pelvis when pelvis is fixed)
Tensor fasciae latae	Enclosed between fascia layers of thigh	Anterior aspect of iliac crest	Iliotibial band of fascia lata	Flexes, abducts, and medially rotates thigh
Adductors—magnus, longus, and brevis	Large muscle mass forming medial aspect of thigh; arise from front of pelvis and insert at various levels on femur	Ischial and pubic rami (magnus); pubis near pubic symphysis (longus); and inferior ramus of pubis (brevis—not shown in figure)	Linea aspera and adductor tubercle of femur (magnus); linea aspera above magnus (longus); linea aspera above longus (brevis)	Adduct and laterally rotate and flex thigh; posterior part of magnus is also a synergist in thigh extension
Pectineus	Overlies adductor brevis on proximal thigh	Pectineal line of pubis	Inferior to lesser trochanter of femur	Adducts, flexes, and laterally rotates thigh
Gracilis	Straplike superficial muscle of medial thigh	Inferior ramus of pubis	Medial surface of head of tibia	Adducts thigh; flexes and medially rotates leg

(continued on p. 116)

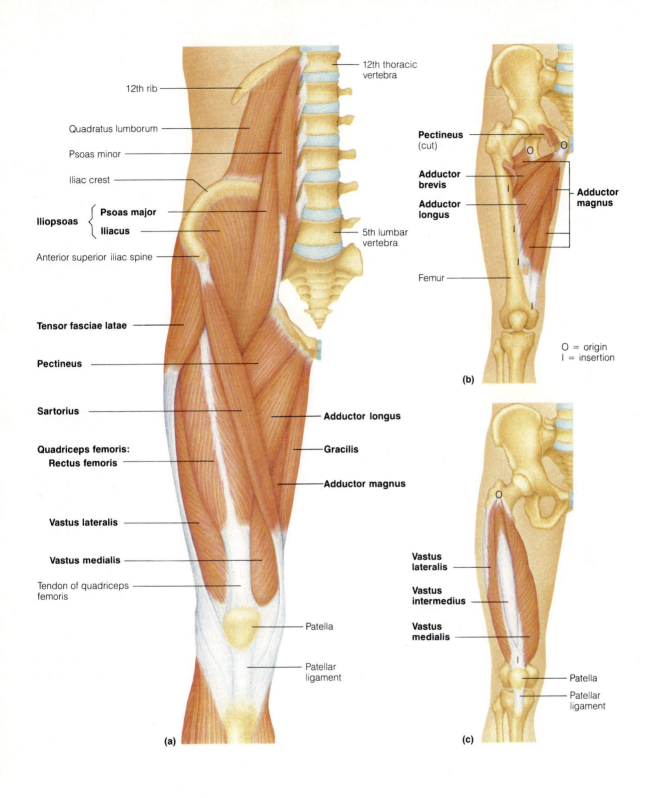

12th thoracic vertebra

12th rib

Quadratus lumborum

Psoas minor

Iliac crest

Iliopsoas { **Psoas major** **Iliacus** }

Anterior superior iliac spine

5th lumbar vertebra

Tensor fasciae latae

Pectineus

Sartorius

Quadriceps femoris:
  Rectus femoris

Vastus lateralis

Vastus medialis

Tendon of quadriceps femoris

Patella

Patellar ligament

Adductor longus

Gracilis

Adductor magnus

(a)

Pectineus (cut)

Adductor brevis

Adductor longus

Adductor magnus

O

O

I

I

I

I

Femur

O = origin
I = insertion

(b)

O

Vastus lateralis

Vastus intermedius

Vastus medialis

I

Patella

Patellar ligament

(c)

Figure 15.9

Anterior view of the deep muscles of the pelvis and superficial muscles of the right thigh.

TABLE 15.7 Muscles Acting on Human Thigh and Leg, Anterior Aspect (*continued*)

Muscle	Comments	Origin	Insertion	Action
Sartorius	Straplike superficial muscle running obliquely across anterior surface of thigh to knee	Anterior superior iliac spine	By an aponeurosis into medial aspect of proximal tibia	Flexes and laterally rotates thigh; flexes knee; known as "tailor's muscle" because it helps bring about cross-legged position in which tailors are often depicted
Quadriceps*				
Rectus femoris	Superficial muscle of thigh; runs straight down thigh; only muscle of group to cross hip joint; arises from two heads	Anterior inferior iliac spine and superior margin of acetabulum	Tibial tuberosity	Extends knee and flexes thigh at hip
Vastus lateralis	Forms lateral aspect of thigh	Greater trochanter and linea aspera	Tibial tuberosity	Extends knee
Vastus medialis	Forms medial aspect of thigh	Linea aspera	Tibial tuberosity	Extends knee
Vastus intermedius	Obscured by rectus femoris; lies between vastus lateralis and vastus medialis on anterior thigh	Anterior and lateral surface of femur (not shown in figure)	Tibial tuberosity	Extends knee

*The quadriceps form the flesh of the anterior thigh and have a common insertion in the tibial tuberosity via the patellar tendon. They are powerful leg extensors enabling humans to kick a football, for example.

TABLE 15.8 Muscles Acting on Human Thigh and Leg, Posterior Aspect (see Figure 15.10)

Muscle	Comments	Origin	Insertion	Action
Gluteus maximus	Largest and most superficial of gluteal muscles (which form buttock mass)	Ilium, sacrum, and coccyx	Gluteal tuberosity of femur and iliotibial tract*	Complex, powerful hip extensor (most effective when hip is flexed, as in climbing stairs—but not as in walking); antagonist of iliopsoas; laterally rotates thigh
Gluteus medius	Partially covered by gluteus maximus	Upper lateral surface of ilium	Greater trochanter of femur	Abducts and medially rotates thigh
Gluteus minimus	Smallest and deepest gluteal muscle	Inferior surface of ilium (not shown in figure)	Greater trochanter of femur	Abducts and medially rotates thigh

(continued on p. 118)

*The iliotibial tract, a thickened lateral portion of the fascia lata, ensheaths all the muscles of the thigh. It extends as a tendinous band from the iliac crest to the knee.

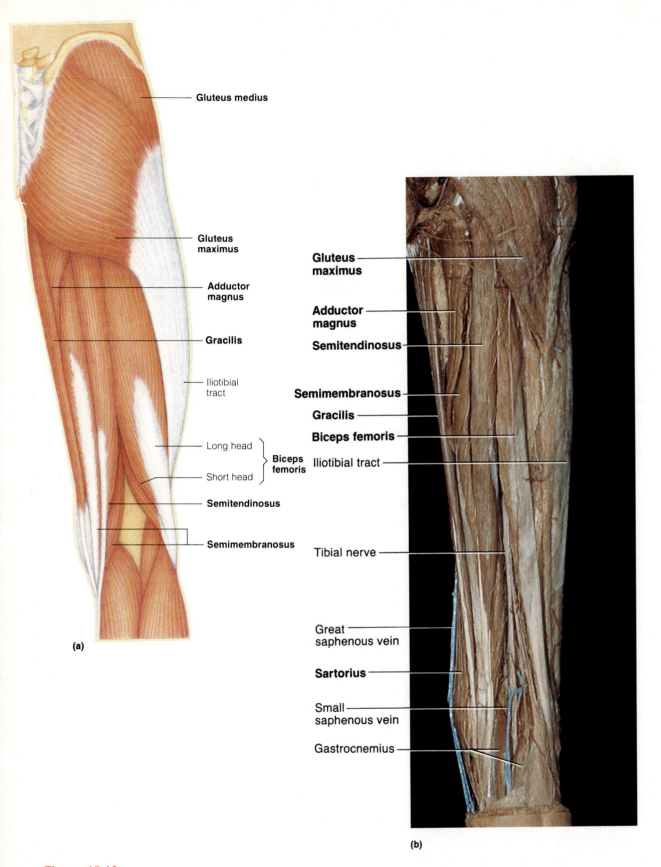

Gluteus medius

Gluteus maximus

Adductor magnus

Gracilis

Iliotibial tract

Long head
Short head
} Biceps femoris

Semitendinosus

Semimembranosus

(a)

Gluteus maximus

Adductor magnus

Semitendinosus

Semimembranosus

Gracilis

Biceps femoris

Iliotibial tract

Tibial nerve

Great saphenous vein

Sartorius

Small saphenous vein

Gastrocnemius

(b)

Figure 15.10

Muscles of right hip and thigh. (a) Superficial view showing the gluteus muscles of the buttock and hamstring muscles of the thigh. (b) Photo of muscles of the posterior thigh.

TABLE 15.8   Muscles Acting on Human Thigh and Leg, Posterior Aspect (*continued*)

Muscle	Comments	Origin	Insertion	Action
Hamstrings*				
Biceps femoris	Most lateral muscle of group; arises from two heads	Ischial tuberosity (long head); linea aspera and distal femur (short head)	Tendon passes laterally to insert into head of fibula and tibia	Extends hip; laterally rotates leg on thigh; flexes knee
Semitendinosus	Medial to biceps femoris	Ischial tuberosity	Medial aspect of upper tibial shaft	Extends thigh; flexes knee; medially rotates leg
Semimembranosus	Deep to semitendinosus	Ischial tuberosity	Medial condyle of tibia	As for semitendinosus

*The hamstrings are the fleshy muscles of the posterior thigh. The name comes from the butchers' practice of using the tendons of these muscles to hang hams for smoking. As a group, they are powerful extensors of the hip; they counteract the powerful quadriceps by stabilizing the knee joint when standing.

TABLE 15.9   Muscles Acting on Human Foot and Ankle (see Figures 15.11 and 15.12)

Muscle	Comments	Origin	Insertion	Action
Superficial posterior (**Figure 15.11(a,b)**):				
Triceps surae	Muscle pair that shapes posterior calf		Via common tendon (Calcaneal Achilles) into heel	Plantarflex foot
Gastrocnemius	Superficial muscle of pair; two prominent bellies	By two heads from medial and lateral condyles of femur	Calcaneus via calcaneal tendon	Crosses knee joint; thus also can flex knee (when foot is dorsiflexed)
Soleus	Deep to gastrocnemius	Proximal portion of tibia and fibula	Calcaneus via calcaneal tendon	

(*continued on p. 121*)

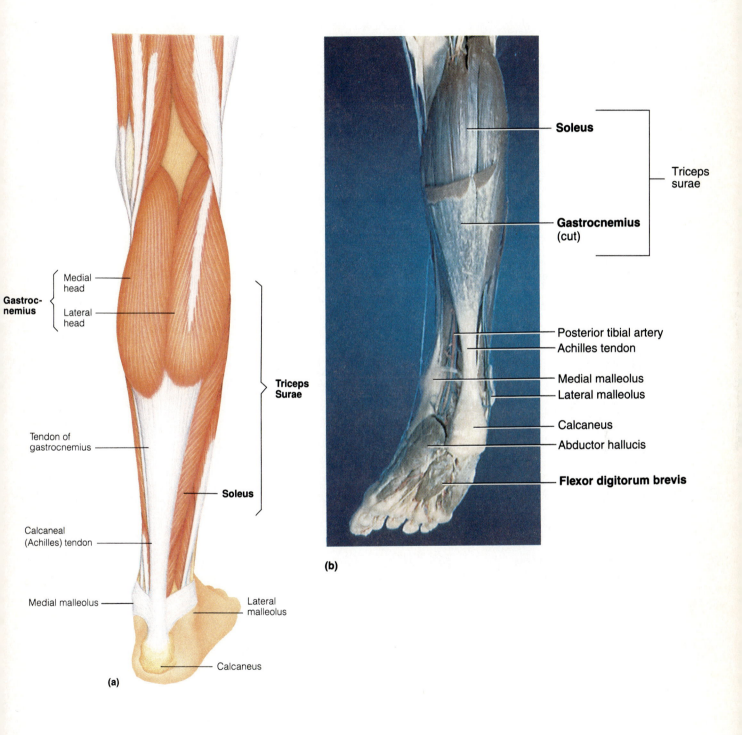

Figure 15.11

Muscles of the right leg. (a) Superficial view of the posterior leg. (b) Photo of posterior aspect of right leg. The gastrocnemius has been transected and its superior part removed.

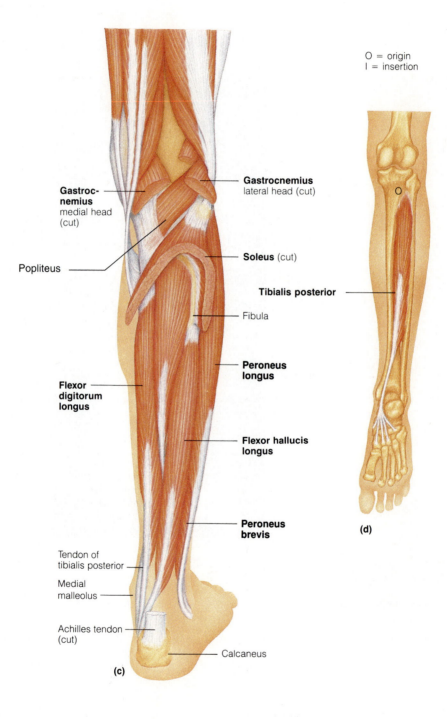

O = origin
I = insertion

Gastroc-
nemius
medial head
(cut)

Gastrocnemius
lateral head (cut)

Popliteus

Soleus (cut)

Tibialis posterior

Fibula

Flexor
digitorum
longus

Peroneus
longus

Flexor hallucis
longus

Peroneus
brevis

Tendon of
tibialis posterior

Medial
malleolus

Achilles tendon
(cut)

Calcaneus

(c)

(d)

Figure 15.11 *(continued)*

Muscles of the right leg. (c) The triceps surae has been removed to show the deep muscles of the posterior compartment. (d) Tibialis posterior shown in isolation so that its origin and insertion may be visualized.

TABLE 15.9 Muscles Acting on Human Foot and Ankle (*continued*)

Muscle	Comments	Origin	Insertion	Action
Deep posterior (Figure 15.11(c,d))				
Popliteus	Thin muscle at posterior aspect of knee	Lateral condyle of femur	Proximal tibia	Flexes and rotates leg medially to "unlock" extended knee when flexion begins
Tibialis posterior	Deep to soleus	Superior portion of tibia and fibula and interosseous membrane	Tendon passes obliquely to medial side of ankle and under arch of foot; inserts into several tarsals and metatarsals, 2–4	Prime mover of foot inversion; plantar flexes foot
Flexor digitorum longus	Runs medial to and partially overlies tibialis posterior	Posterior surface of tibia	Distal phalanges of second through fifth toes	Flexes toes; plantar flexes and inverts foot
Flexor hallucis longus	Lies lateral to inferior aspect of tibialis posterior	Middle portion of fibula shaft	Tendon runs under foot to insert on distal phalanx of great toe	Flexes great toe; plantar flexes and inverts foot
Lateral aspect (Figure 15.11(c) and Figure 15.12(a,b,c))				
Peroneus longus	Superficial lateral muscle; overlies fibula	Head and upper portion of fibula	By long tendon under foot to first metatarsal and medial cuneiform	Plantar flexes and everts foot, helps keep foot flat on ground of peronei group
Peroneus brevis	Smaller muscle; deep to peroneus longus	Distal portion of fibula shaft	By tendon running behind lateral malleolus to insert on proximal end of fifth metatarsal	Plantar flexes and everts foot, as part of peronei group
Peroneus tertius	Small muscle; deep to peroneus brevis	Distal anterior surface of fibula	Tendon passes anteriorly to lateral malleolus; inserts on dorsum of fifth metatarsal	Dorsiflexes and everts foot
Anterior aspect (Figure 15.12(a,b,c))				
Tibialis anterior	Superficial muscle of anterior leg; parallels sharp anterior margin of tibia	Lateral condyle and upper 2/3 of tibia; interosseous membrane	By tendon into inferior surface of first cuneiform and metatarsal	Dorsiflexes and inverts foot
Extensor digitorum longus	Anterolateral surface of leg; lateral to tibialis anterior	Lateral condyle of tibia; proximal 3/4 of fibula; interosseous membrane	Tendon divides into four parts; insert into middle and distal phalanges, toes 2–5	Prime mover of toe extension; dorsiflexes and everts foot
Extensor hallucis longus	Deep to extensor digitorum longus and tibialis anterior	Anteromedial shaft of fibula and interosseous membrane	Tendon inserts on distal phalanx of great toe	Extends great toe and dorsiflexes foot

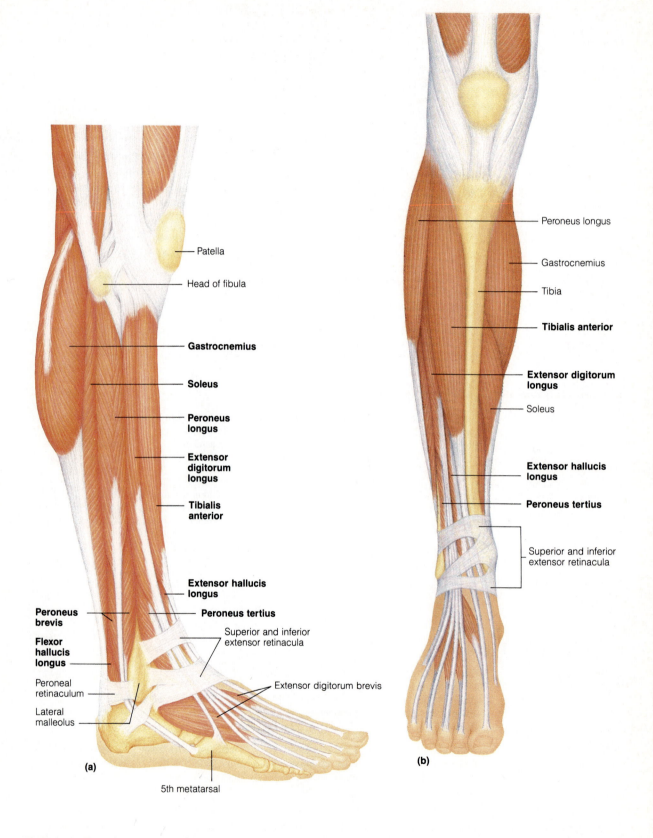

Patella

Head of fibula

Gastrocnemius

Soleus

Peroneus longus

Extensor digitorum longus

Tibialis anterior

Extensor hallucis longus

Peroneus brevis

Flexor hallucis longus

Peroneal retinaculum

Lateral malleolus

Peroneus tertius

Superior and inferior extensor retinacula

Extensor digitorum brevis

(a)

5th metatarsal

Peroneus longus

Gastrocnemius

Tibia

Tibialis anterior

Extensor digitorum longus

Soleus

Extensor hallucis longus

Peroneus tertius

Superior and inferior extensor retinacula

(b)

Figure 15.12

Muscles of right leg. (a) Superficial view of lateral aspect of the leg, illustrating the positioning of the lateral compartment muscles (peroneus longus and brevis) relative to anterior and posterior leg muscles. (b) Superficial view of anterior leg muscles.

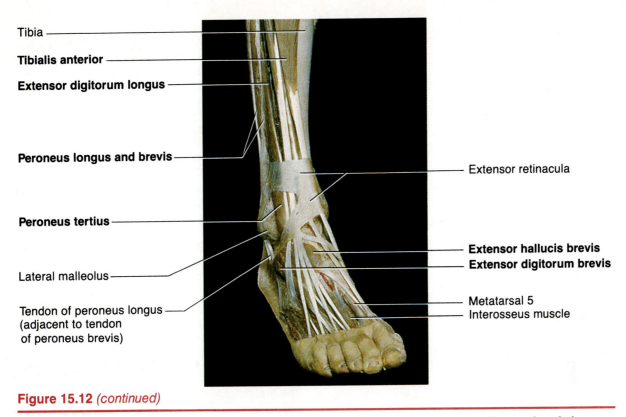

Tibia

Tibialis anterior

Extensor digitorum longus

Peroneus longus and brevis

Peroneus tertius

Lateral malleolus

Tendon of peroneus longus
(adjacent to tendon
of peroneus brevis)

Extensor retinacula

Extensor hallucis brevis
Extensor digitorum brevis

Metatarsal 5
Interosseus muscle

Figure 15.12 *(continued)*

Muscles of the right leg. (c) Photo of anteroinferior aspect of right leg, foot, and extensor retinacula, anterolateral view.

DISSECTION AND IDENTIFICATION OF CAT MUSCLES

The skeletal muscles of all mammals are named in a similar fashion. However, some muscles that are separate in lower animals are fused in humans, and some muscles present in lower animals are lacking in humans. This exercise involves dissection of the cat musculature in conjunction with the study of human muscles, to enhance your knowledge of the human muscular system. Since the aim is to become familiar with the muscles of the human body, you should pay particular attention to the similarities between cat and human muscles. However, pertinent differences will be pointed out as they are encountered.

Preparing the Cat for Dissection

The preserved laboratory animals purchased for dissection have been embalmed with a solution (usually containing formalin) to prevent deterioration of the tissues. The animals are generally delivered in plastic bags that contain a small amount of the embalming fluid. Do not dispose of this fluid when you remove the cat; the fluid prevents the cat from drying out. It is very important to keep the cat's tissues moist, because you will probably use the same cat from now until the end of the course. The embalming fluid may cause your eyes to smart and may dry your skin, but these small irritants are preferable to working with a cat that has become hard and odoriferous due to bacterial action. You can alleviate the skin irritation by using disposable gloves or applying skin cream before dissection.

1.   Obtain a cat, dissection tray, dissection instruments, and a name tag. Mark the name tag with the names of the members of your group and set it aside. The name tag will be attached to the plastic bag at the end of the dissection so that you may identify your animal in subsequent laboratories.

2.   To begin removing the skin, place the cat ventral side down on the dissecting tray. With a scalpel, make a short and shallow incision in the midline of the neck, just to penetrate the skin. From this point on, use scissors. Continue the cut the length of the back to the sacrolumbar region, stopping at the tail (Figure 15.13).

3.   From the dorsal surface of the tail region, make an incision around the tail, encircling the anus and genital organs. The skin will not be removed from this region. Beginning again at the dorsal tail region, make an incision through the skin down each hind leg nearly to the ankle. Continue the cut completely around the ankle.

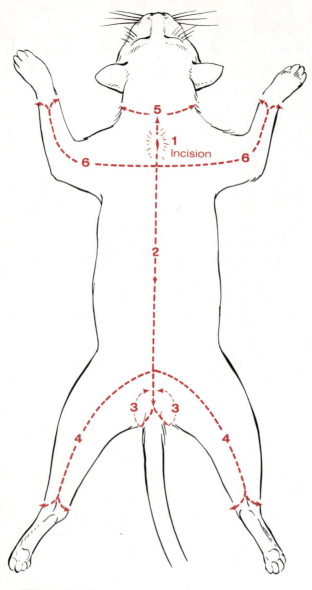

Figure 15.13

Incisions to be made in skinning a cat; numbers indicate sequence.

4.   Return to the neck. Cut the skin around the neck, and then cut down each foreleg to the wrist. Completely cut the skin around the wrists.

5.   Now loosen the skin from the loose connective tissue (superficial fascia) that binds it to the underlying structures. With one hand, grasp the skin on one side of the midline dorsal incision; using your fingers or a blunt probe, break through the "cottony" connective tissue fibers to release the skin from the muscle beneath. Work toward the ventral surface and then toward the neck. As you pull the skin from the body, you should see small white cordlike structures extending from the skin to the muscles at fairly regular intervals. These are the cutaneous nerves that serve the skin. You will also see (particularly as you approach the ventral surface) that a thin layer of muscle fibers remains adhered to the skin. This is the

cutaneous maximus muscle, which enables the cat to move its skin rather like our facial muscles allow us to express emotion. Where the cutaneous maximus fibers cling to those of the deeper muscles, they should be carefully cut free.

6.   You will note as you start to free the skin in the neck region that it is more difficult to remove. Take extra care and time in this region. The large flat **platysma** muscle in the ventral neck region (a skin muscle like the cutaneous maximus) will remain attached to the skin. The skin will not be removed from the head since the cat's muscles are not sufficiently similar to human head muscles to merit study.

7.   Complete the skinning process by freeing the skin from the forelimbs, lower torso, and hindlimbs in the same manner.

8.   Inspect your skinned cat. Note that it is very difficult to see any cleavage lines between the muscles because of the overlying connective tissue, which is white or yellow. If time allows, carefully remove as much of the fat and fascia from the surface of the muscles as possible, using forceps or your fingers. The muscles, when exposed, look grainy or thread-like and are light brown. If this clearing process is done carefully and thoroughly, you will be ready to begin your identification of the superficial muscles.

9.   If the muscle dissection exercises are to be done at a later laboratory session, carefully rewrap the cat's skin around its body. Then dampen several paper towels in embalming fluid and wrap these around the animal. Return the cat to the plastic bag (add more embalming fluid to the bag if necessary), seal the bag, and attach your name tag. Prepare your cat for storage in this way every time the cat is used. Place the cat in the storage container designated by your instructor.

10.   Before leaving the laboratory, dispose of any tissue remnants in the organic debris container, wash the dissecting tray and instruments with soapy water, rinse and dry, and wash down the laboratory bench.

Dissection of Cat Trunk and Neck Muscles

The proper dissection of muscles involves careful separation of one muscle from another and transection of superficial muscles in order to study those lying deeper. In general, when directions are given to transect a muscle, it should be completely freed from all adherent connective tissue and then cut through about halfway between its origin and insertion points.

As a rule, all the fibers of one muscle are held together by a connective tissue sheath (epimysium)

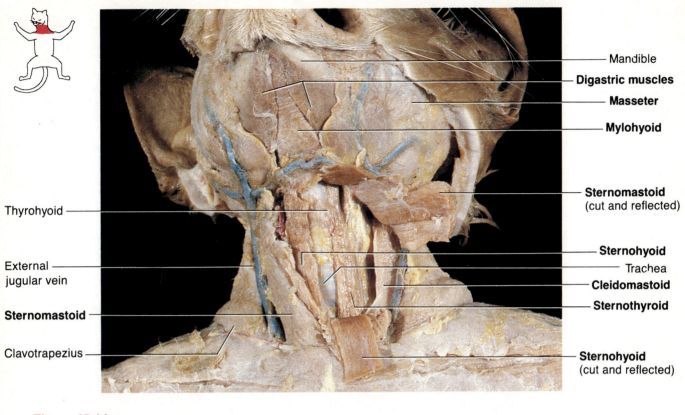

Thyrohyoid

External jugular vein

Sternomastoid

Clavotrapezius

Mandible
Digastric muscles
Masseter
Mylohyoid

Sternomastoid
(cut and reflected)

Sternohyoid
Trachea
Cleidomastoid
Sternothyroid

Sternohyoid
(cut and reflected)

Figure 15.14

Anterior neck muscles of the cat. On the left side of the neck, the sternomastoid and the sternohyoid are transected and reflected to reveal the cleidomastoid and sternothyroid muscles.

and run in the same general direction. Before you begin dissection, observe your skinned cat. If you look carefully, you can see changes in the direction of the muscle fibers, which will help you to locate the muscle borders. Pulling in slightly different directions on two adjacent muscles will usually enable you to expose the normal cleavage line between them. Once cleavage lines have been identified, use a blunt probe to break the connective tissue between muscles and to separate them. If the muscles separate as clean, distinct bundles, your procedure is probably correct. If they appear ragged or chewed up, you are probably tearing a muscle apart rather than separating it from adjacent muscles. Only the muscles that are most easily identified and separated out will be identified in this exercise because of time considerations.

ANTERIOR NECK MUSCLES

1. Using Figure 15.14 as a guide, examine the anterior neck surface of the cat and identify the following superficial neck muscles. (The platysma belongs in this group but was probably removed during the skinning process.) The **sternomastoid** muscle and the more lateral and deeper **cleidomastoid** muscle are joined in humans to form the sternocleidomastoid. The large external jugular vein, which drains the head, should be obvious crossing the anterior

aspect of these muscles. The **mylohyoid** muscle parallels the bottom aspect of the chin, and the **digastric** muscle forms a V over the mylohyoid muscle. Although it is not one of the neck muscles, you can now identify the fleshy **masseter** muscle, which flanks the digastric muscle laterally. Finally, the **sternohyoid** is a narrow muscle between the mylohyoid (superiorly) and the inferior sternomastoid.

2. The deeper muscles of the anterior neck of the cat are small and straplike and hardly worth the effort of dissection. However, one of these deeper muscles can be seen with a minimum of extra effort. Transect the sternomastoid and sternohyoid muscles approximately at mid-belly. Reflect the cut ends to reveal the **sternothyroid** muscle, which runs downward on the anterior surface of the throat just deep and lateral to the sternohyoid muscle. The cleidomastoid muscle, which lies deep to the sternomastoid, is also more easily identified now.

SUPERFICIAL CHEST MUSCLES   In the cat, the chest or pectoral muscles adduct the arm, just as they do in humans. However, humans have only two pectoral muscles, and cats have four—the pectoralis major, pectoralis minor, xiphihumeralis, and pectoantebrachialis (see Figure 15.15). Because of their relatively great degree of fusion, the cat's pectoral

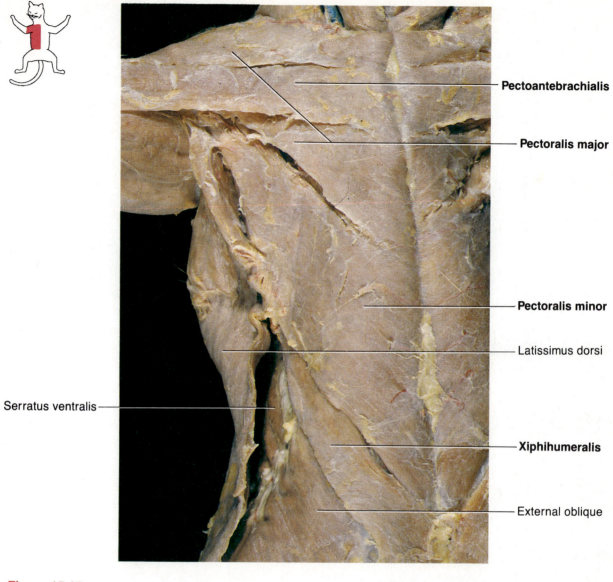

— Pectoantebrachialis

— Pectoralis major

— Pectoralis minor

— Latissimus dorsi

Serratus ventralis —

— **Xiphihumeralis**

— External oblique

Figure 15.15

Superficial thorax muscles, anterior view. Latissimus dorsi is reflected away from the thorax.

muscles give the appearance of a single muscle. The pectoral muscles are rather difficult to dissect and identify, as they do not separate from one another easily.

The **pectoralis major** is 2 to 3 inches wide and can be seen arising on the manubrium just inferior to the sternomastoid muscle of the neck and running to the humerus. Its fibers run at right angles to the longitudinal axis of the cat's body.

The **pectoralis minor** lies beneath the pectoralis major and extends posterior to it on the abdominal surface. It originates on the sternum and inserts on the humerus. Its fibers run obliquely to the long axis of the body, which helps to distinguish it from the pectoralis major. Contrary to what its name implies, the pectoralis minor is a larger, thicker muscle than the pectoralis major.

The **xiphihumeralis** can be distinguished from the posterior edge of the pectoralis minor only by virtue of the fact that its origin is lower—on the xiphoid process of the sternum. Its fibers run parallel to and are fused with those of the pectoralis minor.

The **pectoantebrachialis** is a thin, straplike muscle, about ½ inch wide, lying over the pectoralis major. It originates from the manubrium, passes laterally over the pectoralis major, and merges with the muscles of the arm approximately halfway down the humerus. It has no homologue in humans.

Identify, free, and trace out the origin and insertion of the cat's chest muscles. Refer to Figure 15.15 as you work.

MUSCLES OF THE ABDOMINAL WALL The superficial anterior trunk muscles include those of

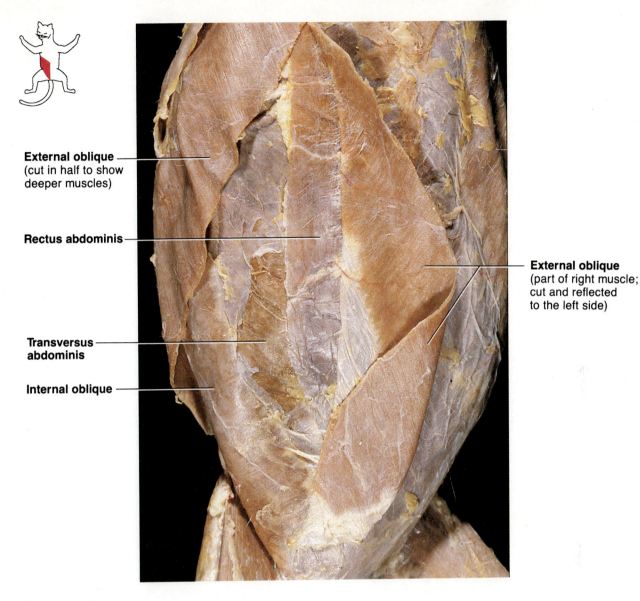

External oblique
(cut in half to show
deeper muscles)

Rectus abdominis

**Transversus
abdominis**

Internal oblique

External oblique
(part of right muscle;
cut and reflected
to the left side)

Figure 15.16

Muscles of the abdominal wall of the cat.

the abdominal wall (see Figure 15.16). Cat musculature in this area is quite similar in function to that
of humans.

1. Complete the dissection of the more superficial
anterior trunk muscles of the cat by identifying the
origins and insertions of the muscles of the abdominal wall. Work carefully here; these muscles are very
thin, and it is easy to miss their boundaries. Begin
with the **rectus abdominis,** a long band of muscle
approximately 1 inch wide running immediately lateral to the midline of the body on the abdominal
surface. Humans have four transverse tendinous
intersections in the rectus abdominis, but they are
absent or difficult to identify in the cat. Identify the
linea alba, which separates the rectus abdominis
muscles. Note the relationship of the rectus abdominis to the other abdominal muscles and their fascia.

2. The **external oblique** is a sheet of muscle immediately beside (and running beneath) the rectus
abdominis (see Figure 15.16). Carefully free and then
transect the external oblique to reveal the anterior
attachment of the rectus abdominis and the deeper
internal oblique. Reflect the external oblique; observe the deeper muscle. Note which way the fibers run.

How does the fiber direction of the internal oblique

compare to the external oblique? _____

_____

3. Free and then transect the internal oblique muscle to reveal the fibers of the **transversus abdominis,**
whose fibers run tranversely across the abdomen.

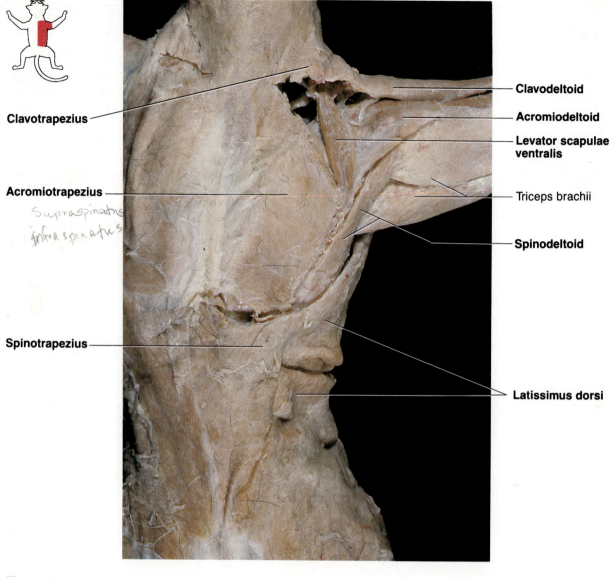

Clavotrapezius

Acromiotrapezius

Supraspinatus
infraspinatus

Spinotrapezius

Clavodeltoid

Acromiodeltoid

Levator scapulae ventralis

Triceps brachii

Spinodeltoid

Latissimus dorsi

Figure 15.17

Superficial muscles of the posterior aspect of the right shoulder, trunk, and neck of the cat.

SUPERFICIAL MUSCLES OF THE SHOULDER, AND POSTERIOR TRUNK AND NECK

Refer to Figure 15.17 as you dissect the following superficial muscles of the posterior surface of the trunk.

1. Turn your cat on its ventral surface and start your observations with the **trapezius group.** Humans have a single large trapezius muscle, but the cat has three separate muscles—the clavotrapezius, acromiotrapezius, and spinotrapezius. The prefix (clavo-, acromio-, and spino-) in each case reveals the muscle's site of insertion. The **clavotrapezius,** the most superior muscle of the group, is homologous to that part of the human trapezius that inserts into the clavicle. Slip a probe under this muscle and follow it to its apparent origin.

Where does the clavotrapezius appear to originate?

_____

Is this similar to its origin in humans? _____

The fibers of the clavotrapezius are continuous inferiorly with those of the clavicular part of the cat's deltoid muscle (clavodeltoid), and the two muscles work together to bring about the extension of the humerus. Release the clavotrapezius muscle from the adjoining muscles. The **acromiotrapezius** is a large, nearly square muscle easily identified by its aponeurosis, which passes over the vertebral border of the scapula. It originates from the cervical and T_1 vertebrae and inserts into the scapular spine. The triangular **spinotrapezius** runs from the thoracic vertebrae to the scapular spine. This is the most posterior of the trapezius muscles in the cat. Now that you know where they are located, pull on the three trapezius muscles.

Do they appear to have the same functions in cats as

in humans? _____

2. The **levator scapulae ventralis,** a flat, straplike muscle, can be located in the triangle created by the division of the fibers of the clavotrapezius and acromiotrapezius. Its anterior fibers run underneath the clavotrapezius from its origin at the base of the skull (occipital bone), and it inserts on the vertebral border of the scapula. In the cat it helps to hold the upper edges of the scapulae together and draws them toward the head.

What is the function of the levator scapulae in humans?

_____

_____

3. The **deltoid group.** Like the trapezius, the human deltoid muscle is represented by three separate muscles in the cat—the clavodeltoid, acromiodeltoid, and spinodeltoid. The **clavodeltoid,** the most superficial muscle of the shoulder, is a continuation of the clavotrapezius below the clavicle, which is this muscle's point of origin (see Figure 15.17). Follow its course down the arm to the point where it merges along a white line with the pectoantebrachialis. Separate it from the pectoantebrachialis, and then transect it and pull it back.

Where does the clavodeltoid insert? _____

What do you think the function of this muscle is?

_____

The **acromiodeltoid** lies posterior to the clavodeltoid and over the top of the shoulder. It is a small tri-

angular muscle that originates on the acromion process of the scapula and inserts into the spinodeltoid posterior to it. The **spinodeltoid** is posterior to the acromiodeltoid and approximately the same size. This muscle is covered with fascia near the anterior end of the scapula. Its tendon extends under the acromiodeltoid muscle and inserts on the humerus. Note that its fibers run obliquely to those of the acromiodeltoid. The acromiodeltoid and clavodeltoid muscles in the cat are synergists that, like the human deltoid muscle, raise and rotate the humerus.

4. The **latissimus dorsi** is a large flat muscle covering most of the lateral surface of the posterior trunk. Its upper edge is covered by the spinotrapezius. As in humans, it inserts into the humerus. But before inserting, its fibers merge with the fibers of many other muscles, among them the xiphihumeralis of the pectoralis group.

DEEP MUSCLES OF THE TRUNK AND POSTERIOR NECK

1. In preparation, transect the latissimus dorsi, the muscles of the pectoralis group, and the spinotrapezius and reflect them back. Be careful not to damage the large brachial nerve plexus, which lies in the axillary space beneath the pectoralis group.

2. The **serratus ventralis,** homologous to the serratus anterior of humans, arises deep to the pectoral muscles and covers the lateral surface of the rib cage. It is easily identified by its fingerlike muscular origins, which arise on the first 9 or 10 ribs. It inserts into the scapula. The anterior portion of the serratus ventralis, which arises from the cervical vertebrae, is homologous to the levator scapulae in humans; both pull the scapula toward the sternum. Trace this muscle to its insertion. In general, in the cat, this muscle acts to pull the scapula posteriorly and downward. Refer to Figure 15.18.

3. Reflect the upper limb to reveal the **subscapularis,** which occupies most of the ventral surface of the scapula. Humans have a homologous muscle.

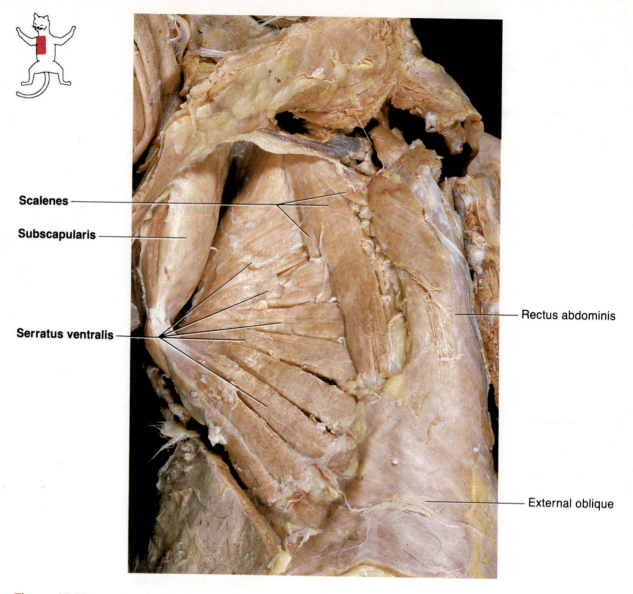

Figure 15.18

Deep muscles of right anterolateral thorax of the cat.

4.   Locate the anterior, posterior, and medial **scalene** muscles on the lateral surface of the cat's neck and trunk. The most prominent and longest of these muscles is the scalenus medius, which lies between the anterior and posterior members. The scalenes originate on the ribs and run upward over the serratus anterior to insert in common on the cervical vertebrae. These muscles draw the ribs anteriorly and bend the neck downward; thus they are homologous to the human scalene muscles, which elevate the ribs and flex the neck. (Note that the difference is only one of position. Humans walk erect, but cats are quadrupeds.)

5.   Transect and reflect the latissimus dorsi, spinodeltoid, acromiodeltoid, and levator scapulae ven-

tralis. The **splenius** is a large flat muscle occupying most of the side of the neck close to the vertebrae. As in humans, it originates on the ligamentum nuchae at the back of the neck and inserts into the occipital bone. It functions to raise the head. Refer to Figure 15.19.

6.   To successfully view the rhomboid muscles, lay the cat on its side and hold its forelegs together. This should spread the scapulae apart. The rhomboid muscles lie between the scapulae and beneath the acromiotrapezius. All the rhomboid muscles originate on the vertebrae and insert on the scapula.
   There are three rhomboids in the cat. The **rhomboid capitis,** the most anterolateral muscle of the group, has no counterpart in the human body. The

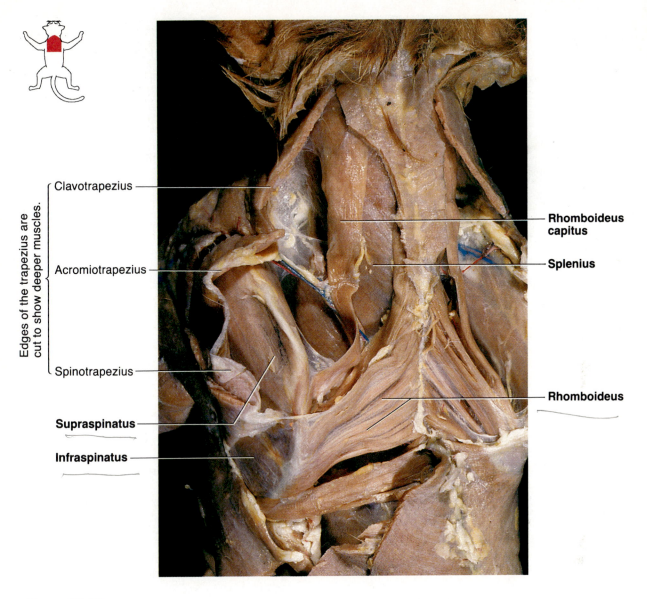

Clavotrapezius

Edges of the trapezius are cut to show deeper muscles.

Acromiotrapezius

Spinotrapezius

Supraspinatus

Infraspinatus

Rhomboideus capitus

Splenius

Rhomboideus

Figure 15.19

Deep muscles of the posterior thorax of the cat.

rhomboid minor, located posteriorly to the rhomboid capitus, is much larger. Its fibers run transversely to those of the rhomboid capitis. The most posterior muscle of the group is the **rhomboid major.** This muscle is so closely fused to the rhomboid minor that many consider them one muscle—the **rhomboideus,** which is homologous to human rhomboid muscles.

7.   The **supraspinatus** and **infraspinatus** muscles are similar to the same muscles in humans. The supraspinatus can be found under the acromiotrapezius, and the infraspinatus under the spinotrapezius. Both originate on the lateral scapular surface and insert on the humerus.

Dissection of Cat Forelimb Muscles

Cat forelimb muscles fall into the same three categories as human upper extremity muscles, but in this section the muscles of the entire forelimb are considered together. Refer to Figure 15.20 as you study these muscles.

MUSCLES OF THE LATERAL SURFACE

1.   The triceps muscle (**triceps brachii**) of the cat can easily be identified if the cat is placed on its side. It is a large fleshy muscle covering the posterior aspect and much of the side of the humerus. As in humans, this muscle arises from three heads, which originate from the humerus and scapula and insert

Figure 15.20

Lateral surface of right forelimb of the cat. The lateral head of the triceps brachii has been transected and reflected.

jointly into the olecranon process of the ulna. Remove the fascia from the upper region of the lateral arm surface to identify the lateral and long heads of the triceps. The long head is approximately twice as long as the lateral head and lies medial to it on the posterior surface of the arm. The medial head can be exposed by transecting the lateral head and pulling it aside. Now pull on the triceps muscle.

How does the function of the triceps muscle compare

in cats and in humans? _____

Anterior and distal to the medial head of the triceps is the tiny anconeus muscle, sometimes called the fourth head of the triceps muscle. Notice its darker color and the manner in which it wraps the tip of the elbow.

2. The **brachialis** can be located anterior to the lateral head of the triceps muscle. Identify its origin on the humerus, and trace its course as it crosses the elbow laterally and inserts on the ulna. It flexes the foreleg of the cat.

Identification of the forearm muscles is very difficult because of the tough fascia sheath that encases them.

3. Remove as much of the connective tissue as possible and cut through the ligaments that secure the tendons at the wrist (transverse carpal ligaments) so that you will be able to follow the muscles to their insertions. Begin your identification of the forearm muscles by examining the lateral surface of the forearm. The muscles of this region are very much

alike in appearance and are difficult to identify accurately unless a definite order is followed. Thus you will begin with the most anterior muscles and proceed to the posterior aspect. Remember to check carefully the tendons of insertion to verify your muscle identification.

4. The ribbonlike muscle on the lateral surface of the humerus is the **brachioradialis.** Note how it passes down the forearm to insert on the styloid process of the radius. (If your removal of the fascia was not very careful, this muscle may have been removed.)

5. The **extensor carpi radialis longus** has a broad origin and is larger than the brachioradialis. It extends down the anterior surface of the radius (see Figure 15.20). Transect this muscle to view the **extensor carpi radialis brevis,** which is partially covered by and sometimes fused with the extensor carpi radialis longus. Both muscles have similar origins, insertions, and actions as their human counterparts.

6. You can see the entire **extensor digitorum communis** along the lateral surface of the forearm. Trace it to its four tendons, which insert on the second to fifth digits. This muscle extends these digits. The **extensor digitorum lateralis** (which is not present in humans) also extends the digits. This muscle lies immediately posterior to the extensor digitorum communis.

7. Follow the **extensor carpi ulnaris** from the lateral epicondyle of the humerus to the ulnar side of the fifth metacarpal. Very often this muscle has a shiny tendon, which helps in its identification.

Figure 15.21

Medial surface of the right forelimb of cat.

MUSCLES OF THE MEDIAL SURFACE

1.   The **biceps brachii** (refer to Figure 15.21) is a large spindle-shaped muscle medial to the brachialis on the anterior surface of the humerus. Pull back the cut ends of the pectoral muscles to get a good view of the biceps. Although this muscle is much more prominent in humans, its origin, insertion, and action are very similar in cats and in humans. Follow the muscle to its origin.

Does the biceps have two heads in the cat? _____

2.   The broad, flat, exceedingly thin muscle on the medioposterior surface of the arm is the **epitrochlearis.** Its tendon originates from the fascia of the latissimus dorsi, and the muscle inserts into the olecranon process of the ulna. This muscle extends the forearm of the cat; it is not found in humans.

3.   The **coracobrachialis** of the cat is very insignificant (approximately ½ in. long) and can be seen as a very small muscle crossing the ventral aspect of the shoulder joint. It runs beneath the biceps brachii to insert on the humerus and has the same function as the human coracobrachialis.

4.   Referring again to Figure 15.21, turn the cat so that the ventral forearm muscles (mostly flexors and pronators) can easily be observed. As in humans, most of these muscles arise from the medial epicondyle of the humerus. The **pronator teres** runs from the medial epicondyle of the humerus and declines in size as it approaches its insertion on the radius. Do not bother to trace it to its insertion.

5.   The **flexor carpi radialis** runs from the medial epicondyle of the humerus to insert into the second and third metacarpals, as in humans.

6.   The large flat muscle in the center of the medial surface is the **palmaris longus.** Its origin on the medial epicondyle of the humerus abuts that of the pronator teres and is shared with the flexor carpi radialis. The palmaris longus extends down the forearm to terminate in four tendons on the digits. Comparatively speaking, this muscle is much larger in cats than in humans. The **flexor carpi ulnaris** arises from a two-headed origin (medial epicondyle of the humerus and olecranon of the ulna). Its two bellies (fleshy parts) pass downward to the wrist, where they are united by a single tendon that inserts into the carpals of the wrist. As in humans, this muscle flexes the wrist.

Dissection of Cat Hindlimb Muscles

Remove the fat and fascia from all the thigh surfaces, but do not cut through or remove the **fascia lata,** which is a tough white aponeurosis covering the anterolateral surface of the thigh from the hip to the leg. If the cat is a male, the cordlike sperm duct will be embedded in the fat near the pubic symphysis. Carefully clear around, but not in, this region.

POSTEROLATERAL HINDLIMB MUSCLES

1.   Turn the cat on its ventral surface and identify the following superficial muscles of the hip and thigh, referring to Figure 15.22. (The deeper hip muscles of the cat will not be identified.) Viewing the lateral aspect of the hindlimb, you will identify these muscles in sequence from the anterior to the posterior aspects of the hip and thigh. Most anteriorly is the **sartorius.** Approximately 1½ in. wide, it extends

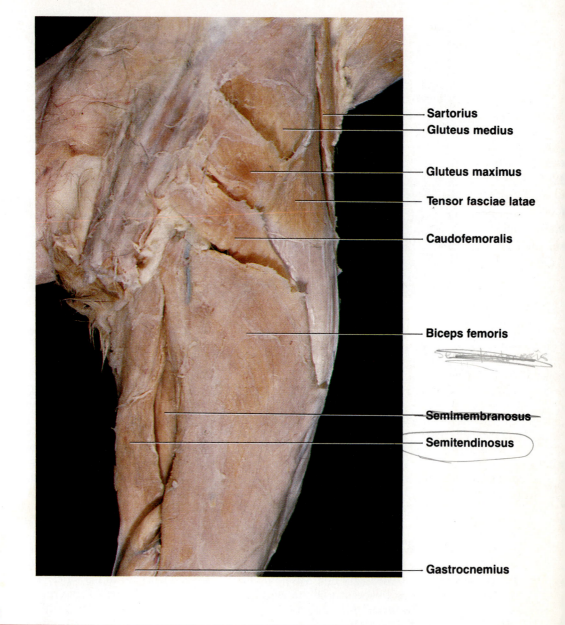

- Sartorius
- Gluteus medius
- Gluteus maximus
- Tensor fasciae latae
- Caudofemoralis
- Biceps femoris
- Semimembranosus
- Semitendinosus
- Gastrocnemius

Figure 15.22

Muscles of the right posterolateral thigh in the cat; superficial view.

around the lateral aspect of the leg to the anterior surface, where the major portion of it lies. Free it from the adjacent muscles and pass a blunt probe under it to trace its origin and insertion. Homologous to the sartorius muscle in humans, it adducts and rotates the thigh, but in addition, the cat sartorius acts as a knee extensor. Transect this muscle.

2. The **tensor fasciae latae** is posterior to the sartorius. It is wide at its superior end, where it originates on the iliac crest, and narrows as it approaches its insertion into the fascia lata (or iliotibial band), which runs to the proximal tibial region. Transect its superior end and pull it back to expose the **gluteus medius** lying beneath it. This is the largest of the gluteus muscles in the cat. It originates on the ilium and inserts on the greater trochanter of the femur. The gluteus medius overlays and obscures the glu-

teus minimus, pyriformis, and gemellus muscles (which will not be identified here).

3. The **gluteus maximus** is a small triangular hip muscle posterior to the superior end of the tensor fascia lata and paralleling it. In humans the gluteus maximus is a large fleshy muscle forming most of the buttock mass, but in the cat it is only about ½ in. wide and 2 in. long, smaller than the gluteus medius. The gluteus maximus covers part of the gluteus medius as it extends from the sacral region and the end of the femur. It abducts the thigh.

4. Posterior to the gluteus maximus, identify the triangular **caudofemoralis,** which originates on the caudal vertebrae and inserts into the patella via an aponeurosis. There is no homologue to this muscle

in humans; in cats it abducts the thigh and flexes the vertebral column.

5.   The **hamstring muscles,** a group in the lower extremity, includes the biceps femoris, the semitendinosus, and the semimembranosus muscles. The **biceps femoris** is a large, powerful muscle that covers most of the posterolateral surface of the thigh. It is 1½ to 2 in. wide throughout its length. Trace it from its origin on the ischial tuberosity to its insertion on the tibia. The **semitendinosus** can be partially seen beneath the posterior border of the biceps femoris. Transect and reflect the biceps muscle to reveal the whole length of the semitendinosus. Contrary to what its name implies ("half-tendon"), this muscle is muscular and fleshy except at its insertion. It is uniformly about ¾ in. wide as it runs down the thigh from the ischial tuberosity to the medial side of the ulna. It acts to bend the knee. The **semimembranosus,** a large muscle lying medial to the semitendinosus and largely obscured by it, is best seen in an anterior view of the thigh (see Figure 15.24b). If desired, however, the semitendinosus can be transected to view it from the posterior aspect. The semimembranosus is larger and broader than the semitendinosus. Like the other hamstrings, it originates on the ischial tuberosity and inserts on the medial epicondyle of the femur and the medial tibial surface.

How does the semimembranosus compare with its

human homologue? _____

_____

6.   Remove the heavy fascia covering the lateral surface of the shank. Moving from the posterior to the anterior aspect, identify the following muscles on the posterolateral shank (leg). (Refer to Figure 15.23.) First reflect the lower portion of the biceps femoris to see the origin of the **triceps surae,** the large composite muscle of the calf. Humans also have a triceps surae. The **gastrocnemius** is a portion of the triceps surae and the largest muscle on the shank. As in humans, it has two heads and inserts via the Achilles tendon into the calcaneus. Run a probe beneath this muscle and then transect it to reveal the **soleus,** which is deep to the gastrocnemius.

7.   Another important group of muscles in the leg are the **peroneus muscles** (see Figure 15.23), which appear as a slender, evenly shaped superficial muscle lying anterior to the triceps surae. Originating on the fibula and inserting on the digits and metatarsals, the peroneus muscles flex the foot.

8.   The **extensor digitorum longus** is anterior to the peroneus muscles. Its origin, insertion, and action in cats are similar to the homologous human muscle. The **tibialis anterior** is anterior to the extensor digitorum longus. The tibialis anterior is roughly tri-

angular in cross section and heavier at its proximal end. Locate its origin on the proximal fibula and tibia and its insertion on the first metatarsal. You can see the sharp edge of the tibia at the anterior border of this muscle. As in humans, it is a flexor of the foot.

ANTEROMEDIAL HINDLIMB MUSCLES

1.   Turn the cat onto its dorsal surface to identify the muscles of the anteromedial hindlimb (refer to Figure 15.24). Note once again the straplike sartorius at the surface of the thigh, which you have already identified and transected. It originates on the ilium and inserts on the medial region of the tibia.

2.   Reflect the cut ends of the sartorius to identify the **quadriceps** muscles. The **vastus medialis** lies just beneath the sartorius. Resting close to the femur, it arises from the ilium and inserts into the patellar tendon. The small cylindrical muscle anterior and lateral to the vastus medialis is the **rectus femoris.** In cats this muscle originates entirely from the femur.

What is the origin of the rectus femoris in humans?

_____

_____

Free the rectus femoris from the large fleshy **vastus lateralis,** the most lateral muscle of this group which lies deep to the tensor fascia latae. The vastus lateralis arises from the lateral femoral surface and inserts, with the other vasti muscles, into the patellar tendon. Transect this muscle to identify the deep **vastus intermedius,** the smallest of the vasti muscles. It lies medial to the vastus lateralis and merges superiorly with the vastus medialis. (The vastus intermedius is not shown in the figure.)

3.   The **gracilis** is a broad muscle that covers the posterior portion of the medial aspect of the thigh (see Figure 15.24a.) It originates on the pubic symphysis and inserts on the media proximal tibial surface. In cats the gracilis adducts the leg and draws it posteriorly.

How does this compare with the human gracilis?

_____

_____

4.   Free and transect the gracilis to view the adductor muscles below it. The **adductor femoris** is a large muscle that lies beneath the gracilis and abuts the semimembranosus medially. Its origin is the pubic ramus and the ischium, and its fibers pass downward to insert on most of the length of the femoral shaft. The adductor femoris is homologous to the human adductor magnus, brevis, and longus. Its function is

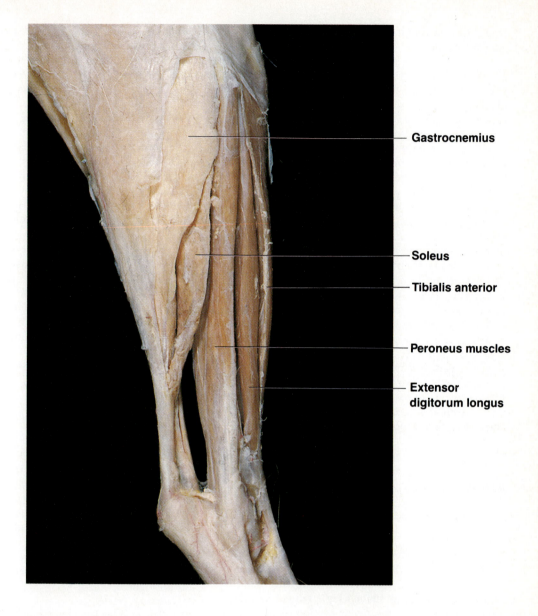

Gastrocnemius

Soleus

Tibialis anterior

Peroneus muscles

Extensor digitorum longus

Figure 15.23

Superficial muscles of the posterolateral aspect of the right leg.

to extend the thigh after it has been drawn forward and to adduct the thigh. A small muscle about an inch long—the **adductor longus**—touches the superior margin of the adductor femoris. It originates on the pubic bone and inserts on the proximal surface of the femur.

5.   Before continuing your dissection, note the **femoral triangle** (Scarpa's triangle), an important area bordered by the proximal edge of the sartorius and the adductor muscles. It is usually possible to identify the femoral artery (injected with red latex) and the femoral vein (injected with blue latex), which span the triangle. (You will identify these vessels again in your study of the circulatory system.) If

your instructor wishes you to identify the pectineus and iliopsoas, remove these vessels and continue with steps 6 and 7.

6.   Examine the superolateral margin of the adductor longus to locate the small **pectineus.** It is normally covered by the gracilis (which you have cut and reflected). The pectineus, which originates on the pubis and inserts on the proximal end of the femur, is similar in all ways to its human homologue.

7.   Just lateral to the pectineus you can see a small portion of the **iliopsoas,** a long and cylindrical muscle. Its origin is on the transverse processes of T_1 through T_{12} and the lumbar vertebrae, and it passes posteriorly toward the body wall to insert on the medial

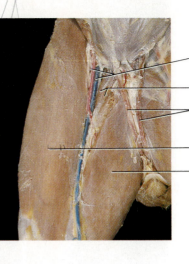

(handwritten annotations: Sartorius, Coracobra..., Pectineus, Adductor longus, Adductor femoris, Semimembranosus, Semitendinosus, V. medialis, Rectus femoris, V. lateralis)

(a)

Femoral artery and vein
Adductors
Spermatic cords
Sartorius
Gracilis

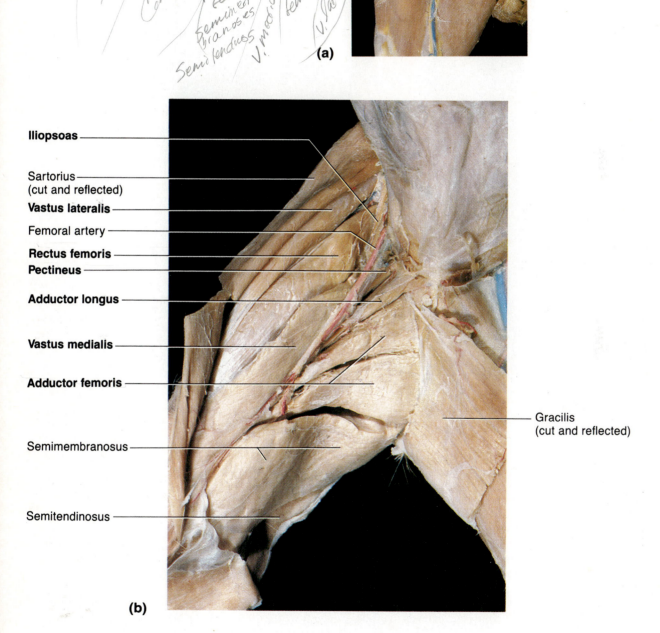

Iliopsoas

Sartorius (cut and reflected)

Vastus lateralis

Femoral artery

Rectus femoris

Pectineus

Adductor longus

Vastus medialis

Adductor femoris

Semimembranosus

Semitendinosus

Gracilis (cut and reflected)

(b)

Figure 15.24

Superficial muscles of the anteromedial thigh: (a) gracilis and sartorius are intact in this superficial view of the right thigh.
(b) The gracilis and sartorius are transected and reflected to show deeper muscles.

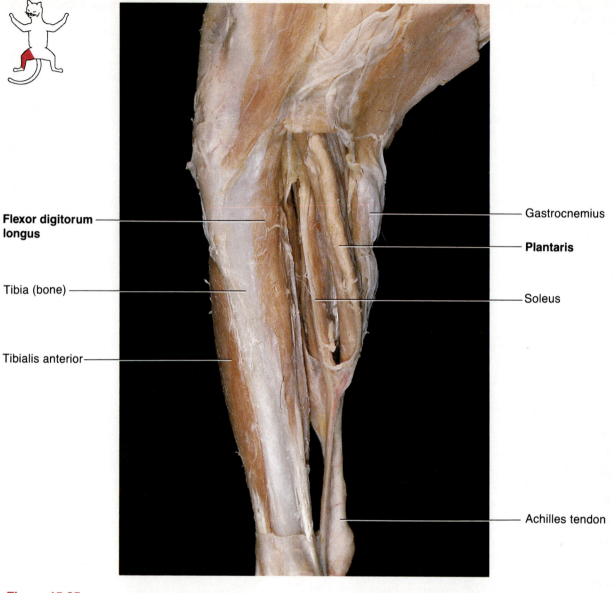

Flexor digitorum longus

Tibia (bone)

Tibialis anterior

Gastrocnemius

Plantaris

Soleus

Achilles tendon

Figure 15.25

Superficial muscles of the right anteromedial leg of cat.

aspect of the proximal femur. The iliopsoas flexes and laterally rotates the thigh. It corresponds to the human iliopsoas and psoas minor.

8.   Reidentify the gastrocnemius of the shank and then the **plantaris,** which is fused with the lateral head of the gastrocnemius. (See Figure 15.25.) It originates from the lateral aspect of the femur and patella, and its tendon passes around the calcaneus to insert on the second phalanx. Working with the triceps surae, it flexes the digits and extends the foot.

9.   Anterior to the plantaris is the **flexor digitorum longus,** a long, tapering muscle with two heads. It originates on the lateral surfaces of the proximal fibula

and tibia and inserts via four tendons into the terminal phalanges. As in humans, it flexes the toes.

10.   The **tibialis posterior** is a long, flat muscle lateral and deep to the flexor digitorum longus. Its origin is on the medial surface of the head of the fibula and the ventral tibia; it then merges with a flat, shiny tendon to insert into the tarsals.

11.   The **flexor hallucis longus** is a long muscle lateral to the tibialis posterior. It originates from the posterior tibia and passes downward to the ankle. It is a uniformly broad muscle in the cat. As in humans, it is a flexor of the great toe.

16

Histology of Nervous Tissue

OBJECTIVES

1. To differentiate between the functions of neurons and neuroglia and to list the functional characteristics of each.

2. To list four types of neuroglia cells.

3. To identify the following anatomic characteristics of a neuron on an appropriate diagram or projected slide:

   Cell body (perikaryon)
   Nucleus and nucleolus
   Neuron fibers (axons, dendrites)
   Neurilemma
   Nodes of Ranvier
   Nissl bodies
   Neurofibrils
   Myelin sheath
   Axonal terminals
   Axon hillock

4. To state the functions of axons, dendrites, axonal terminals, neurofibrils, and myelin sheaths.

5. To explain how a nerve impulse is transmitted from one neuron to another.

6. To explain the role of Schwann cells in the formation of the myelin sheath.

7. To classify neurons according to structure and function.

8. To distinguish between a nerve and a tract and between a ganglion and a nucleus.

9. To describe the structure of a nerve, identifying the connective tissue coverings (endoneurium, perineurium, and epineurium) and citing their functions.

MATERIALS

Model of a "typical" neuron (if available)
Compound microscope
Histologic slides of an ox spinal cord smear and teased myelinated nerve fibers
Prepared slides of Purkinje cells (cerebellum), pyramidal cells (cerebrum), and a dorsal root ganglion
Prepared slide of a nerve (cross section)

The nervous system is the master integrating and coordinating system, continuously monitoring and processing sensory information both from the external environment and from within the body. Every thought, action, and sensation is a reflection of its activity. Like a computer, it processes and integrates new "inputs" with information previously fed into it ("programmed") to produce an appropriate response ("readout"). However, no manmade computer can possibly compare in complexity and scope to the human nervous system.

Despite its complexity, nervous tissue is made up of just two principal cell populations: the **neurons** and the **neuroglia** (glial cells). The *neuroglia,* literally "nerve glue," include astrocytes, oligodendrocytes, microglia, and ependymal cells (Figure 16.1). These cells serve the needs of the neurons by acting as supportive and protective cells, myelinating cells, and phagocytes. In addition, they probably serve a nutritive function by acting as a selective barrier between the capillary blood supply and the neurons. Although neuroglia resemble neurons in some ways (they have fibrous cellular extensions), they are not capable of generating and transmitting nerve impulses, a capability that is highly developed in neurons.

NEURON ANATOMY

The delicate **neurons** are the structural units of nervous tissue. They are highly specialized to exhibit irritability and conductivity, both essential for the effective transmission of messages (nerve impulses) from one part of the body to another.

Although neurons differ structurally, they have many identifiable features in common (Figure 16.2). All have a **cell body** from which slender processes or fibers extend. Neuron cell bodies, which are found only in the CNS (brain or spinal cord) or in **ganglia** (collections of nerve cell bodies outside the CNS), make up the gray matter of the nervous system. The neuron processes running through the CNS form **tracts** of white matter; outside the CNS they form the peripheral **nerves.**

The neuron cell body contains a large round nucleus surrounded by cytoplasm (neuroplasm). The

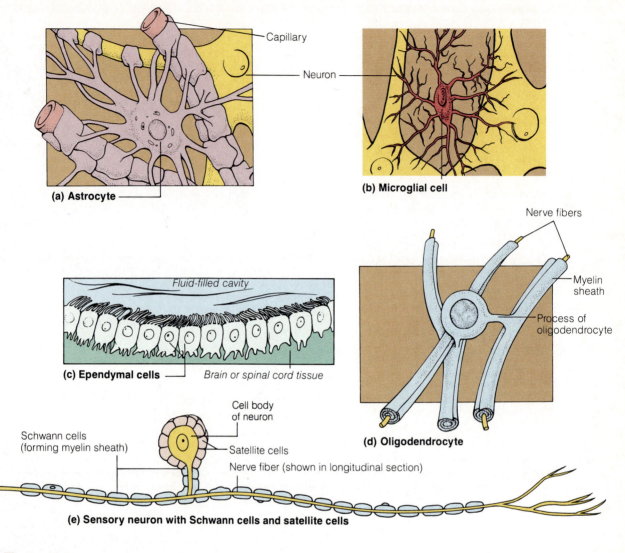

Figure 16.1

Supporting cells of nervous tissue: (a) astrocyte; (b) microglial cell; (c) ependymal cells; (d) oligodendrocyte; (e) nerve cell with Schwann cells and satellite cells.

cytoplasm is riddled with neurofibrils and with darkly staining structures called Nissl bodies. **Neurofibrils** are believed to have a support and nutritive function; **Nissl bodies,** an elaborate type of rough endoplasmic reticulum, are involved in the metabolic activities of the cell.

Neuron processes that conduct electrical signals *toward* the nerve cell body are called **dendrites;** those that conduct *away from* the nerve cell body are called **axons.** Typically, neurons have only one axon (which may branch into **collaterals**) but may have many dendrites, depending on the neuron type.

In general, a neuron is excited by other neurons when their axons release neurotransmitters close to its dendrites or cell body. The impulse generated travels across the nerve cell body and down the axon. As Figure 16.2 shows, the axon begins at a slightly

enlarged cell body structure called the **axon hillock** and ends in many small structures called **axonal terminals,** or synaptic knobs. These terminals store the neurotransmitter chemical in tiny vesicles. Each axonal terminal is separated from the cell body or dendrites of the next neuron by a tiny gap called the **synaptic cleft.** Thus, although they are close, there is no actual physical contact between neurons. When an impulse reaches the axonal terminals, the synaptic vesicles rupture and release the neurotransmitter into the synapse. The neurotransmitter then diffuses across the synapse to bind to membrane receptors on the next neuron, initiating the action potential.*

*Specialized synapses in skeletal muscle are called myoneural junctions. They are discussed in Exercise 14.

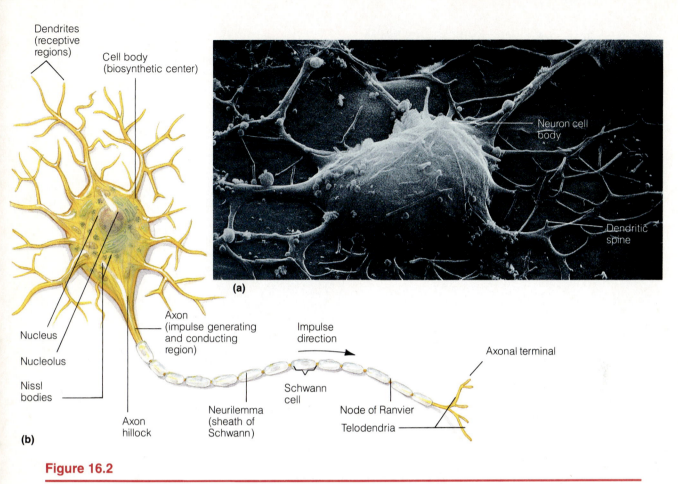

Dendrites
(receptive
regions)

Cell body
(biosynthetic center)

Neuron cell
body

Dendritic
spine

Nucleus

Nucleolus

Nissl
bodies

Axon
(impulse generating
and conducting
region)

Impulse
direction

Axonal terminal

Schwann
cell

Axon
hillock

Neurilemma
(sheath of
Schwann)

Node of Ranvier

Telodendria

(a)

(b)

Figure 16.2

Structure of a typical motor neuron: (a) photomicrograph showing the neuron cell body and dendrites with obvious dendritic spines (synapse sites), 5000X; (b) diagrammatic view.

Many drugs can influence the transmission of impulses and synapses. Some, like caffeine, are stimulants which decrease the receptor neuron's threshold and make it more irritable. Others block transmission by binding competitively with the receptor sites or by interfering with the release of neurotransmitter by the axonal terminals. As might be anticipated, some of these drugs are used as pain-killers or tranquilizers. ■

Most long nerve fibers (both axons and dendrites) are covered with a fatty material called myelin, and such fibers are referred to as **myelinated fibers.** Axons leaving the CNS are typically heavily myelinated by special cells called **Schwann cells,** which wrap themselves tightly around the axon jelly-roll fashion (Figure 16.3). During the wrapping process, the cytoplasm is squeezed from between adjacent layers of the Schwann cell membranes, so that when the process is completed a tight core of plasma membrane material (protein-lipoid material) encompasses the axon. This wrapping is the **myelin sheath.** The bulk of the Schwann cell cytoplasm ends up just beneath the outermost portion of its plasma membrane. The peripheral part of the Schwann cell and its membrane is referred to as the **neurilemma.** Since the myelin sheath is formed by many individual Schwann cells, it has gaps or is a discontinuous sheath; the indentations in the sheath are called **nodes of Ranvier** (see Figure 16.2).

Within the CNS, myelination is accomplished by glial cells called **oligodendrocytes** (see Figure 16.1d). These CNS sheaths do not exhibit the neurilemma seen in fibers myelinated by Schwann cells. Because of its chemical composition, myelin acts to insulate the fibers and greatly increases the speed of neurotransmission by neuron fibers.

1.   Study the typical neuron shown in Figure 16.2, noting the structural details described above, and then identify these structures on a neuron model.

2.   Obtain a prepared slide of the ox spinal cord smear, which has large, easily identifiable neurons. Study one representative neuron under oil immersion and identify the cell body, the nucleus,

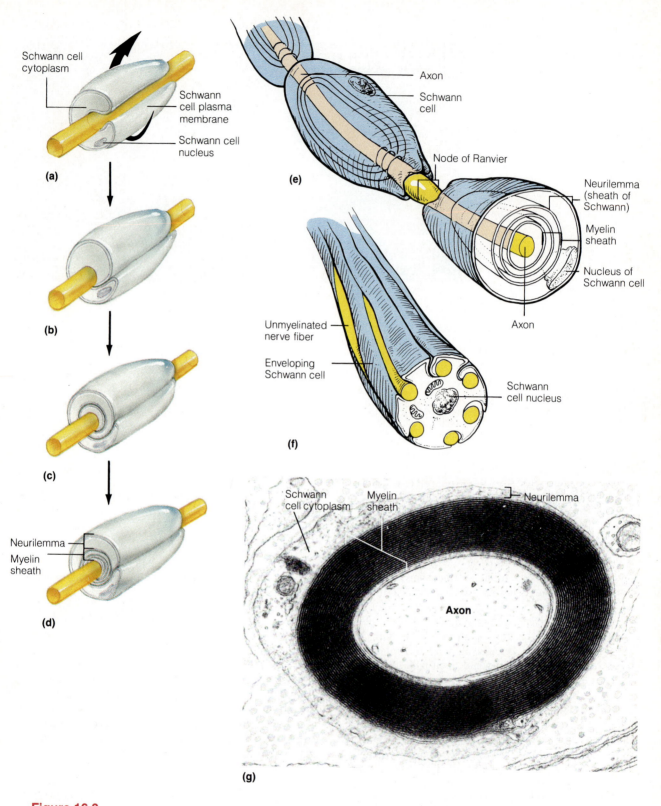

Schwann cell cytoplasm

Schwann cell plasma membrane

Schwann cell nucleus

(a)

(b)

Neurilemma

Myelin sheath

(c)

Neurilemma

Myelin sheath

(d)

Axon

Schwann cell

Node of Ranvier

Neurilemma (sheath of Schwann)

Myelin sheath

Nucleus of Schwann cell

Axon

(e)

Unmyelinated nerve fiber

Enveloping Schwann cell

Schwann cell nucleus

(f)

Schwann cell cytoplasm

Myelin sheath

Neurilemma

Axon

(g)

Figure 16.3

Myelination of neuron processes by individual Schwann cells. (a)-(d) A Schwann cell becomes apposed to an axon and envelops it in a trough. It then begins to rotate around the axon, wrapping it loosely in successive layers of its plasma membrane. Eventually, the Schwann cell cytoplasm is forced from between the membranes and comes to lie peripherally just beneath the exposed portion of the Schwann cell membrane. The tight membrane wrappings surrounding the axon form the myelin sheath. The area of Schwann cell cytoplasm and its exposed membrane are referred to as the neurilemma or sheath of Schwann. (e) Longitudinal view of a myelinated axon showing portions of adjacent Schwann cells and the node of Ranvier between them; (f) unmyelinated fibers. Schwann cells may associate loosely with several axons, which they partially invest. In such cases, Schwann cell coiling around the axons does not occur. (g) Photomicrograph of a myelinated axon, cross-sectional view.

the large prominent "owl's eye" nucleoli, and the granular Nissl bodies. If possible, distinguish the axon from the many dendrites. Sketch the cell in the space provided here, and label the important anatomic details you observe.

Sketch a portion of a myelinated nerve fiber in the space provided here, illustrating two or three nodes of Ranvier. Label the axon, myelin sheath, nodes, and neurilemma.

3.   Obtain a prepared slide of teased myelinated nerve fibers and identify the following: nodes of Ranvier, neurilemma, axis cylinder (the axon itself), Schwann cell nuclei, and myelin sheath.

Do the nodes seem to occur at consistent intervals,

or are they irregularly distributed? _____

Explain the significance of this finding: _____

_____

_____

_____

_____

NEURON CLASSIFICATION

Neurons may be classified on the basis of structure or of function. Figure 16.4 depicts both schemes.

Basis of Structure

Neurons may be differentiated according to the number of processes attached to the nerve cell body. In **unipolar neurons,** one very short process extends from the nerve cell body. However, this process soon splits into proximal and distal parts—the axon and dendrite, respectively. Nearly all neurons that conduct impulses toward the CNS are unipolar.

Bipolar neurons have two processes—one axon and one dendrite—attached to the nerve cell body. This neuron type is quite rare, typically found only as part of the receptor apparatus of the eye, ear, and olfactory mucosa.

Many processes issue from the cell body of **multipolar neurons,** all classified as dendrites except for a single axon. Most neurons in the brain and spinal cord (CNS neurons) and those whose axons carry impulses away from the CNS fall into this last category.

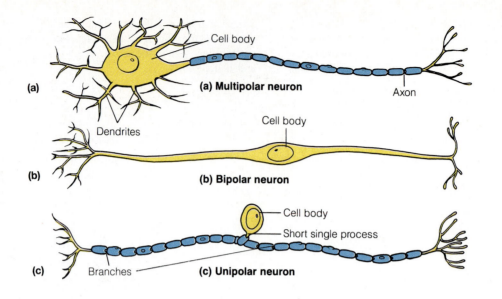

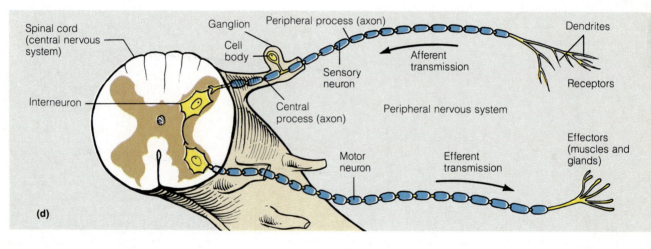

Figure 16.4

Classification of neurons. (a-c) On the basis of structure: (a) multipolar; (b) bipolar; (c) unipolar. (d) On the basis of function, there are sensory and motor neurons, and interneurons. Sensory (afferent) neurons conduct impulses from the body's sensory receptors to the central nervous system; most are unipolar neurons with their nerve cell bodies in ganglia in the PNS. Motor (efferent) neurons transmit impulses from the CNS to effectors such as muscles and glands. Association neurons (interneurons) complete the communication line between sensory and motor neurons. They are typically multipolar and their nerve cell bodies reside in the CNS.

Obtain prepared slides of Purkinje cells of the cerebellar cortex, pyramidal cells of the cerebral cortex, and a dorsal root ganglion. As you observe them under the microscope, try to pick out the anatomic details depicted in Figure 16.5. Note that the viewed neurons of the cerebral and cerebellar tissues (both brain tissues) are extensively branched; in contrast, the neurons of the dorsal root ganglion are more rounded. You may also be able to identify astrocytes (a type of neuroglia) in the brain tissue slides if you examine them closely.

Which of these neuron types would be classified as

multipolar neurons? _____

Which as unipolar? _____

Basis of Function

In general, neurons carrying impulses from the sensory receptors in the internal organs (viscera) or in the skin are termed **sensory,** or **afferent, neurons** (refer back to Figure 16.4). The dendritic endings of sensory neurons are often equipped with specialized

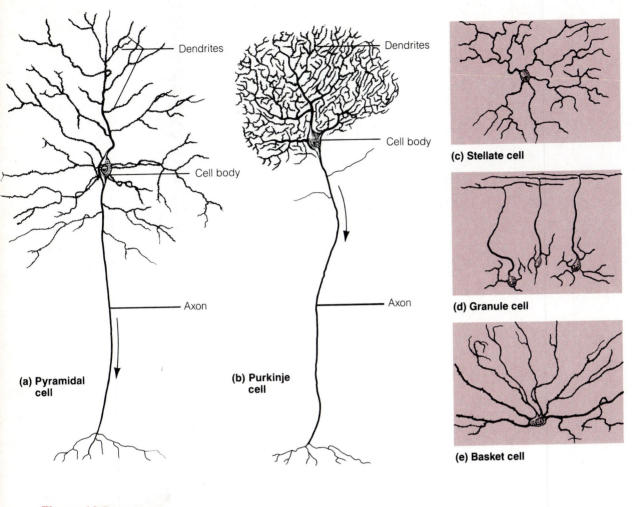

Dendrites

Dendrites

Cell body

Cell body

Axon

Axon

(a) Pyramidal cell

(b) Purkinje cell

(c) Stellate cell

(d) Granule cell

(e) Basket cell

Figure 16.5

Structure of selected neurons: (a) pyramidal cell of the cerebral cortex and (b) Purkinje cells of the cerebellum have extremely long axons, but they are easily distinguished by their dendrite-branching patterns, whereas cerebellar neurons such as (c) stellate cells, (d) granule cells, and (e) basket cells have short (or even absent) axons and profuse dendrites.

receptors that are stimulated by specific changes in their immediate environment. The cell bodies of sensory neurons are always found in a ganglion outside the CNS.

Neurons carrying activating impulses from the CNS to the viscera and/or body muscles and glands are termed **motor, or efferent, neurons.** The cell bodies of motor neurons are always located in the CNS.

The third functional category of neurons are the **interneurons, or association neurons,** which connect sensory and motor neurons. Their nerve cell bodies, like those of motor neurons, are always located within the CNS.

STRUCTURE OF A NERVE

A nerve is a bundle of neuron fibers or processes wrapped in connective tissue coverings that extends to and/or from the CNS and visceral organs or structures of the body periphery (such as skeletal muscles, glands, and skin).

Within a nerve, each fiber is surrounded by a delicate connective tissue sheath called an **endoneurium,** which insulates it from the other neuron processes adjacent to it. (The endoneurium is often mistaken for the myelin sheath; it is instead an additional sheath that surrounds the myelin sheath.) Groups of fibers are bound by a coarser connective tissue, called the **perineurium,** to form bundles of fibers called **fascicles.** Finally, all the fascicles are bound

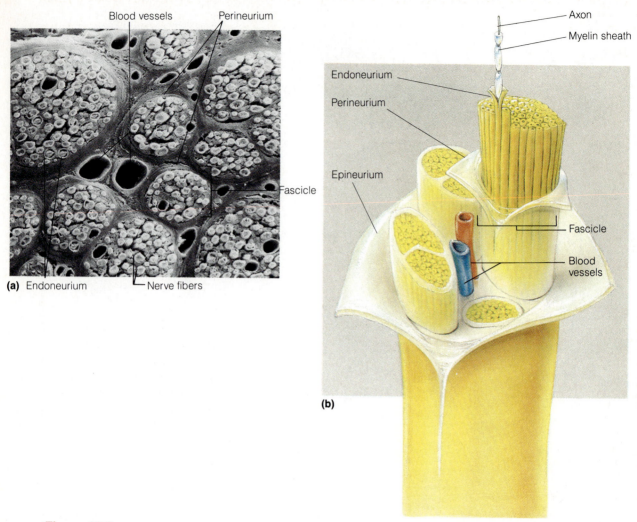

Figure 16.6

Structure of a nerve: (a) scanning electron micrograph of a cross section of a portion of a nerve (400X); (b) three-dimensional view of a portion of a nerve, showing connective tissue wrappings.

together by a tough, white fibrous connective tissue sheath called the **epineurium** to form the cordlike nerve (Figure 16.6). In addition to the connective tissue wrappings, blood vessels and lymphatic vessels serving the fibers also travel within a nerve.

Like neurons, nerves are classified according to the direction in which they transmit impulses. Nerves carrying both sensory (afferent) and motor (efferent) fibers are called **mixed nerves;** all spinal nerves are mixed nerves. Nerves that carry only sensory processes and conduct impulses only toward the CNS are referred to as **sensory,** or **afferent, nerves.** A few of the cranial nerves are pure sensory nerves, but the majority are mixed nerves. The ventral roots of the spinal cord, which carry only motor fibers, can be considered **motor,** or **efferent, nerves.**

Examine under the compound microscope a prepared cross section of a peripheral nerve. Sketch the nerve, identifying and labeling nerve fibers, myelin sheaths, fascicles, and endoneurium, perineurium, and epineurium sheaths.

17

EXERCISE 17

Gross Anatomy of the Brain and Cranial Nerves

OBJECTIVES

1. To identify or locate the following brain structures on a dissected specimen, human brain model (or slices), or appropriate diagram, and to state their functions:

   - *cerebral hemisphere structures:* lobes, important fissures, lateral ventricles, basal nuclei, corpus callosum, fornix, septum pellucidum
   - *diencephalon structures:* thalamus, intermediate mass, hypothalamus, optic chiasma, pituitary gland, mammillary bodies, pineal body, choroid plexus of the third ventricle, interventricular foramen
   - *brain stem structures:* corpora quadrigemina, cerebral aqueduct, cerebral peduncles of the midbrain, pons, medulla, fourth ventricle
   - *cerebellum structures:* cerebellar hemispheres, vermis, arbor vitae

2. To describe the composition of the gray and white matter.

3. To locate the well-recognized functional areas of the human cerebral hemispheres

4. To define *gyri,* and *fissures* (sulci).

5. To identify the three meningeal layers and state their function, and to locate the falx cerebri, falx cerebelli, and tentorium cerebelli.

6. To state the function of the arachnoid villi and dural sinuses.

7. To discuss the formation, circulation, and drainage of cerebrospinal fluid.

8. To identify at least four pertinent anatomic differences between the human brain and that of the sheep (or other mammal).

9. To identify the cranial nerves by number and name on an appropriate model or diagram, stating the origin and function of each.

MATERIALS

Human brain model (dissectible)
Preserved human brain (if available)
Coronally sectioned human brain slice (if available)
Preserved sheep brain (meninges and cranial nerves intact)
Dissecting tray and instruments
Materials for cranial nerve tests (as desired by the instructor)

When viewed alongside all nature's animals, humans are indeed unique; the key to their uniqueness is found in the brain. Only in humans has the brain region called the cerebrum become so elaborated and grown out of proportion that it overshadows other brain areas. Other animals are primarily concerned with informational input and response for the sake of survival and preservation of the species, but human beings devote considerable time to nonsurvival ends. They are the only animals who manipulate abstract ideas and search for knowledge for its own sake, who are capable of emotional response and artistic creativity, or who can anticipate the future and guide their lives according to ethical and moral values. For all this, humans can thank their overgrown cerebrum.

We can be considered composite reflections of our brain's experience. If all past sensory input could mysteriously and suddenly be "erased," we would be unable to walk, talk, or communicate in any manner. Spontaneous movement would be apparent, as in a fetus, but no voluntary integrated function of any type would be possible. Clearly we would cease to be the same individuals.

Because of the complexity of the nervous system, its anatomic structures are usually considered in terms of two principal divisions: the **central nervous system (CNS)** and the **peripheral nervous system (PNS).** The central nervous system consists of the brain and spinal cord, which primarily interpret incoming sensory information and issue instructions based on past experience. The peripheral nervous system consists of the cranial and spinal nerves, ganglia, and sensory receptors. These structures serve as communication lines as they carry impulses—from the sensory receptors to the CNS and from the CNS to the appropriate glands or muscles.

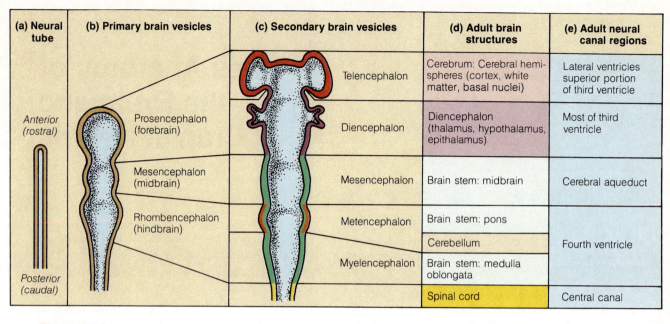

(a) Neural tube	(b) Primary brain vesicles	(c) Secondary brain vesicles		(d) Adult brain structures	(e) Adult neural canal regions
Anterior (rostral)	Prosencephalon (forebrain)	Telencephalon		Cerebrum: Cerebral hemispheres (cortex, white matter, basal nuclei)	Lateral ventricles superior portion of third ventricle
		Diencephalon		Diencephalon (thalamus, hypothalamus, epithalamus)	Most of third ventricle
	Mesencephalon (midbrain)	Mesencephalon		Brain stem: midbrain	Cerebral aqueduct
	Rhombencephalon (hindbrain)	Metencephalon		Brain stem: pons	Fourth ventricle
				Cerebellum	
Posterior (caudal)		Myelencephalon		Brain stem: medulla oblongata	
				Spinal cord	Central canal

Figure 17.1

Embryonic development of the human brain. (a) The neural tube becomes subdivided into (b) the primary brain vesicles, which subsequently form (c) the secondary brain vesicles, which differentiate into (d) the adult brain structures. (e) The adult structures derived from the neural canal.

In this exercise both CNS (brain) and PNS (cranial nerves) structures will be studied because of their close anatomic relationship.

THE HUMAN BRAIN

During embryonic development of all vertebrates, the CNS first makes its appearance as a simple tube-like structure, the **neural tube,** that extends down the dorsal medial plane. By the fourth week, the human brain begins to form as an expansion of the anterior or rostral end of the neural tube (the end toward the head). Shortly thereafter, constrictions appear, dividing the developing brain into three major regions—the forebrain, midbrain, and hindbrain (Figure 17.1). The remainder of the neural tube becomes the spinal cord.

During fetal development, two anterior outpocketings extend from the forebrain and grow rapidly to form the cerebral hemispheres. Because of space restrictions imposed by the skull, the cerebral hemispheres grow posteriorly and inferiorly and finally end up enveloping and obscuring the rest of the forebrain and midbrain structures. Somewhat later in development, the dorsal portion of the hindbrain also enlarges to produce the cerebellum. The central canal of the neural tube, which remains continuous throughout the brain and cord, becomes enlarged in four regions of the brain to form chambers called **ventricles.**

External Anatomy

Generally, in studying the major brain areas in the laboratory, the brain is considered in terms of four major regions: the cerebral hemispheres, diencephalon, brain stem, and cerebellum. The correlation between these four anatomic regions and the structures of the forebrain, midbrain, and hindbrain is also outlined in Figure 17.1.

CEREBRAL HEMISPHERES   The **cerebral hemispheres** are the most superior portion of the brain (Figure 17.2). Their entire surface is thrown into elevated ridges of tissue called **gyri** that are separated by depressed areas called **fissures,** or **sulci.** Many of these fissures and gyri are important anatomic landmarks.

The cerebral hemispheres are divided by a single deep fissure, the **longitudinal fissure.** The **central sulcus** divides the **frontal lobe** from the **parietal lobe,** and the **lateral sulcus** separates the **temporal lobe** from the parietal lobe. The **parieto-occipital sulcus,** which divides the **occipital lobe** from the parietal lobe, is not visible externally. Note that the cerebral hemisphere lobes are named for the cranial bones that lie over them.

The functional areas of the cerebral hemispheres have also been located (Figure 17.2d). The **primary (somatic) sensory area** is located in the **postcentral gyrus** of the parietal lobe. Impulses traveling from the body's sensory receptors (such as those for pressure, pain, and temperature) are localized in this area of the brain. ("This information is from my big toe.") Immediately posterior to the primary sensory area is

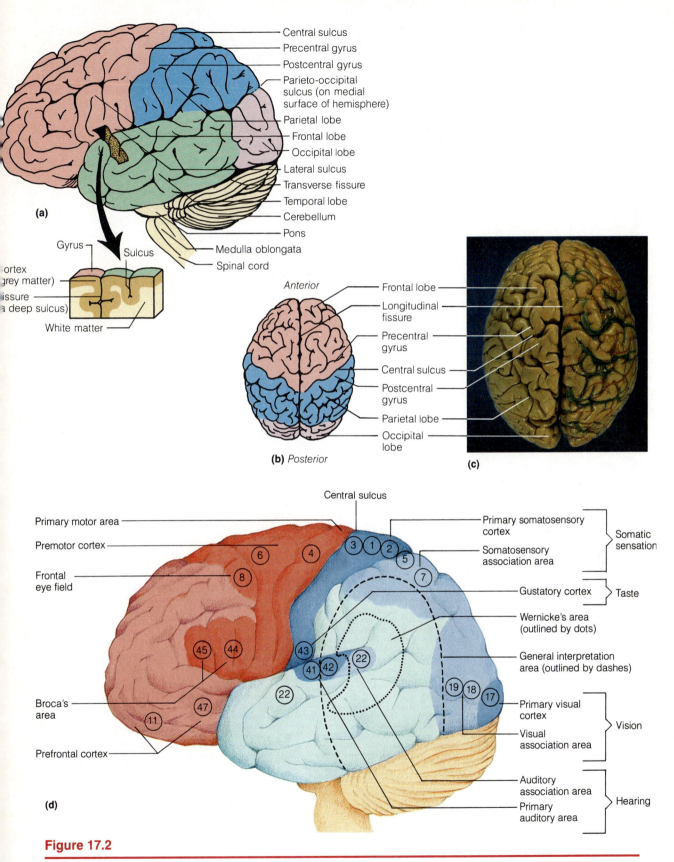

Figure 17.2

External structure (lobes and fissures) of the cerebral hemispheres: (a) left lateral view of the brain; (b) superior view; (c) photograph of the superior aspect of the human brain; (d) functional areas of the left cerebral cortex. The olfactory area, which is deep to the temporal lobe on the medial hemispheric surface, is not identified in (d). Numbers indicate brain regions plotted by the Brodman system.

the **somatosensory association area,** in which the meaning of incoming stimuli is analyzed. ("Ouch! I have a *pain* there.") Thus, the somatosensory association area allows you to become aware of pain, coldness, a light touch, and the like.

Impulses from the special sense organs are interpreted in other specific areas also noted in Figure 17.2(d). For example, the visual area is located in the posterior portion of the occipital lobe and the auditory area is located in the temporal lobe in the gyrus bordering the lateral sulcus. The olfactory area is located deep within the temporal lobe along its medial surface, in a region called the **uncus** (see Figure 17.4b).

The **primary motor area,** which is responsible for conscious or voluntary movement of the skeletal muscles, is located in the **precentral gyrus** of the frontal lobe. A specialized motor speech area called **Broca's area** is found at the base of the precentral gyrus just above the lateral sulcus. Damage to this area (which is located only in one cerebral hemisphere, usually the left) reduces or eliminates the ability to articulate words.

Areas involved in higher intellectual reasoning are believed to lie in the anterior portions of the frontal lobes. A rather poorly defined region in the temporal lobe is Wernicke's area, sometimes called the **speech area,** in which unfamiliar words are sounded out. Wernicke's area, like Broca's area, is located in one cerebral hemisphere only, typically the left.

The cell bodies of cerebral neurons involved in these functions are found only in the outermost gray matter of the cerebrum, the area called the **cerebral cortex.** The balance of cerebral tissue—the deeper **cerebral white matter**—is composed of fiber tracts carrying impulses to or from the cortex.

1. Using a model of the human brain (and a preserved human brain, if available), identify the areas and structures of the cerebral hemispheres described above.

DIENCEPHALON   The **diencephalon,** sometimes considered the most superior portion of the brain stem, is embryologically part of the forebrain, along with the cerebral hemispheres.

1. Turn the brain model so the ventral surface of the brain can be viewed. Using Figure 17.3 as a guide, start superiorly and identify the externally visible structures that mark the position of the floor of the diencephalon. These are the **olfactory bulbs** and **tracts, optic nerves, optic chiasma** (where the fibers of the optic nerves partially cross over), **optic tracts, pituitary gland,** and **mammillary bodies.**

BRAIN STEM

1. Continue inferiorly to identify the **brain stem** structures—the **cerebral peduncles** (fiber tracts

in the **midbrain** connecting the pons below with cerebrum above), the pons, and the medulla oblongata. *Pons* means "bridge," and the **pons** consists primarily of motor and sensory fiber tracts connecting the brain with lower CNS centers. The lowest brain stem region, the **medulla,** is also composed primarily of fiber tracts. You can see the **decussation of pyramids,** a crossover point for the major motor tract (pyramidal tract) descending from the motor areas of the cerebrum to the cord, on the medulla's anterior surface. The medulla also houses many vital autonomic centers involved in the control of heart rate, respiratory rhythm, and blood pressure as well as involuntary centers involved in the initiation of vomiting, swallowing, and so on.

CEREBELLUM

1. Turn the brain model so you can see the dorsal aspect. Identify the large cauliflowerlike **cerebellum,** which projects dorsally from under the occipital lobe of the cerebrum. Note that, like the cerebrum, the cerebellum has two hemispheres and a convoluted surface. It also has an outer cortex made up of gray matter with an inner region of white matter.

2. Remove the cerebellum to view the **corpora quadrigemina,** located on the posterior aspect of the midbrain, a brain stem structure. The two superior prominences are the **superior colliculi** (visual reflex centers), and the two smaller inferior prominences are the **inferior colliculi** (auditory reflex centers).

Internal Anatomy

The deeper structures of the brain have also been well mapped. Like the external structures, these can be studied in terms of the four major regions.

CEREBRAL HEMISPHERES

1. Take the brain model apart so you can see a median sagittal view of the internal brain structures (Figure 17.4). Observe the model closely to see the extent of the outer cortex (gray matter), which contains the nerve cell bodies of cerebral neurons. (The pyramidal cells of the cerebral motor cortex are representative of the neurons seen in the precentral gyrus.)

2. Now observe the deeper area of white matter, which is composed of fiber tracts. The fiber tracts found in the cerebral hemisphere white matter are named *association tracts* if they connect two portions of the same hemisphere, *projection tracts* if they run between the cerebral cortex and the lower brain or spinal cord, and *commissures* if they run from one hemisphere to another. Observe the large **corpus callosum,** the major commissure connecting the cerebral hemispheres. The corpus callosum arches above the structures of the diencephalon. Note also the **fornix,** a band-

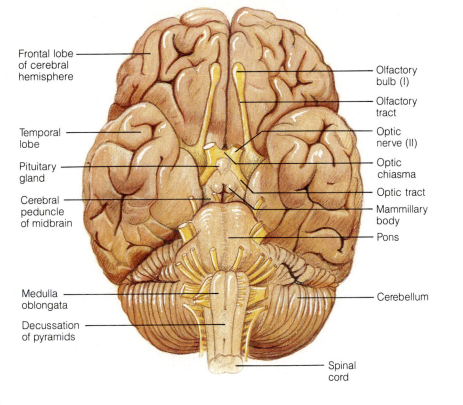

Frontal lobe
of cerebral
hemisphere

Temporal
lobe

Pituitary
gland

Cerebral
peduncle
of midbrain

Medulla
oblongata

Decussation
of pyramids

Olfactory
bulb (I)

Olfactory
tract

Optic
nerve (II)

Optic
chiasma

Optic tract

Mammillary
body

Pons

Cerebellum

Spinal
cord

Figure 17.3

Ventral aspect of the human brain, showing the three regions of the brain stem. Only a small portion of the midbrain can be seen; the rest is surrounded by other brain regions.

like fiber tract concerned with olfaction as well as limbic system functions, and the membranous **septum pellucidum,** which separates the lateral ventricles of the cerebral hemispheres.

3. In addition to the gray matter of the cerebral cortex, there are several "islands" of gray matter, the **basal nuclei,** buried deep within the white matter of the cerebral hemispheres, flanking the lateral and third ventricles. You can see the basal nuclei if you have an appropriate dissectible model or a coronally or cross-sectioned human brain slice. Otherwise, Figure 17.5 will suffice.

The basal nuclei, which are important subcortical motor nuclei (and part of the extra pyramidal system), are involved in the regulation of voluntary motor activities. The most important of them are the arching, comma-shaped **caudate nucleus,** the **claustrum,** the **amygdaloid nucleus** (located at the tip of the caudate nucleus), and the **lentiform nucleus,** which is composed of the **putamen** and **globus pallidus nuclei.** A band of projection fibers coursing down from the precentral (motor) gyrus combines with sensory fibers traveling to the sensory cortex to form a broad band of fibrous material called the **internal capsule.** The internal capsule gives these basal nuclei a striped appearance. This is why the caudate nucleus and the lentiform nucleus are some-

times referred to as the **corpus striatum,** or "striped body."

4. Note the relationship of the lateral ventricles and corpus callosum to the diencephalon structures; that is, hypothalamus, thalamus, and third ventricle—from the cross-sectional viewpoint (see Figure 17.5b).

DIENCEPHALON

1. The major internal structures of the diencephalon are the thalamus, hypothalamus, and epithalamus (Figure 17.4). The **thalamus** consists of two large lobes of gray matter that laterally enclose the shallow third ventricle of the brain. A slender stalk of thalamic tissue, the **massa intermedia,** or **intermediate mass,** connects the two thalamic lobes and bridges the ventricle. The thalamus is a relay station for sensory impulses passing upward to the cortical sensory areas for localization and interpretation. Locate also the **interventricular foramen** or **foramen of Monro,** a tiny orifice connecting the third ventricle with the lateral ventricle on the same side.

2. The **hypothalamus** makes up the floor and a part of the inferior lateral walls of the third ventricle. It is an important autonomic center involved in the regulation of body temperature, water bal-

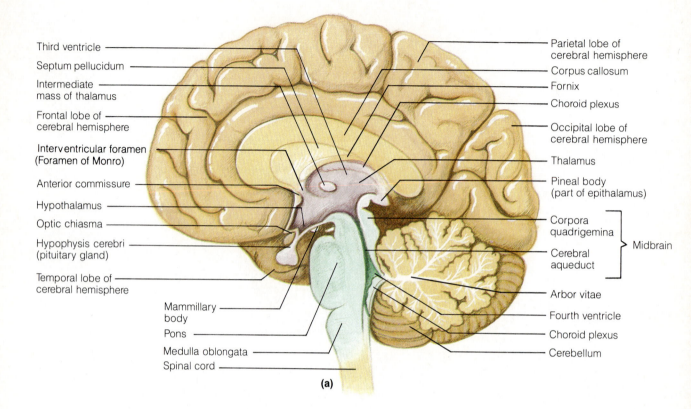

Third ventricle
Septum pellucidum
Intermediate mass of thalamus
Frontal lobe of cerebral hemisphere
Interventricular foramen (Foramen of Monro)
Anterior commissure
Hypothalamus
Optic chiasma
Hypophysis cerebri (pituitary gland)
Temporal lobe of cerebral hemisphere
Mammillary body
Pons
Medulla oblongata
Spinal cord

Parietal lobe of cerebral hemisphere
Corpus callosum
Fornix
Choroid plexus
Occipital lobe of cerebral hemisphere
Thalamus
Pineal body (part of epithalamus)
Corpora quadrigemina
Cerebral aqueduct
Arbor vitae
Fourth ventricle
Choroid plexus
Cerebellum
Midbrain

(a)

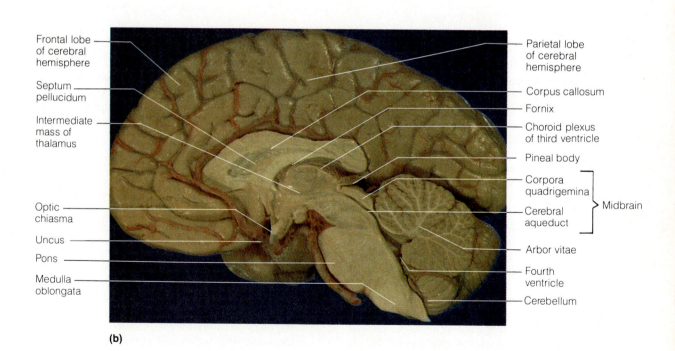

Frontal lobe of cerebral hemisphere
Septum pellucidum
Intermediate mass of thalamus
Optic chiasma
Uncus
Pons
Medulla oblongata

Parietal lobe of cerebral hemisphere
Corpus callosum
Fornix
Choroid plexus of third ventricle
Pineal body
Corpora quadrigemina
Cerebral aqueduct
Arbor vitae
Fourth ventricle
Cerebellum
Midbrain

(b)

Figure 17.4

Diencephalon and brain stem structures as seen in a midsagittal section of the brain. (a) Diagrammatic view; (b) photograph.

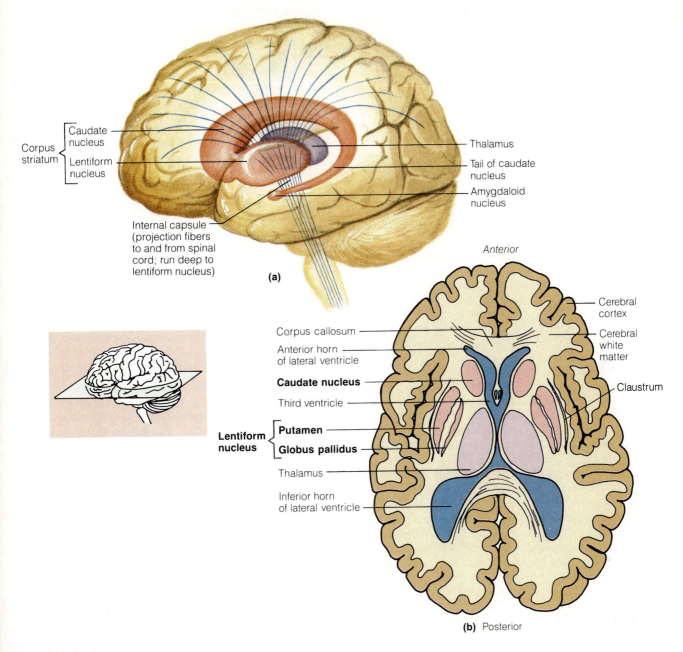

Figure 17.5

Basal (cerebral) nuclei: (a) three-dimensional view of the basal (cerebral) nuclei showing their position within the cerebrum; (b) a transverse section of the cerebrum and diencephalon showing the relationship of the basal nuclei to the thalamus and the lateral and third ventricles.

ance, and fat and carbohydrate metabolism as well as in many other activities and drives. Locate again the pituitary gland, or **hypophysis,** which hangs from the anterior floor of the hypothalamus by a slender stalk, the **infundibulum.** (The pituitary gland is usually not present in preserved brain specimens.) In life, the pituitary rests in the fossa of the sella turcica portion of the sphenoid bone. Its function is discussed in Exercise 22.

Anterior to the pituitary, the optic chiasma portion of the optic pathway to the brain can also be identified. The **mammillary bodies,** hypothalamic reflex centers for olfaction, bulge exteriorly from the floor of the hypothalamus posterior to the pituitary gland.

3. The **epithalamus** forms the roof of the third ventricle and is the most dorsal portion of the diencephalon. Important structures in the epithalamus are the **pineal body** or **gland** (a

neuroendocrine structure), and the **choroid plexus** of the third ventricle. The choroid plexuses, knotlike collections of capillaries within each ventricle, secrete the cerebrospinal fluid.

BRAIN STEM

1. Now trace the short midbrain from the mammillary bodies to the rounded pons below. Continue to refer to Figure 17.4. The **cerebral aqueduct** is a slender canal traveling through the midbrain; it connects the third ventricle to the fourth ventricle in the hindbrain below. The cerebral penduncles and the rounded corpora quadrigemina make up the midbrain tissue anterior and posterior (respectively) to the cerebral aqueduct.

2. Locate the hindbrain structures. Trace the rounded pons to the medulla oblongata below, and identify the fourth ventricle posterior to these structures. Attempt to identify the single median aperture and the two lateral apertures (see page 156), three orifices found in the roof of the fourth ventricle. These apertures serve as conduits for cerebrospinal fluid to circulate into the subarachnoid space from the fourth ventricle.

CEREBELLUM   Examine the cerebellum. Note that it is composed of two lateral hemispheres connected by a midline lobe called the **vermis.** As in the cerebral hemispheres, the cerebellum has an outer cortical area of gray matter and an inner area of white matter. The treelike branching of the cerebellar white matter is referred to as the **arbor vitae,** or tree of life. The cerebellum is concerned with the unconscious coordination of skeletal muscle activity and the control of balance and equilibrium. Fibers converge on the cerebellum from the equilibrium apparatus of the inner ear, visual pathways, proprioceptors of the tendons and skeletal muscles, and from many other areas. Thus the cerebellum remains constantly aware of the position and state of tension of the various body parts.

Meninges of the Brain

The brain (and spinal cord) are covered and protected by three connective tissue membranes called **meninges** (Figure 17.6). The outermost meninx is the leathery **dura mater,** a double-layered membrane. One of its layers is attached to the inner surface of the skull, forming the periosteum; the other forms the outermost brain covering and is continuous with the dura mater of the spinal cord.

The dural layers are fused together except in three areas where the inner membrane extends inward to form a septum that secures the brain to structures inside the cranial cavity. One such extension, the **falx cerebri,** dips into the longitudinal fissure between the cerebral hemispheres to attach to the crista galli of the ethmoid bone of the skull. The cavity created at this point is the large **superior sagittal sinus,** which collects blood draining from the brain tissue. The **falx cerebelli,** separating the two cerebellar

hemispheres, and the **tentorium cerebelli,** separating the cerebrum from the cerebellum below, are two other important inward folds of the inner dural membrane.

The middle meninx, the weblike **arachnoid,** underlies the dura mater and is partially separated from it by the **subdural space.** Threadlike projections bridge the **subarachnoid space** to attach the arachnoid mater to the innermost meninx, the **pia mater.** The delicate pia mater is extensively vascularized and clings tenaciously to the surface of the brain, following its convolutions.

In life, the subarachnoid space is filled with cerebrospinal fluid. Specialized projections of the arachnoid tissue called **arachnoid villi** protrude through the dura mater to allow the cerebrospinal fluid to drain back into the venous circulation via the superior sagittal and other dural sinuses.

Meningitis, an inflammation of the meninges, is a serious threat to the brain because of the intimate association between the brain and meninges. Meningitis is often diagnosed by taking a sample of cerebrospinal fluid from the subarachnoid space. ■

Cerebrospinal Fluid

The cerebrospinal fluid, much like plasma in composition, is continually formed by the **choroid plexuses,** small capillary knots hanging from the roof of the ventricles of the brain. The cerebrospinal fluid in and around the brain forms a watery cushion that protects the delicate brain tissue against blows to the head.

Within the brain, the cerebrospinal fluid circulates from the two lateral ventricles (in the cerebral hemispheres) into the third ventricle via the **interventricular foramina** (foramina of Monro), and then through the cerebral aqueduct of the midbrain into the fourth ventricle in the hindbrain (Figure 17.7). A portion of the fluid reaching the fourth ventricle continues down the central canal of the spinal cord, but the bulk of it circulates into the subarachnoid space, exiting through the three foramina in the walls of the fourth ventricle (the two lateral apertures, also called the foramina of Luschka, and the single median aperture). The fluid returns to the blood in the dural sinuses through the arachnoid villi.

Ordinarily, cerebrospinal fluid forms and drains at a constant rate. However, under certain conditions—for example, obstructed drainage or circulation resulting from tumors or anatomic deviations—the cerebrospinal fluid begins to accumulate and exerts increasing pressure on the brain. ■

CRANIAL NERVES

The cranial nerves are part of the peripheral nervous system and not part of the brain proper, but they are most appropriately identified in conjunction with the study of brain anatomy. The 12 pairs of cranial nerves primarily serve the head and neck. Only one pair (the vagus nerves) extends to the thoracic and

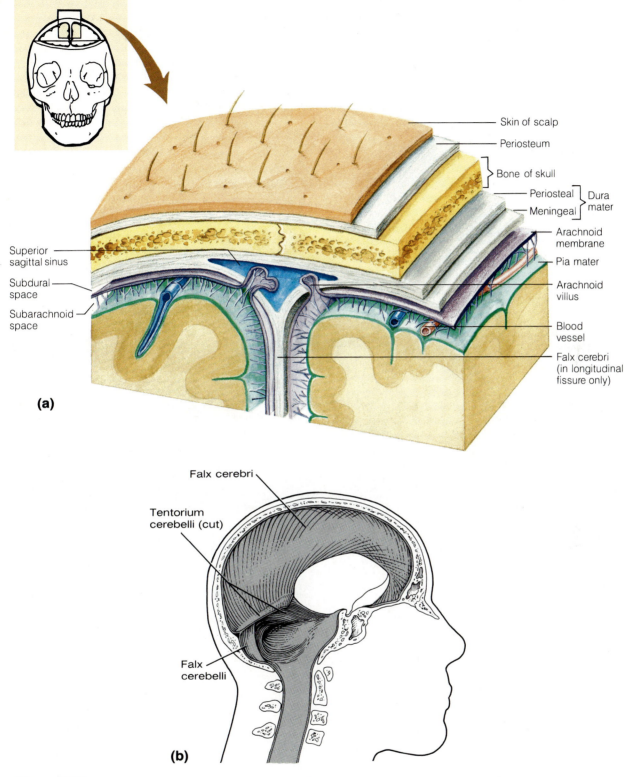

Skin of scalp
Periosteum
Bone of skull
Periosteal ⎤ Dura
Meningeal ⎦ mater
Arachnoid membrane
Pia mater
Arachnoid villus
Blood vessel
Falx cerebri (in longitudinal fissure only)

Superior sagittal sinus
Subdural space
Subarachnoid space

(a)

Falx cerebri
Tentorium cerebelli (cut)
Falx cerebelli

(b)

Figure 17.6

Meninges of the brain: (a) three-dimensional frontal section showing the relationship of the dura mater, arachnoid, and pia mater. The meningeal dura forms the falx cerebri fold, which extends into the longitudinal fissure and attaches the brain to the ethmoid bone of the skull. A dural sinus, the superior sagittal sinus, is enclosed by the dural membranes superiorly. Arachnoid villi, which return cerebrospinal fluid to the dural sinus, are also shown. (b) Position of the dural folds, the falx cerebri, tentorium cerebelli, and falx cerebelli.

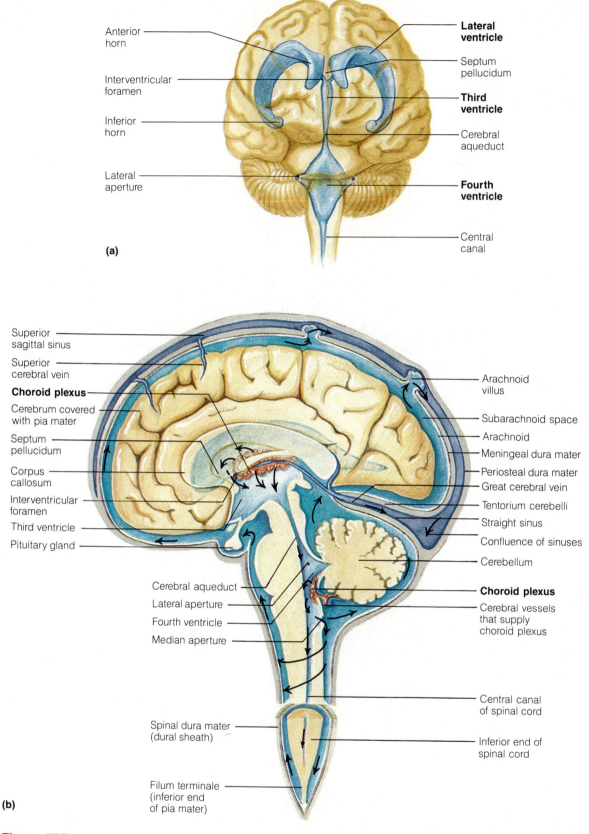

Anterior horn

Interventricular foramen

Inferior horn

Lateral aperture

Lateral ventricle

Septum pellucidum

Third ventricle

Cerebral aqueduct

Fourth ventricle

Central canal

(a)

Superior sagittal sinus

Superior cerebral vein

Choroid plexus

Cerebrum covered with pia mater

Septum pellucidum

Corpus callosum

Interventricular foramen

Third ventricle

Pituitary gland

Cerebral aqueduct

Lateral aperture

Fourth ventricle

Median aperture

Spinal dura mater (dural sheath)

Filum terminale (inferior end of pia mater)

Arachnoid villus

Subarachnoid space

Arachnoid

Meningeal dura mater

Periosteal dura mater

Great cerebral vein

Tentorium cerebelli

Straight sinus

Confluence of sinuses

Cerebellum

Choroid plexus

Cerebral vessels that supply choroid plexus

Central canal of spinal cord

Inferior end of spinal cord

(b)

Figure 17.7

Location and circulatory pattern of cerebrospinal fluid. (a) Anterior view; note that different regions of the large lateral ventricles are indicated by the terms *anterior horn*, *posterior horn*, and *inferior horn*. (b) The cerebrospinal fluid flows from the lateral ventricles, through the interventricular foramina, into the third ventricle, and then into the fourth ventricle via the cerebral aqueduct. (The relative position of the right lateral ventricle is indicated by the pale blue area deep to the corpus callosum and septum pellucidum.)

abdominal cavities. All but the first two pairs (olfactory and optic nerves) arise from the brain stem and pass through foramina in the base of the skull to reach their destination.

The cranial nerves are numbered consecutively, and in most cases their names reflect the major structures they control. The cranial nerves are described by name, number, origin, course, and function in Table 17.1. This information should be committed to memory.

A mnemonic device that might be helpful for remembering the cranial nerves in order is, "*Oh, oh, oh, to touch and feel very good velvet, Ah.*" The first letter of each word, and both letters of the final "*Ah,*" will remind you of the first letter of the cranial nerve name.

1. Observe the anterior surface of the brain model to identify the cranial nerves. Figure 17.8 may also aid you in this study.

Most cranial nerves are mixed nerves (containing both motor and sensory fibers). However, close scrutiny of Table 17.1 will reveal that three pairs of cranial nerves (optic, olfactory, and vestibulocochlear) are purely sensory in function.

The last column of the table describes techniques for testing cranial nerves, which is an important part of any neurologic examination. This infor-

mation may help you understand cranial nerve function, especially as it pertains to some aspects of brain function. If materials are provided for cranial nerve testing, conduct the tests following the directions given in the "Testing" column of Table 17.1.

2. You may recall that the nerve cell bodies of neurons are always located within the central nervous system or in specialized collections of nerve cell bodies (ganglia) outside the CNS. Nerve cell bodies of the sensory cranial nerves are located in ganglia; those of the mixed cranial nerves are found both within the brain and in peripheral ganglia. Several cranial nerve ganglia are named in the chart below. *Using your textbook or an appropriate reference,* name the cranial nerve the ganglion is associated with and state its location.

Cranial nerve ganglion	Cranial nerve	Site of ganglion
gasserian		
geniculate		
inferior		
superior		
spiral		
vestibular		

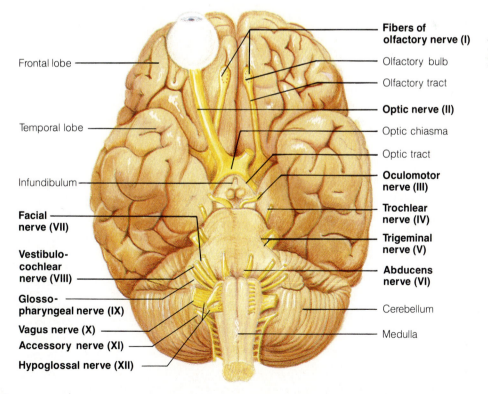

Figure 17.8

Ventral aspect of the human brain, showing the cranial nerves.

TABLE 17.1 The Cranial Nerves (see Figure 17.8)

Name and number	Origin and course	Function	Testing
I. Olfactory	Fibers arise from olfactory mucosa and run through cribriform plate of ethmoid bone to synapse with olfactory bulbs.	Purely sensory—carries impulses associated with sense of smell.	Person is asked to sniff aromatic substances, such as oil of cloves and vanilla, and to identify each.
II. Optic	Fibers arise from retina of eye and pass through optic foramen in sphenoid bone. Fibers of two optic nerves then take part in forming optic chiasma (with partial crossover of fibers) after which they continue on to thalamus as the optic tracts. Final fibers of this pathway travel from the thalamus to the optic cortex as the optic radiation.	Purely sensory—carries impulses associated with vision.	Vision and visual field are determined with eye chart and by testing the point at which the person first sees an object (finger) moving into the visual field. Fundus of eye viewed with ophthalmoscope to detect papilledema (swelling of optic disk), or point at which optic nerve leaves the eye and to observe blood vessels.
III. Oculomotor	Fibers emerge from midbrain and exit from skull via superior orbital fissure to run to eye.	Somatic motor fibers to inferior oblique and superior, inferior, and medial rectus muscles, which direct eyeball, and to levator palpebrae muscles of eyelid; parasympathetic fibers to iris and smooth muscle controlling lens shape (control reflex responses to varying light intensity and focusing of eye for near vision); contains proprioceptive sensory fibers carrying impulses from extrinsic eye muscles.	Pupils are examined for size, shape, and equality. Pupillary reflex is tested with penlight (pupils should constrict when illuminated). Convergence for near vision is tested, as is subject's ability to follow objects up, down, side to side, and diagonally.
IV. Trochlear	Fibers emerge from midbrain and exit from skull via superior orbital fissure to run to eye.	Provides somatic motor fibers to superior oblique muscle (an extrinsic eye muscle); conveys proprioceptive impulses from same muscle to brain.	Tested in common with cranial nerve III.
V. Trigeminal	Fibers emerge from pons and form three divisions, which exit separately from skull: mandibular division through foramen ovale in sphenoid bone, maxillary division via foramen rotundum in sphenoid bone, and ophthalmic division through superior orbital fissure of eye socket.	Major sensory nerve of face; conducts sensory impulses from skin of face and anterior scalp, from mucosae of mouth and nose, and from surface of eyes; mandibular division also contains motor fibers that innervate muscles of mastication and muscles of floor of mouth.	Sensations of pain, touch, and temperature are tested with safety pin and hot and cold objects. Corneal reflex tested with wisp of cotton. Motor branch assessed by asking person to clench his teeth, open mouth against resistance, and move jaw side to side.
VI. Abducens	Fibers leave inferior region of pons and exit from skull via superior orbital fissure to run to eye.	Carries motor fibers to lateral rectus muscle of eye and proprioceptive fibers from same muscle to brain.	Tested in common with cranial nerve III.

TABLE 17.1   *(continued)*

Name and number	Origin and course	Function	Testing
VII. Facial	Fibers leave pons and travel through temporal bone via internal acoustic meatus, exiting via stylomastoid foramen to reach facial region.	Mixed—supplies somatic motor fibers to muscles of facial expression and parasympathetic motor fibers to lacrimal and salivary glands; carries sensory fibers from taste receptors of anterior portion of tongue.	Anterior two-thirds of tongue is tested for ability to taste sweet (sugar), salty, sour (vinegar), and bitter (quinine) substances. Symmetry of face is checked. Subject is asked to close eyes, smile, whistle, and so on. Tearing is assessed with ammonia fumes.
VIII. Vestibulocochlear	Fibers run from inner-ear equilibrium and hearing apparatus, housed in temporal bone, through internal acoustic meatus to enter the pons.	Purely sensory—vestibular branch transmits impulses associated with sense of equilibrium from vestibular apparatus and semicircular canals; cochlear branch transmits impulses associated with hearing from cochlea.	Hearing is checked by air and bone conduction using tuning fork.
IX. Glossopharyngeal	Fibers emerge from medulla and leave skull via jugular foramen to run to throat region.	Mixed—somatic motor fibers serve pharyngeal muscles, and parasympathetic motor fibers serve salivary glands; sensory fibers carry impulses from pharynx, tonsils, posterior tongue (taste buds), and pressure receptors of carotid artery.	Position of the uvula is checked. Gag and swallowing reflexes are checked. Subject is asked to speak and cough. Posterior third of tongue may be tested for taste.
X. Vagus	Fibers emerge from medulla and pass through jugular foramen and descend through neck region into thorax and abdominal regions.	Fibers carry somatic motor impulses to pharynx and larynx and sensory fibers from same structures; very large portion is composed of parasympathetic motor fibers, which supply heart and smooth muscles of abdominal visceral organs; transmits sensory impulses from viscera.	As for cranial nerve IX (IX and X are tested in common, since they both innervate muscles of throat and mouth.)
XI. Accessory	Fibers arise from medulla and superior aspect of spinal cord and travel through jugular foramen to reach muscles of neck and back.	Provides somatic motor fibers to sternocleidomastoid and trapezius muscles and to muscles of soft palate, pharynx, and larynx (spinal and medullary fibers respectively); proprioceptive impulses are conducted from these muscles to brain.	Sternocleidomastoid and trapezius muscles are checked for strength by asking person to rotate head and shrug shoulders against resistance.
XII. Hypoglossal	Fibers arise from medulla and exit from skull via hypoglossal canal to travel to tongue.	Carries somatic motor fibers to muscles of tongue and proprioceptive impulses from tongue to brain.	Person is asked to protrude and retract tongue. Any deviations in position are noted.

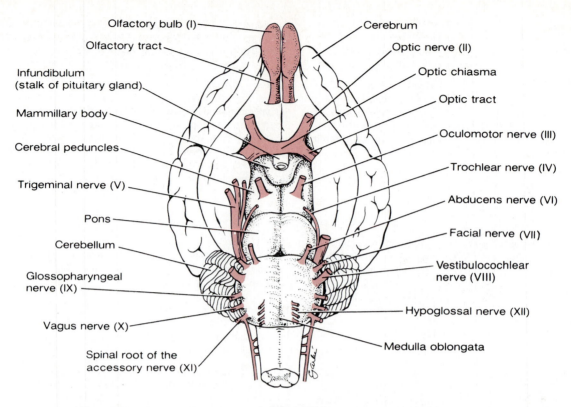

Olfactory bulb (I)
Olfactory tract
Infundibulum (stalk of pituitary gland)
Mammillary body
Cerebral peduncles
Trigeminal nerve (V)
Pons
Cerebellum
Glossopharyngeal nerve (IX)
Vagus nerve (X)
Spinal root of the accessory nerve (XI)

Cerebrum
Optic nerve (II)
Optic chiasma
Optic tract
Oculomotor nerve (III)
Trochlear nerve (IV)
Abducens nerve (VI)
Facial nerve (VII)
Vestibulocochlear nerve (VIII)
Hypoglossal nerve (XII)
Medulla oblongata

(a)

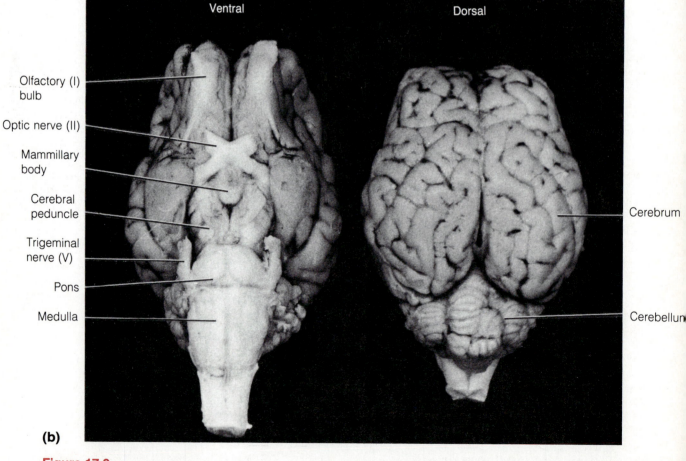

Ventral Dorsal

Olfactory (I) bulb
Optic nerve (II)
Mammillary body
Cerebral peduncle
Trigeminal nerve (V)
Pons
Medulla

Cerebrum
Cerebellum

(b)

Figure 17.9

Intact sheep brain: (a) ventral view; (b) photograph showing ventral and dorsal views, courtesy of Ann Allworth.

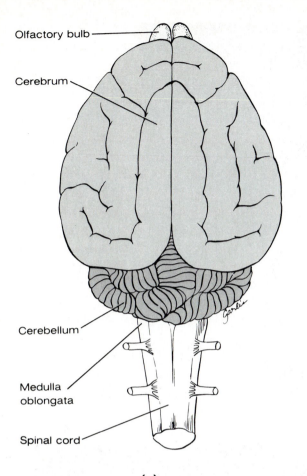

Olfactory bulb

Cerebrum

Cerebellum

Medulla oblongata

Spinal cord

(c)

Figure 17.9 *(continued)*

Intact sheep brain: (c) dorsal view.

DISSECTION OF THE SHEEP BRAIN

The brain of any mammal is enough like the human brain to warrant comparison. Obtain a sheep brain, dissecting pan, and instruments, and bring them to your laboratory bench.

1.   Place the intact sheep brain ventral surface down on the dissecting pan and observe the dura mater. Feel its consistency and note its toughness. Cut through the dura mater along the line of the longitudinal fissure (which separates the cerebral hemispheres) to enter the sagittal sinus. Gently force the cerebral hemispheres apart laterally to expose the corpus callosum deep to the longitudinal fissure.

2.   Carefully remove the dura mater and examine the superior surface of the brain. Note that its surface is thrown into convolutions (fissures and gyri), just as the human brain is. Locate the arachnoid

meninx, which appears on the brain surface as a delicate "cottony" material spanning the fissures. In contrast, the innermost meninx, the pia mater, closely follows the cerebral contours.

Dorsal Structures

1.   Refer to Figure 17.9 (b) and (c) as a guide in identifying the following structures. The cerebral hemispheres should be easy to locate. How do the size of the sheep's cerebral hemispheres and the depth of the fissures compare to those

in the human brain? _____

_____

_____

_____

2.   Carefully examine the cerebellum. Note that it is not divided longitudinally, in contrast to the human cerebellum, and that its fissures are oriented differently. What dural falx is missing that

is present in humans? _____

3.   Locate the three pairs of cerebellar peduncles, fiber tracts that connect the cerebellum to other brain structures, by lifting the cerebellum dorsally away from the brain stem. The most posterior pair, the inferior cerebellar peduncles, connect the cerebellum to the medulla. The middle cerebellar peduncles attach the cerebellum to the pons, and the superior cerebellar peduncles run from the cerebellum to the midbrain.

4.   To expose the dorsal surface of the midbrain, gently spread the cerebrum and cerebellum apart, as shown in Figure 17.10. Identify the corpora quadrigemina, which appear as four rounded prominences on the dorsal midbrain surface. What is the function of the corpora quadrigemina?

_____

_____

_____

_____

Also locate the pineal body, which appears as a small oval protrusion in the midline just anterior to the corpora quadrigemina.

Ventral Structures

Figure 17.9 (a) and (b) show the important features of the ventral surface of the brain.

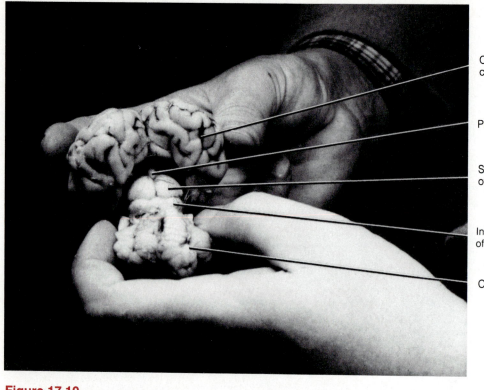

Occipital lobe of
cerebral hemisphere

Pineal body

Superior colliculi
of corpora quadrigemina

Inferior colliculi
of corpora quadrigemina

Cerebellum

Figure 17.10

Means of exposing the dorsal midbrain structures of the sheep brain. (Photo courtesy of Ann Allworth.)

1. Look for the clublike olfactory bulbs, anteriorly on the inferior surface of the frontal lobes of the cerebral hemispheres. Axons of olfactory neurons run from the nasal mucosa through the perforated cribriform plate of the ethmoid bone to synapse with the olfactory bulbs.

    How does the size of these olfactory bulbs compare with those of humans? _____

    _____

    _____

    Is the sense of smell more important as a protective and a food-getting sense in sheep *or*

    in humans? _____

2. The optic nerve (II) carries sensory impulses from the retina of the eye. Thus this cranial nerve is involved in the sense of vision. Identify the optic nerves, optic chiasma, and optic tracts.

3. Posterior to the optic chiasma, two structures protrude from the ventral aspect of the hypothalamus—the infundibulum (stalk of the pituitary gland) immediately posterior to the optic chiasma and the mammillary body. Notice that

the mammillary body is a single rounded eminence; in humans it is a double structure.

4. Identify the cerebral peduncles on the ventral aspect of the midbrain just posterior to the mammillary body of the hypothalamus. The cerebral penducles are fiber tracts connecting the cerebrum and medulla. Identify the large oculomotor nerves (III), which arise from the ventral midbrain surface, and the tiny trochlear nerves (IV), which can be seen at the junction of the midbrain and pons. Both these cranial nerves provide motor fibers to extrinsic muscles of the eyeball.

5. Move posteriorly from the midbrain to identify first the pons and then the medulla oblongata, both hindbrain structures composed primarily of ascending and descending fiber tracts.

6. Return to the junction of the pons and midbrain and proceed posteriorly to identify the following cranial nerves, all arising from the pons: the trigeminal nerves (V), which are involved in chewing and sensations of the head and face; the abducens nerves (VI), which abduct the eye (and thus work in conjunction with cranial nerves III and IV); and the large facial nerves (VII), which are involved in taste sensation, gland function (salivary and lacrimal glands), and facial expression.

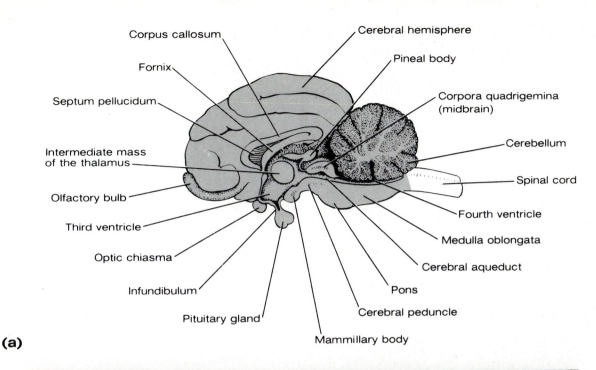

Corpus callosum

Fornix

Septum pellucidum

Intermediate mass of the thalamus

Olfactory bulb

Third ventricle

Optic chiasma

Infundibulum

Pituitary gland

Mammillary body

Cerebral hemisphere

Pineal body

Corpora quadrigemina (midbrain)

Cerebellum

Spinal cord

Fourth ventricle

Medulla oblongata

Cerebral aqueduct

Pons

Cerebral peduncle

(a)

Corpus callosum

Fornix

Optic chiasma

Cerebral hemisphere

Cerebellum

Cerebral peduncle

Pons   Medulla

(b)

Figure 17.11

Sagittal section of the sheep brain showing internal structures. (a) Diagrammatic view; (b) photograph, courtesy of Ann Allworth.

7. Continue posteriorly to identify the purely sensory vestibulocochlear nerves (VIII), which are involved in the sensations of hearing and equilibrium; the glossopharyngeal nerves (IX), which contain motor fibers innervating throat structures and sensory fibers transmitting taste stimuli (in conjunction with cranial nerve VII); the vagus nerves (X), often called "wanderers," which serve many organs of the head, thorax, and abdominal cavity; the accessory nerves (XI), which serve muscles of the neck, larynx, and shoulder; and the hypoglossal nerves (XII), which stimulate tongue and neck muscles. Note that the accessory nerves arise from both the medulla and the spinal cord.

Internal Structures

1. The internal structure of the brain can only be examined after further dissection. Place the brain ventral side down on the dissecting pan and make a cut completely through it in a superior to inferior direction. Cut through the longitudinal fissure, corpus callosum, and midline of the cerebellum. Refer to Figure 17.11 as you work.

2. The thin nervous tissue membrane immediately ventral to the corpus callosum that separates the lateral ventricles is the septum pellucidum. Pierce this

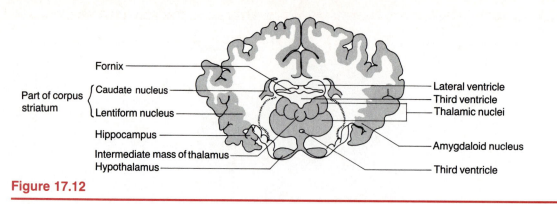

Figure 17.12

Frontal section of a sheep brain. Major structures revealed are the location of major basal nuclei deep in the interior, the thalamus, hypothalamus, and the lateral and third ventricles.

membrane and probe the lateral ventricle cavity. The fiber tract ventral to the septum pellucidum and anterior to the third ventricle is the fornix.

How does the size of the fornix in this brain compare

with the human fornix? _____

_____

Why do you suppose this is so? (Hint: What is the

function of this band of fibers?) _____

_____

_____

_____

3.   Identify the thalamus, which forms the walls of the third ventricle and is located posterior and ventral to the fornix. The intermediate mass spanning the ventricular cavity appears as an oval protrusion of the thalamic wall. Anterior to the intermediate mass, locate the interventricular foramen, a canal connecting the lateral ventricle on the same side with the third ventricle.

4.   The hypothalamus forms the floor of the third ventricle. Identify the optic chiasma, infundibulum, and mammillary body on its exterior surface. You can see the pineal body at the superior–posterior end of the third ventricle just beneath the junction of the corpus callosum and fornix.

5.   The midbrain can be located by identifying the corpora quadrigemina that form its dorsal roof. Follow the cerebral aqueduct (the narrow canal connecting the third and fourth ventricles) through the midbrain tissue to the fourth ventricle. Identify the cerebral peduncles, which form its anterior walls.

6.   The pons and medulla can be found anterior to the fourth ventricle. The medulla continues into the spinal cord without any obvious anatomic change, but the point at which the fourth ventricle narrows to a small canal is generally accepted as the beginning of the spinal cord.

7.   The cerebellum can be seen posterior to the fourth ventricle. Note its internal treelike arrangement of white matter, the arbor vitae.

8.   If time allows, obtain another sheep brain and section it along the frontal plane so that the cut passes through the infundibulum. Compare your specimen to the diagrammatic view in Figure 17.12, and attempt to identify all the structures shown in the figure.

9.   Check with your instructor to determine if cow spinal cord sections (preserved) are available for the spinal cord studies in Exercise 18. If not, save the small portion of the spinal cord from your brain specimen. Otherwise, dispose of all the organic debris in the appropriate laboratory containers and clean the dissecting instruments and tray before leaving the laboratory.

Spinal Cord, Spinal Nerves, and the Autonomic Nervous System

OBJECTIVES

1. To identify on a spinal cord model or appropriate diagram gray and white matter, anterior median fissure, posterior median sulcus, central canal, dorsal, ventral, and lateral horns of the gray matter, ventral and dorsal roots, dorsal root ganglia, and posterior, lateral, and anterior funiculi, and to cite the neuron type found in these areas (where applicable).

2. To name two major areas where the spinal cord is enlarged, and to explain the reasons for this anatomic characteristic.

3. To define *conus medullaris, cauda equina,* and *filum terminale*.

4. To locate on a diagram the fiber tracts in the spinal cord white matter and to state their functional importance.

5. To list two major functions of the spinal cord.

6. To name the meningeal coverings of the spinal cord and state their function.

7. To describe the origin, fiber composition, and distribution of the spinal nerves, differentiating between ventral and dorsal roots, the spinal nerve proper, and ventral and dorsal rami, and to discuss the result of transecting these structures.

8. To discuss the distribution of the dorsal rami and ventral rami of the spinal nerves.

9. To identify the four major nerve plexuses, the major nerves of each, and their distribution.

10. To identify on a dissected animal the musculocutaneous, radial, median, and ulnar nerves of the upper limb and the femoral, saphenous, sciatic, common peroneal, and tibial nerves of the lower limb.

11. To identify the site of origin and the function of the sympathetic and parasympathetic divisions of the autonomic nervous system, and to state how the autonomic nervous system differs from the somatic nervous system.

MATERIALS

Spinal cord model (cross section)
Laboratory charts of the spinal cord and spinal nerves and sympathetic chain
Red and blue pencils
Preserved cow spinal cords with meninges and nerve roots intact (or spinal cord segment saved from the brain dissection in Exercise 18)
Dissecting tray and instruments
Dissecting microscope
Histologic slide of spinal cord (cross section)
Compound microscope
Animal specimen from previous dissections

ANATOMY OF THE SPINAL CORD

The cylindrical spinal cord, a continuation of the brain stem, is an association and communication center. It plays a major role in spinal reflex activity and provides neural pathways to and from higher nervous centers. Enclosed within the vertebral canal of the spinal column, the spinal cord extends from the foramen magnum of the skull to the first or second lumbar vertebra, where it terminates in the cone-shaped **conus medullaris** (Figure 18.1). Like the brain,

it is cushioned and protected by the meninges. The dural and arachnoid meningeal coverings extend beyond the conus medullaris, approximately to the level of S_2. A fibrous extension of the pia mater extends even farther, i.e., into the coccygeal canal as the **filum terminale.** Because of this anatomic characteristic, cerebrospinal fluid can easily be obtained by a *lumbar* puncture (or lumbar tap) from the subarachnoid space below L_3 without endangering the delicate spinal cord.

In humans, 31 pairs of spinal nerves arise from the spinal cord and pass through intervertebral foramina to serve the body area at their approximate

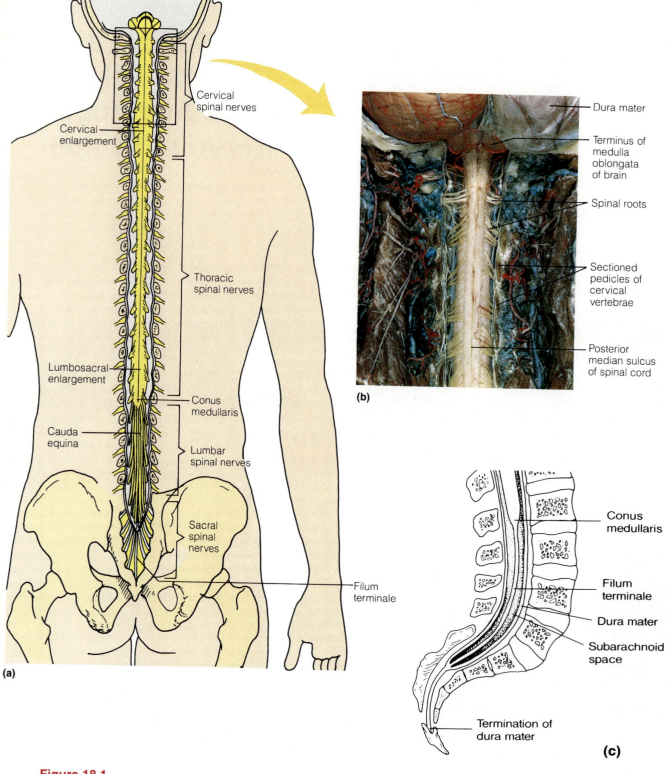

Cervical spinal nerves

Cervical enlargement

Thoracic spinal nerves

Lumbosacral enlargement

Conus medullaris

Cauda equina

Lumbar spinal nerves

Sacral spinal nerves

Filum terminale

(a)

Dura mater

Terminus of medulla oblongata of brain

Spinal roots

Sectioned pedicles of cervical vertebrae

Posterior median sulcus of spinal cord

(b)

Conus medullaris

Filum terminale

Dura mater

Subarachnoid space

Termination of dura mater

(c)

Figure 18.1

Structure of the spinal cord. (a) The vertebral arches have been removed to show the dorsal aspect of the spinal cord (and its nerve roots). The dura mater is cut and reflected laterally. The various regions of the spinal cord are indicated in relation to the vertebral column as cervical, thoracic, lumbar, and sacral. (b) Photograph of the cervical region of the spinal cord; the meningeal coverings have been removed to show the spinal roots that give rise to the spinal nerves. (c) Lateral view, showing the extent of the filum terminale.

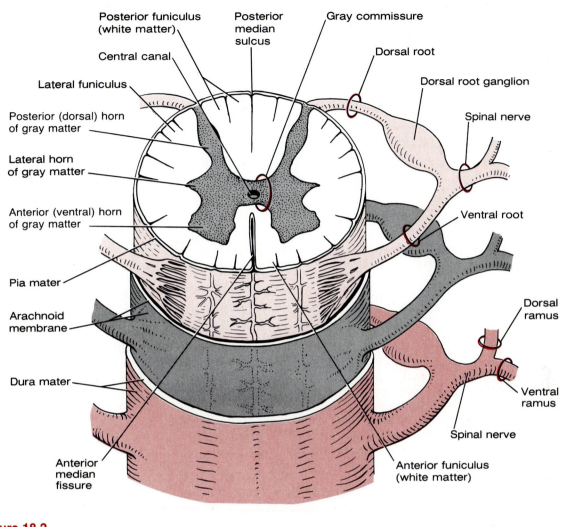

Figure 18.2

Anatomy of the human spinal cord (cross section).

level of emergence. The cord is about the size of a thumb in circumference for most of its length, but there are obvious enlargements in the cervical and lumbar areas where the nerves serving the upper and lower limbs issue from the cord.

Because the cord does not extend to the end of the spinal column, the spinal nerves emerging from the inferior end of the cord must travel through the vertebral canal for some distance before exiting at the appropriate intervertebral foramina. This collection of spinal nerves traversing the inferior end of the vertebral canal is called the **cauda equina** because of its similarity to a horse's tail (the literal translation of *cauda equina*).

 Obtain a model of a cross section of a spinal cord and identify its structures as they are described next.

Gray Matter

In cross section, the gray matter of the spinal cord looks like a butterfly or the letter H (Figure 18.2). The two posterior projections are called the **posterior**, or **dorsal**, **horns;** the two anterior projections are the **anterior**, or **ventral**, **horns.** The tips of the anterior horns are broader and less tapered than those of the posterior horns. In the thoracic and lumbar regions of the cord, there is also a lateral outpocketing of gray matter on each side referred to as the **lateral horn.** The central area of gray matter connecting the two vertical regions is the **gray commissure.** The gray commissure surrounds the **central canal** of the cord, which contains cerebrospinal fluid.

Neurons with specific functions can be localized in the gray matter. The posterior horns, for instance, contain association neurons and sensory fibers that enter the cord from the body periphery via the **dorsal root.** The cell bodies of these sensory neurons are found in an enlarged area of the dorsal root called the **dorsal root ganglion.** The anterior horns contain

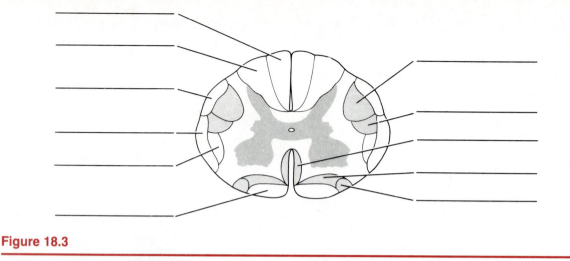

Figure 18.3

Cross section of spinal cord showing the relative positioning of its major tracts.

nerve cell bodies of motor neurons of the somatic nervous system (voluntary system), which send their axons out via the **ventral root** of the cord to enter the adjacent spinal nerve. The **spinal nerves** are formed from the fusion of the dorsal and ventral roots. The lateral horns, where present, contain cell bodies of sympathetic motor neurons of the autonomic nervous system (sympathetic division). Their axons also leave the cord via the ventral roots, with those of the motor neurons of the anterior horns.

White Matter

The white matter of the spinal cord is nearly bisected by fissures (see Figure 18.2). The more open anterior fissure is the **anterior median fissure,** and the posterior one is the **posterior median sulcus.** The white matter is composed of myelinated fibers—some running to higher centers, some traveling from the brain to the cord, and some conducting impulses from one side of the cord to the other.

Because of the irregular shape of the gray matter, the white matter on each side of the cord can be divided into three primary regions or white columns: the **posterior, lateral,** and **anterior funiculi.** Each of the funiculi contains a number of fiber **tracts** composed of axons with the same origin, terminus, and function. Tracts conducting sensory impulses to the brain are called ascending, or sensory, tracts; those carrying impulses from the brain to the skeletal muscles are descending, or motor, tracts.

Because it serves as the transmission pathway between the brain and the body periphery, the spinal cord is an extremely important functional area. Even though it is protected by meninges and cerebrospinal fluid in the vertebral canal, it is highly vulnerable to traumatic injuries, such as might occur in an automobile accident.

When the cord is transected (or severely traumatized), both motor and sensory functions are lost in body areas normally served by that (and lower) regions of the spinal cord. Injury to certain cord areas may even result in a permanent paralysis of both legs (paraplegia) or of all four limbs (quadraplegia). ■

With the help of your textbook or a laboratory chart showing the tracts of the spinal cord, label Figure 18.3 with the tract names that follow. Since each tract is represented on both sides of the cord, for clarity you can label the motor tracts on the right side of the diagram and the sensory tracts on the left side of the diagram. *Color ascending tracts red and descending tracts blue.* Then fill in the functional importance of each tract beside its name below. As you work, try to be aware of how the naming of the tracts is related to their anatomic distribution.

Fasciculus gracilis _____

Fasciculus cuneatus _____

Dorsal spinocerebellar _____

Ventral spinocerebellar _____

Lateral spinothalamic _____

Ventral spinothalamic _____

Lateral corticospinal _____

Ventral corticospinal _____

Rubrospinal _____

Tectospinal _____

Olivospinal _____

Vestibulospinal _____

Spinal Cord Dissection

1.   Obtain a dissecting tray and instruments and a segment of preserved spinal cord (from a cow or saved from the brain specimen used in Exercise 17). Identify the tough outer meninx (dura mater) and the weblike arachnoid membrane.

What name is given to the third meninx, and where

is it found? _____

_____

Peel back the dura mater and observe the fibers making up the dorsal and ventral roots. If possible, identify a dorsal root ganglion.

2.   Cut a thin cross section of the cord and identify the anterior and posterior horns of the gray matter with the naked eye or with the aid of a dissecting microscope.

How can you be certain that you are correctly iden-

tifying the anterior and posterior horns? _____

_____

_____

Also identify the central canal, white matter, anterior median fissure, posterior median sulcus, and posterior, anterior, and lateral funiculi.

3.   Obtain a prepared slide of the spinal cord (cross section) and a compound microscope. Examine the slide carefully under low power. Observe the shape of the central canal.

Is it basically circular or oval? _____

What would you expect to find in this canal in the

living animal? _____

Can any neuron cell bodies be seen? _____

Where? _____

_____

What type of neurons would these most likely be—

motor, association, or sensory? _____

_____

SPINAL NERVES AND NERVE PLEXUSES

The 31 pairs of human spinal nerves arise from the fusions of the ventral and dorsal roots of the spinal cord. (Figure 18.4 shows how the nerves are named according to their point of issue.) Since the ventral roots contain the myelinated axons of motor neurons located in the cord and the dorsal roots carry sensory fibers entering the cord, all spinal nerves are **mixed nerves.** The first pair of spinal nerves leaves the vertebral canal between the base of the occiput and the atlas, but all the rest exit via the intervertebral foramina. The second through seventh pairs of cervical nerves emerge *above* the vertebra for which they are named. C_8 emerges between C_7 and T_1. The remaining spinal nerves emerge from the spinal cord below the same-numbered vertebra.

Almost immediately after emerging, each nerve divides into **dorsal** and **ventral rami.** (Thus each spinal nerve is only about 1 or 2 cm long.) The rami, like the spinal nerves, contain both motor and sensory fibers. The smaller dorsal rami serve the skin and musculature of the posterior body trunk at their approximate level of emergence. The ventral rami of spinal nerves, T_2-T_{12}, pass anteriorly as the **intercostal nerves** to supply the muscles of intercostal spaces and the skin and muscles of the anterior and lateral trunk. The ventral rami of all other spinal nerves form complex networks of nerves called **plexuses.** These plexuses serve the motor and sensory needs of the muscles and skin of the limbs. The fibers of the ventral rami unite in the plexuses and then diverge again to form peripheral nerves, which contain fibers from more than one spinal nerve. The four major nerve plexuses and their chief peripheral nerves are illustrated in Figures 18.4 and 18.5 and are described below. Their names and site of origin should be committed to memory.

The **cervical plexus** arises from the ventral rami of C_1 through C_5 to supply muscles of the shoulder and neck. The major motor branch of this plexus is the **phrenic nerve,** which passes into the thoracic cavity in front of the first rib to innervate the diaphragm. The primary danger of a broken neck is that the phrenic nerve may be severed, leading to paralysis of the diaphragm and cessation of breathing.

The **brachial plexus** is complex, arising from the ventral rami of C_5 through C_8 and T_1. The plexus becomes subdivided into five major peripheral nerves. The **axillary nerve** which serves the muscles and skin of the shoulder, has the most limited distribution. The large **radial nerve** passes down the posterolateral surface of the arm and forearm, supplying all the

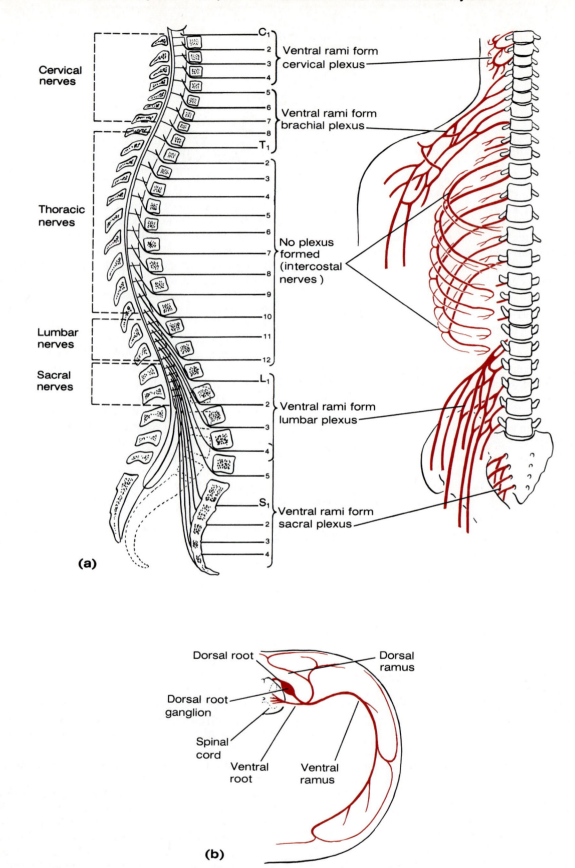

Cervical nerves

Ventral rami form cervical plexus

Ventral rami form brachial plexus

Thoracic nerves

No plexus formed (intercostal nerves)

Lumbar nerves

Sacral nerves

Ventral rami form lumbar plexus

Ventral rami form sacral plexus

(a)

Dorsal root

Dorsal ramus

Dorsal root ganglion

Spinal cord

Ventral root

Ventral ramus

(b)

Figure 18.4

Human spinal nerves: (a) relationship of spinal nerves to vertebrae (areas of plexuses formed by the ventral rami are indicated; (b) relative distribution of the ventral and dorsal rami of a spinal nerve (cross section of left trunk).

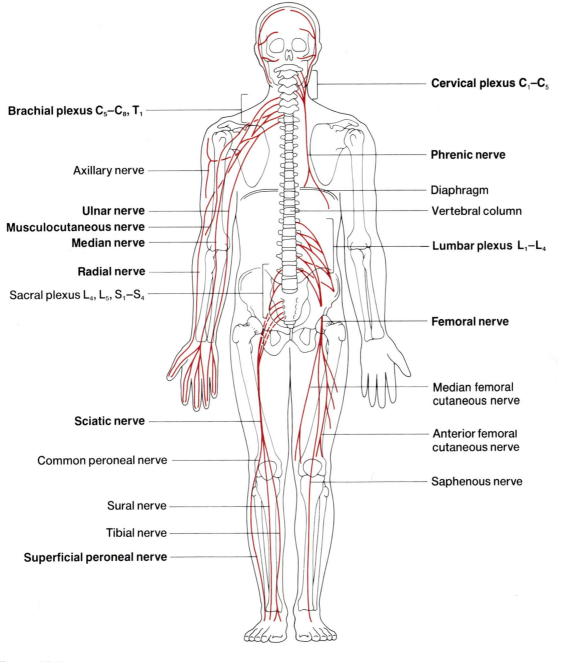

Cervical plexus C_1–C_5

Brachial plexus C_5–C_8, T_1

Axillary nerve

Phrenic nerve

Diaphragm

Ulnar nerve

Vertebral column

Musculocutaneous nerve

Median nerve

Lumbar plexus L_1–L_4

Radial nerve

Sacral plexus L_4, L_5, S_1–S_4

Femoral nerve

Median femoral cutaneous nerve

Sciatic nerve

Anterior femoral cutaneous nerve

Common peroneal nerve

Saphenous nerve

Sural nerve

Tibial nerve

Superficial peroneal nerve

Figure 18.5

Nerve plexuses and the major nerves arising from each. For clarity, each plexus is illustrated only on one side of the body.

extensor muscles of the arm, forearm, and hand and the skin along its course. The radial nerve is often injured in the axillary region by the pressure of a crutch or by hanging one's arm over the back of a chair. The **median nerve** passes down the antero-medial surface of the arm to supply most of the flexor muscles in the forearm and several muscles in the hand (plus the skin of the lateral surface of the palm of the hand). The **musculocutaneous nerve** supplies the flexor muscles of the anterior arm and skin of the lateral surface of the forearm. The **ulnar nerve** travels down the posteromedial surface of the arm.

It courses around the medial epicondyle of the humerus to supply the flexor carpi ulnaris, the ulnar head of the flexor digitorum profundus of the forearm, and all intrinsic muscles of the hand not served by the median nerve. It supplies the skin of the medial third of the hand, both its anterior and posterior surfaces. Trauma to the ulnar nerve, which often occurs when the elbow is hit, produces a smarting sensation commonly referred to as "hitting the funny bone."

The **lumbosacral plexus,** which serves the pelvic region of the trunk and the lower limbs, is actually a complex of two plexuses, the lumbar plexus and

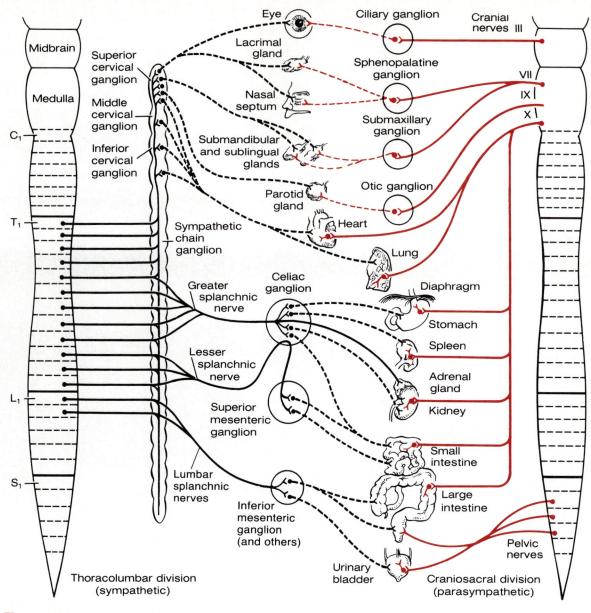

Figure 18.6

The autonomic nervous system. Solid lines indicate preganglionic nerve fibers; dotted lines indicate postganglionic nerve fibers.

the sacral plexus (see Figure 18.5). The **lumbar plexus** arises from ventral rami of L_1 through L_4 (and sometimes T_{12}). Its nerves serve the lower abdominopelvic region and the anterior thighs. The largest nerve of this plexus is the **femoral nerve,** which passes beneath the inguinal ligament to innervate the anterior thigh muscles. The cutaneous branches of the femoral nerve (median and anterior femoral cutaneous and the saphenous nerves) supply the skin of the anteromedial surface of the entire lower limb.

Arising from L_4 through L_5 and S_1 through S_4, the nerves of the **sacral plexus** supply the buttock, the posterior surface of the thigh, leg, and foot. The major peripheral nerve of this plexus is the **sciatic**

nerve, which is the largest nerve in the body. The sciatic nerve leaves the pelvis through the greater sciatic notch and travels down the posterior thigh, serving its flexor muscles and skin. In the popliteal region, the sciatic nerve divides into the **common peroneal nerve** and the **tibial nerve,** which together supply the balance of the leg muscles and skin, both directly and via several branches.

Identify each of the four major nerve plexuses (and its major nerves) shown in Figure 18.5 on a large laboratory chart. Trace the course of the nerves.

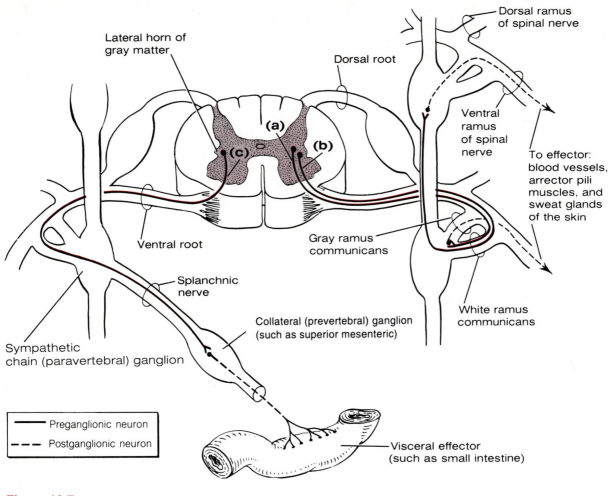

Figure 18.7

Sympathetic pathways: (a) synapse in a sympathetic chain ganglion at the same level; (b) synapse in a sympathetic chain ganglion at a different level; (c) synapse in a collateral ganglion.

THE AUTONOMIC NERVOUS SYSTEM

The **autonomic nervous system** is the subdivision of the PNS that regulates body activities that are generally not under conscious control. It is composed of a special group of motor neurons serving cardiac muscle (the heart), smooth muscle (found in the walls of the visceral organs and blood vessels), and internal glands. Because these structures function without conscious control, this system has often been referred to as the involuntary nervous system.

There is a basic anatomic difference between the motor pathways of the **somatic** (voluntary) **nervous system,** which innervates the skeletal muscles, and those of the autonomic nervous system. In the somatic division, the cell bodies of the motor neurons reside in the CNS (spinal cord or brain), and their axons, sheathed in spinal nerves, extend all the way to the skeletal muscles they serve. However, the autonomic nervous system consists of chains of two motor neu-rons. The first motor neuron of each pair, called the *preganglionic neuron,* resides in the brain or cord. Its axon leaves the CNS to synapse with the second motor neuron (*postganglionic neuron*), which is located in a ganglion outside the CNS. The axon of the postganglionic neuron then extends to the organ it serves.

The autonomic nervous system has two major functional subdivisions (Figure 18.6). These, the sympathetic and parasympathetic divisions, serve most of the same organs but generally cause opposing or antagonistic effects.

Sympathetic Division

The preganglionic neurons of the sympathetic (or thoracolumbar) division are in the lateral horns of the gray matter of the spinal cord from T_1 through L_2. The preganglionic axons leave the cord via the ventral root (in conjunction with the axons of the somatic motor neurons), enter the spinal nerve, and then travel briefly in the ventral ramus (Figure 18.7).

From the ventral ramus, they pass through a small branch called the **white ramus communicans** to enter a **paravertebral** ganglion in the **sympathetic chain,** or **trunk,** which lies alongside the vertebral column.

Having reached the ganglion, an axon may take one of three main courses (see Figure 18.7). First, it may synapse with a postganglionic neuron in *that* sympathetic chain ganglion. Second, the axon may travel upward or downward through the sympathetic chain to synapse with a postganglionic neuron at another level. In either of these two instances, the postganglionic axons then reenter the ventral or dorsal ramus of a spinal nerve via a **gray ramus communicans** and travel in the ramus to innervate skin structures (sweat glands, arrector pili muscles attached to hair follicles, and the smooth muscles of blood vessel walls). Third, the axon may pass through the ganglion without synapsing and form part of the **splanchnic nerves,** which travel to the viscera to synapse with a postganglionic neuron in a **collateral** or **prevertebral ganglion.** (The major collateral ganglia—the celiac, superior mesenteric, and inferior mesenteric ganglia—supply the abdominal and pelvic visceral organs.) The postganglionic axon then leaves the collateral ganglion and travels to a nearby visceral organ.

Parasympathetic Division

The preganglionic neurons of the parasympathetic (or craniosacral) division are located in brain nuclei of cranial nerves III, VII, IX, X and in the S_2 through the S_4 level of the spinal cord. The axons of the preganglionic neurons of the cranial region travel in their respective cranial nerves to the immediate area of the head and neck organs to be stimulated. There they synapse with the postganglionic neuron in a **ter-**

minal or **intramural ganglion.** The postganglionic neuron then sends out a very short axon to the organ it serves. In the sacral region, the preganglionic axons leave the ventral roots of the spinal cord and collectively form the **pelvic nerves,** which travel to the pelvic cavity. In the pelvic cavity, the preganglionic axons synapse with the postganglionic neurons in terminal ganglia located on or close to the organs served.

Locate the sympathetic chain on the spinal nerve chart.

Autonomic Functioning

As noted earlier, body organs served by the autonomic nervous system receive fibers from both the sympathetic and parasympathetic divisions. The only exceptions are the structures of the skin, some glands, most blood vessels, and the adrenal medulla, all of which receive sympathetic innervation only. When both divisions serve an organ, they have antagonistic effects. This is because their postganglionic axons release different neurotransmitters. The parasympathetic fibers, called **cholinergic fibers,** release acetylcholine; the sympathetic postganglionic fibers, called **adrenergic fibers,** release norepinephrine; and the preganglionic fibers of both divisions release acetylcholine.

The parasympathetic division is often referred to as the "resting–digesting" system because it maintains the visceral organs in a state most suitable for normal functions and internal homeostasis; that is,

Organ or function	Parasympathetic effect	Sympathetic effect
Heart		
Bronchioles of lungs		
Digestive tract activity		
Urinary bladder		
Iris of the eye		
Blood vessels of skeletal muscles		
Blood vessels of visceral organs		
Penis		
Sweat glands		
Adrenal medulla		
Pancreas		

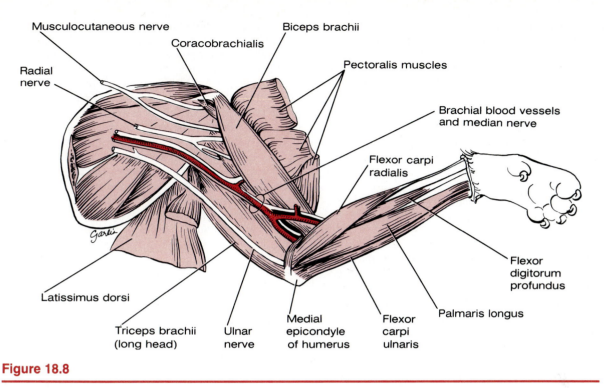

Figure 18.8

Brachial plexus and major blood vessels of the left forelimb of the cat (ventral aspect). See dissection photo, Color Plate C.

it promotes normal digestion and elimination. In contrast, activation of the sympathetic division is referred to as the "fight or flight" response because it readies the body to cope with situations that threaten homeostasis. Under such emergency conditions, the sympathetic nervous system induces an increase in heart rate and blood pressure, dilation of the bronchioles of the lungs, an increase in blood sugar levels, and many other effects that help the individual cope with the stressor.

As we grow older, our sympathetic nervous system gradually becomes less and less efficient, particularly in vasoconstriction of blood vessels. When elderly people stand up quickly after sitting or lying down, they often become light-headed or faint because the sympathetic nervous system is not able to react quickly enough to counteract the pull of gravity by activating the vasoconstrictor fibers; and so blood pools in the feet. This condition, *orthostatic hypotension,* is a type of low blood pressure resulting from changes in body position as described. ■

Several body organs are listed in the chart on page 174. *Using your textbook as a reference,* list the effect of the sympathetic and parasympathetic divisions on each.

Dissection of Cat Spinal Nerves

The cat has 38 or 39 pairs of spinal nerves (as compared to 31 in humans). Of these, 8 are cervical, 13 thoracic, 7 lumbar, 3 sacral, and 7 or 8 caudal.

A complete dissection of the cat's spinal nerves would be extraordinarily time-consuming and exacting and is not warranted in a basic anatomy and physiology course. However, it is desirable for you to have some dissection work to complement your study of the anatomic charts. Thus at this point you will perform a partial dissection of the brachial plexus and lumbosacral plexus and some of the major nerves.

NERVES OF THE BRACHIAL PLEXUS

 1. Place your cat specimen on the dissecting tray, dorsal side down. Reflect the cut ends of the left pectoralis muscles to expose the large brachial plexus in the axillary region (Figure 18.8 and Color Plate C). Carefully clean the exposed nerves as far back toward their points of origin as possible.

2. The musculocutaneous nerve is the most superior nerve of this group. It splits into two subdivisions that run under the margins of the coracobrachialis and biceps brachii muscles. Trace its fibers into the ventral muscles of the arm it serves.

3. Locate the large radial nerve inferior to the musculocutaneous nerve. The radial nerve serves the dorsal muscles of the arm and forearm. Follow it into the three heads of the triceps brachii muscle.

4. In the cat, the median nerve is closely associated with the brachial artery and vein. It courses through the upper arm to supply the ventral muscles of the

forearm (with the exception of the flexor carpi ulnaris and the ulnar head of the flexor digitorum profundus). It also innervates some of the intrinsic hand muscles, as in humans.

5.   The ulnar nerve is the most posterior of the large brachial plexus nerves. Follow it as it travels down the arm, passing over the medial epicondyle of the humerus, to supply the flexor carpi ulnaris and the ulnar head of the flexor digitorum profundus (and the hand muscles).

NERVES OF THE LUMBOSACRAL PLEXUS

1.   To locate the femoral nerve arising from the lumbar plexus, first identify the right femoral triangle, which is bordered by the sartorius and adductor muscles of the anterior thigh (Figure 18.9). The large femoral nerve travels through this region after emerging from the psoas major muscle in close association with the femoral artery and vein. Follow the nerve into the muscles and skin of the anterior thigh, which it supplies. Note also its cutaneous branch in the cat, the saphenous nerve, which continues down the anterior medial surface of the leg (with the greater saphenous artery and vein) to supply the skin of the anterior shank and foot.

2.   Turn the cat ventral side down so you can view the posterior aspect of the lower limb (Figure 18.10, p. 178). Reflect the ends of the transected biceps femoris muscle to view the large cordlike sciatic nerve. The sciatic nerve arises from the sacral plexus and serves the dorsal thigh muscles and all the muscles of the leg and foot. Follow the nerve as it travels down the posterior thigh lateral to the semimembranosus muscle. Note that it divides just superior to the gastrocnemius muscle of the calf into its two major branches, which serve the leg.

3.   Identify the tibial nerve medially and the common peroneal nerve, which curves over the lateral surface of the gastrocnemius.

4.   When you have finished making your observations, wrap the cat for storage and clean all dissecting tools and equipment before leaving the laboratory.

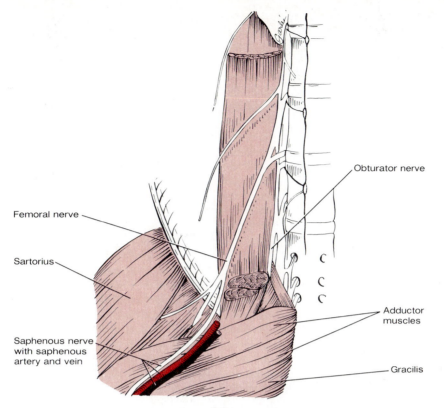

Femoral nerve

Sartorius

Saphenous nerve
with saphenous
artery and vein

Obturator nerve

Adductor
muscles

Gracilis

(a)

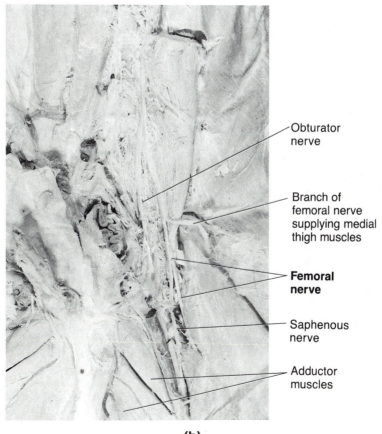

Obturator
nerve

Branch of
femoral nerve
supplying medial
thigh muscles

**Femoral
nerve**

Saphenous
nerve

Adductor
muscles

(b)

Figure 18.9

Lumbar plexus of the cat (ventral view): (a) art; (b) photo.

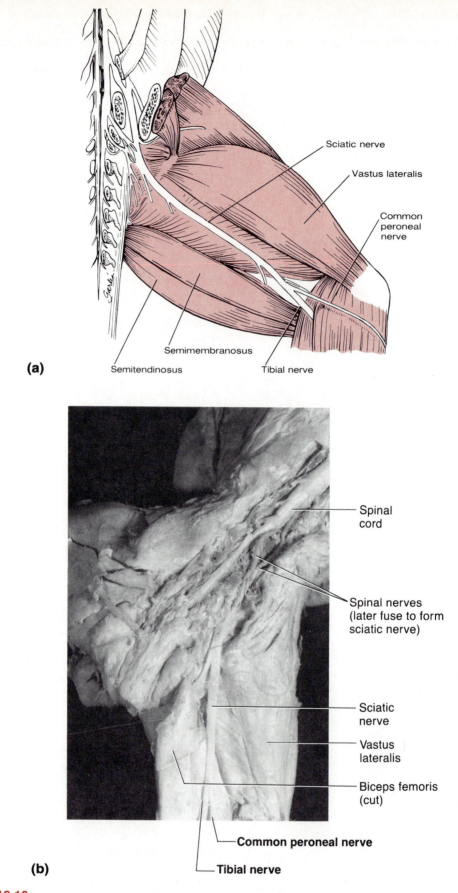

(a)

Sciatic nerve

Vastus lateralis

Common peroneal nerve

Semimembranosus

Semitendinosus

Tibial nerve

Spinal cord

Spinal nerves (later fuse to form sciatic nerve)

Sciatic nerve

Vastus lateralis

Biceps femoris (cut)

Common peroneal nerve

Tibial nerve

(b)

Figure 18.10

Sacral plexus of the cat (dorsal view): (a) art; (b) photo.

Special Senses: Vision

OBJECTIVES

1. To describe the structure and function of the accessory visual structures.
2. To identify the structural components of the eye when provided with a model, an appropriate diagram, or a preserved sheep or cow eye, and list the function(s) of each.
3. To describe the cellular makeup of the retina.
4. To discuss the mechanism of image formation on the retina.
5. To trace the visual pathway to the optic cortex and note the effects of damage to various parts of this pathway.
6. To define the following terms:

   refraction   myopia
   accommodation   hyperopia
   convergence   cataract
   astigmatism   glaucoma
   emmetropia   conjunctivitis

7. To discuss the importance of the pupillary and convergence reflexes.
8. To explain the difference between the rods and cones with respect to visual perception and retinal localization.
9. To state the importance of an ophthalmoscopic examination.

MATERIALS

Dissectible eye model
Chart of eye anatomy
Preserved cow or sheep eye
Dissecting pan and instruments
Laboratory lamp or penlight
Histologic section of an eye showing retinal layers
Compound microscope
Snellen eye chart (floor marked with chalk to indicate 20-ft distance from posted Snellen chart)
Ishihara's color-blindness plates
Ophthalmoscope
1-inch-diameter disks of colored paper (white, red, blue, green)
White, red, blue, and green chalk
Metric ruler; meter stick
Test tubes
Common straight pins

ANATOMY OF THE EYE

External Anatomy and Accessory Structures

The adult human eye is a sphere measuring about 1 inch (2.5 cm) in diameter. Only about one-sixth of the eye's anterior surface is observable, the remainder is enclosed and protected by a cushion of fat and the walls of the bony orbit.

Six **external,** or **extrinsic, eye muscles** attached to the exterior surface of each eyeball control eye movement and make it possible for the eye to follow a moving object. The names and positioning of these extrinsic muscles are noted in Figure 19.1. Their actions are given in the chart on page 181.

The anterior surface of each eye is protected by the **eyelids,** or **palpebrae.** (See Figure 19.2.) The medial and lateral junctions of the upper and lower eyelids are referred to as the **medial** and **lateral canthus** (respectively). A mucous membrane, the **conjunctiva,** lines the internal surface of the eyelids and

continues over the anterior surface of the eyeball to its junction with the cornea epithelium. The conjunctiva secretes mucus, which aids in lubricating the eyeball. Inflammation of the conjunctiva, often accompanied by redness of the eye, is called **conjunctivitis.**

Projecting from the border of each eyelid is a row of short hairs, the **eyelashes.** The **ciliary glands,** a type of sebaceous gland, are associated with the eyelash hair follicles and help lubricate the eyeball by producing an oily secretion. An inflammation of one of these glands is called a **sty.** The **meibomian glands,** located posterior to the eyelashes, also secrete an oily substance.

The **lacrimal apparatus** consists of the **lacrimal gland, lacrimal canals, lacrimal sac,** and the **nasolacrimal duct.** The lacrimal glands are situated superior to the lateral aspect of each eye. They continually liberate a dilute salt solution (tears) that flows onto the anterior surface of the eyeball through several small ducts. The tears flush across the eyeball into the lacrimal canals medially, then into the lacrimal sac, and finally into the nasolacrimal duct, which empties into the nasal cavity. The lacrimal secretion

Name	Innervation (cranial nerve)	Action
Lateral rectus	VI	Moves eye horizontally (laterally)
Medial rectus	III	Moves eye horizontally (medially)
Superior rectus	III	Elevates eye
Inferior oblique	III	Elevates eye and turns it laterally
Inferior rectus	III	Depresses eye
Superior oblique	IV	Depresses eye and turns it laterally

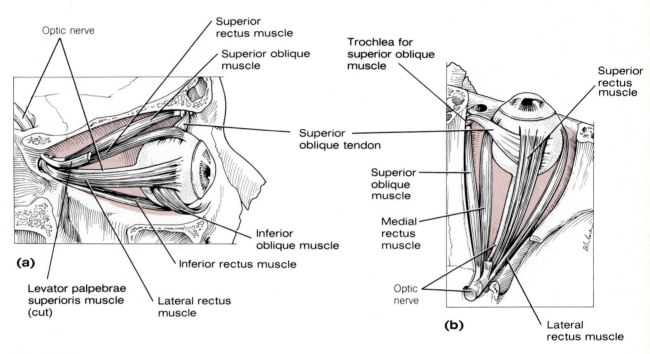

Figure 19.1

Extrinsic muscles of the eye: (a) lateral view of right eye; (b) superior view of right eye.

also contains **lysozyme,** an antibacterial enzyme. Because it constantly flushes the eyeball, lysozyme cleanses and protects the eye surface as it moistens and lubricates it. As we age, our eyes tend to become dry due to decreased lacrimation, and thus are more vulnerable to bacterial invasion and irritation.

Observe the eyes of another student and identify as many of the accessory structures as possible. Ask the student to look to the left. What extrinsic eye muscles are responsible for this action?

Right eye _____

Left eye _____

Internal Anatomy of the Eye

Obtain a dissectible eye model and identify its internal structures as they are described below. As you work, also refer to Figure 19.3.

Anatomically, the wall of the eye is constructed of three tunics, or coats. The outermost **fibrous tunic** is primarily composed of the **sclera,** thick white connective tissue observable anteriorly as the "white of the eye." Its anteriormost portion is modified to form the transparent **cornea,** through which light enters the eye.

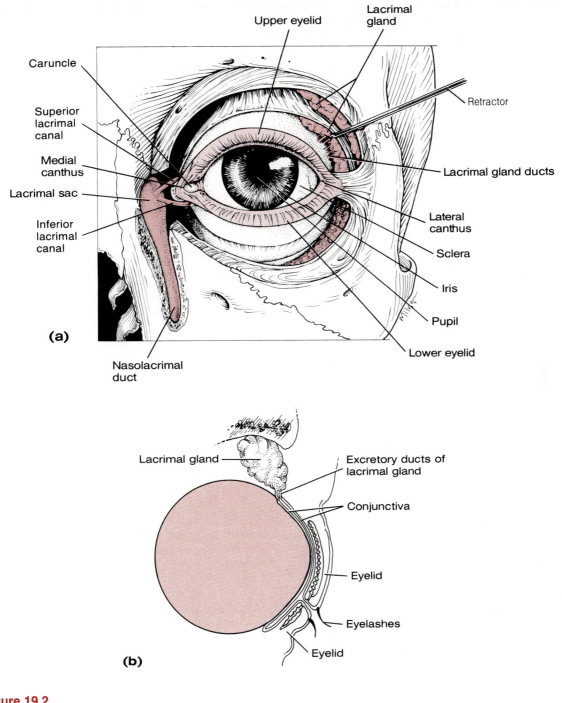

External anatomy of the eye and accessory structures: (a) anterior view; (b) sagittal section.

The middle **vascular tunic** consists primarily of the **choroid,** a richly vascular nutritive layer that contains a dark pigment. Anteriorly, the choroid is modified to form the **ciliary body,** to which the lens is attached, and then the pigmented **iris.** The iris is incomplete, resulting in a rounded opening, the **pupil,** through which light passes.

The iris is composed of circularly and a radially arranged smooth muscle fibers and acts as a reflexly activated diaphragm in regulating the amount of light entering the eye. In close vision and bright light, the circular muscles of the iris contract, and the pupil becomes more constricted. In distant vision and in dim light, the radial fibers contract to enlarge (dilate) the pupil which allows more light to enter the eye.

The innermost **sensory tunic** of the eye is the **retina,** which extends anteriorly only to the ciliary body. The retina contains the photoreceptor cells, the

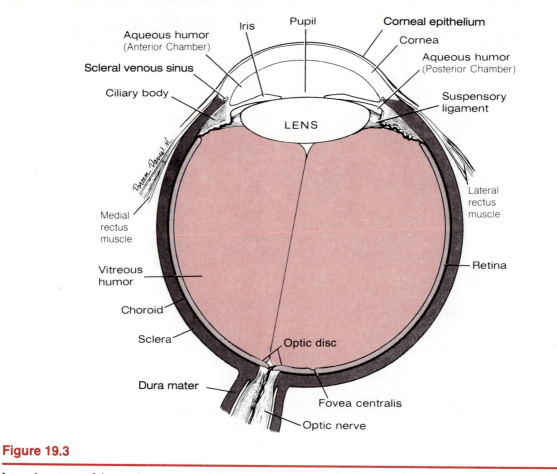

Iris
Pupil
Corneal epithelium
Aqueous humor (Anterior Chamber)
Cornea
Scleral venous sinus
Aqueous humor (Posterior Chamber)
Ciliary body
Suspensory ligament
LENS
Lateral rectus muscle
Medial rectus muscle
Vitreous humor
Retina
Choroid
Sclera
Optic disc
Dura mater
Fovea centralis
Optic nerve

Figure 19.3

Internal anatomy of the eye (transverse section).

rods and **cones,** which act as transducers to convert light energy into nerve impulses that are transmitted to the optic cortex of the brain. Vision is the result. The photoreceptor cells are distributed over the entire retina, except where the optic nerve leaves the eyeball. This site is called the **optic disc,** or blind spot. Lateral to each blind spot is an area called the **macula lutea** (yellow spot), an area of high cone density. In its center is the **fovea centralis,** a minute pit about ½ mm in diameter, which contains only cones and is the area of greatest visual acuity. Focusing for discriminative vision occurs in the fovea centralis.

Light entering the eye is focused on the retina by the **lens,** a flexible crystalline structure held vertically in the eye's interior by **suspensory ligaments** attached to the ciliary body. Activity of the ciliary muscle of the ciliary body changes lens thickness to allow light to be properly focused on the retina. In the elderly the lens becomes increasingly hard and opaque. **Cataracts,** which are the result of this process, cause vision to become hazy or entirely obstructed. ■

The lens divides the eye into two segments: the anterior segment anterior to the lens, which contains a clear watery fluid called the **aqueous humor,** and the posterior segment behind the lens, filled with a gellike substance, the **vitreous humor,** or **vitreous**

body. The anterior segment is further divided into **anterior** and **posterior chambers,** located before and after the iris, respectively. The aqueous humor is continually formed by the **ciliary processes** of the ciliary body. It helps to maintain the intraocular pressure of the eye and provides nutrients for the avascular lens and cornea. The aqueous humor is resorbed into the **scleral venous sinus.** The vitreous humor provides the major internal reinforcement of the eyeball.

Any interference with aqueous fluid drainage increases intraorbital pressure and compresses the retina and optic nerve, resulting in pain and possible blindness, a condition called **glaucoma.** ■

DISSECTION OF THE COW (SHEEP) EYE

1. Obtain a preserved cow or sheep eye, dissecting instruments, and a dissecting pan.

2. Examine the external surface of the eye, noting the thick cushion of adipose

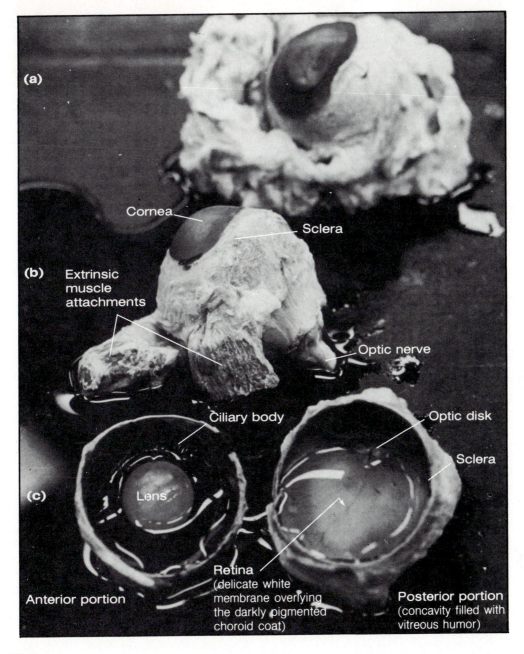

(a)

Cornea

Sclera

(b) Extrinsic
muscle
attachments

Optic nerve

Ciliary body

Optic disk

Sclera

(c)   Lens

Retina
(delicate white
membrane overlying
the darkly pigmented
choroid coat)

Anterior portion

Posterior portion
(concavity filled with
vitreous humor)

Figure 19.4

Anatomy of the cow eye: (a) cow eye (entire) removed from orbit (note the large amount of fat cushioning the eyeball); (b) cow eye (entire) with fat removed to show the extrinsic muscle attachments and optic nerve; (c) cow eye cut along the coronal plane to reveal internal structures. (Photograph courtesy of Jack Scanlon, Holyoke Community College Art Graphics Studio.)

tissue. Identify the optic nerve (cranial nerve II) as it leaves the eyeball, the remnants of the extrinsic eye muscles, the conjunctiva, the sclera, and the cornea. The normally transparent cornea is opalescent or opaque if the eye has been preserved. Refer to Figure 19.4 as you work.

3.   Trim away most of the fat and connective tissue, but leave the optic nerve intact. Holding the eye with the cornea facing downward, make an incision with a sharp scalpel into the sclera about ¼ inch above the cornea. Complete the incision around the circumference of the eyeball parallel to the corneal edge.

4.   Carefully lift the posterior part of the eyeball away from the lens. (Conditions being proper, the vitreous body should remain with the posterior part of the eyeball.)

5. Examine the anterior portion of the eye and identify the following structures:

Ciliary body: black pigmented body that appears to be a halo encircling the lens.
Lens: biconvex structure that appears opaque in preserved specimens.
Suspensory ligaments: delicate fibers attaching the lens to the ciliary body.

Carefully remove the lens and identify the adjacent structures:

Iris: anterior continuation of the ciliary body penetrated by the pupil.
Cornea: more convex anteriormost portion of the sclera; normally transparent but cloudy in preserved specimens.

6. Examine the posterior portion of the eyeball. Remove the vitreous humor, and identify the following structures:

Retina: appears as a delicate white, probably crumpled membrane that separates easily from the pigmented choroid.

Note its point of attachment. What is this point called?

_____

Pigmented choroid coat: appears iridescent in the cow or sheep eye owing to a special reflecting surface called the **tapetum lucidum.** This specialized surface reflects the light within the eye and is found in the eyes of animals that live under conditions of low-intensity light. It is not found in humans.

MICROSCOPIC ANATOMY OF THE RETINA

The retina consists of two types of cells: a pigmented epithelial layer, which abuts the choroid, and an inner cell layer composed of neurons, which is in contact with the vitreous humor (Figure 19.5). The inner nervous layer is composed of three major neuronal populations. These are, from outer to inner aspect, the **photoreceptor layer** (rods and cones), the **bipolar cells,** and the **ganglion cell neuron layer.**

The rods are the specialized receptors for dim light. Visual interpretation of their activity is in gray tones. The cones are color receptors that permit high levels of visual acuity, but they function only under conditions of high light intensity; thus, for example, no color vision is possible in moonlight. Only cones are found in the fovea centralis, and their number decreases as the retinal periphery is approached.

Conversely, the rods are most dense in the periphery, and their number decreases as the macula is approached.

Light must pass through the ganglion cell neuron layer and the bipolar neuron layer to reach and excite the rods and cones, which then initiate electrical signals that pass to the bipolar neurons. These in turn stimulate the ganglion cells, whose axons leave the retina in the tight bundle of fibers known as the optic nerve. The retinal layer is thickest (approximately 4 μm) where the optic nerve attaches to the eyeball because an increasing number of ganglion cell axons converge at this point. It thins as it approaches the ciliary body.

 Obtain a histologic slide of a longitudinal section of the eye. Identify the retinal layers by comparing it to Figure 19.5.

VISUAL PATHWAYS TO THE BRAIN

The axons of the ganglion cells of the retina converge at the posterior aspect of the eyeball and exit from the eye as the optic nerve. At the **optic chiasma,** the fibers from the medial side of each eye cross over to the opposite side. The fiber tracts thus formed are called the **optic tracts.** Each optic tract contains fibers from the lateral side of the eye on the same side and from the medial side of the opposite eye. The optic tract fibers synapse with neurons in the **lateral geniculate nucleus** of the thalamus, whose axons form the **optic radiation,** terminating in the **optic cortex** in the occipital lobe of the brain. Here they synapse with the cortical cells, and visual interpretation occurs.

After examining Figure 19.6, determine what the effects of lesions in the following areas would have on vision:

In the right optic nerve _____

_____

Through the optic chiasma _____

_____

In the left optic tract _____

_____

In the right cerebral cortex (visual area) _____

_____

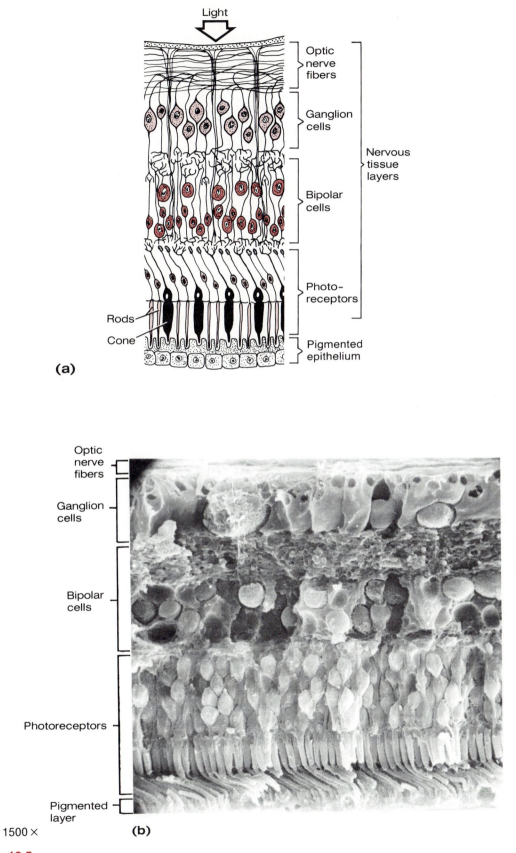

Light

Optic nerve fibers

Ganglion cells

Nervous tissue layers

Bipolar cells

Photo-receptors

Rods

Cone

Pigmented epithelium

(a)

Optic nerve fibers

Ganglion cells

Bipolar cells

Photoreceptors

Pigmented layer

(b)

Figure 19.5

Microscopic anatomy of the cellular layers of the retina: (a) diagrammatic view; (b) scanning electron micrograph. (From *Tissues and Organs: A Text-Atlas of Scanning Electron Microscopy* by Richard G. Kessel and Randy H. Kardon. W.H. Freeman and Company. Copyright © 1979.)

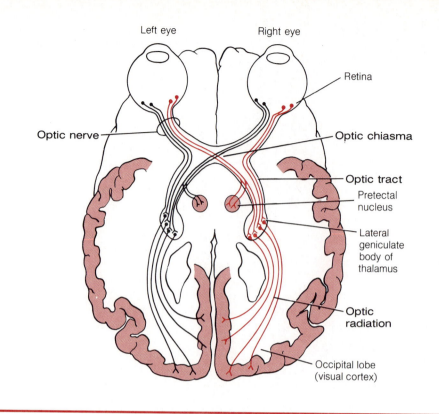

Figure 19.6

Visual pathway to the brain. (Note that fibers from the lateral portion of each retinal field do not cross at the optic chiasma.)

VISUAL TESTS
AND EXPERIMENTS

Demonstration of
the Blind Spot

1.   Hold Figure 19.7 about 18 inches from your eyes. Close your left eye, and focus your right eye on the X, which should be positioned so that it is directly in line with your right eye. Move the figure slowly toward your face, keeping your right eye focused on the X. When the dot focuses on the blind spot, which lacks photoreceptors, it will disappear.

X   ●

Figure 19.7

Blind spot test figure.

2.   Have your laboratory partner record in metric units the distance at which this occurs. The dot will reappear as the figure is moved closer. Distance at which the dot disappears:

Right eye _____

Repeat the test for the left eye, this time closing the right eye and focusing the left eye on the dot. Record the distance at which the X disappears:

Left eye _____

Afterimages

When light from an object strikes **rhodopsin,** the purple pigment contained in the rods of the retina, it triggers a photochemical reaction that causes the rhodopsin to be split into its colorless precursor molecules (a protein called opsin and vitamin A). This event, called *bleaching of the pigment,* initiates a chain of events leading to impulse transmission along fibers of the optic nerve. Once bleaching has occurred in a rod, the photoreceptor pigment must be resynthesized before the rod can be restimulated. This takes a certain period of time. Both phenomena—that is, the stimulation of the photoreceptor cells and their subsequent inactive period—can be demonstrated indirectly in terms of positive and negative afterimages.

1.   Stare at a bright light bulb for a few seconds, and then gently close your eyes for approximately one minute.

2.   Record, in the sequence of occurrence, what you "saw" after closing your

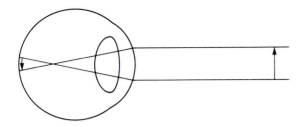

Figure 19.8

Refraction of light in the eye resulting in the production of a real image on the retina.

eyes: _____

_____

_____

The bright image of the light bulb initially seen was a **positive afterimage** caused by the continued firing of the rods. The dark image of the light bulb that subsequently appeared against a lighter background was the **negative afterimage,** an indication that the rhodopsin in the affected photoreceptor cells had been bleached.

Refraction, Tests for Visual Acuity, and Astigmatism

When light rays pass from one medium to another, their velocity, or speed of transmission, changes, and the rays are bent or refracted. Thus the light rays in the visual field are refracted as they encounter the cornea, lens, and vitreous body of the eye.

The refractive index (bending power) of the cornea and vitreous humor are constant; however, the refractive index of the lens can be varied by changing its shape, that is, by making it more or less convex, thus causing the light to be properly converged and focused on the retina. The greater the lens convexity, or bulge, the more the light will be bent. Conversely, the less the lens convexity (the flatter it is), the less it bends the light. In general, light from a distant source (over 20 feet) approaches the eye as parallel rays and no change in lens convexity is necessary for it to focus properly on the retina. However, light from a close source tends to diverge, and the convexity of the lens must increase to make close vision possible. To achieve this, the ciliary muscle contracts, decreasing the tension on the suspensory ligaments attached to the lens. Thus, a lens capable of bringing a *close* object into sharp focus is stronger (more convex) than a lens focusing on a more distant object. The ability of the eye to focus differentially for objects of near vision (less than 20 feet) is called **accommodation.** It should be noted that the image formed on the retina as a result of the refractory activity of the lens (see Figure 19.8) is a **real image** (reversed from left to right, inverted, and smaller than the object).

The normal or **emmetropic eye** is able to accommodate properly; however, visual problems may result from lenses that are too strong or too "lazy" (overconverging and underconverging, respectively) or from structural problems such as an eyeball that is too long or too short to provide for proper focusing by the lens, or a cornea or lens with improper curvatures.

Individuals in whom the image normally focuses in front of the retina are said to have **myopia,** or "nearsightedness"; they can see close objects without difficulty, but distant objects are blurred or seen indistinctly. Correction requires a concave lens, which causes the light reaching the eye to diverge.

If the image focuses behind the retina, the individual is said to have **hyperopia** or farsightedness. Such persons have no problems with distant vision but need glasses with convex lenses to augment the converging power of the lens for close vision.

Irregularities in the curvatures of the lens and/or the cornea lead to a blurred vision problem called **astigmatism.** Cylindrically ground lenses, which compensate for inequalities in the curvatures of the refracting surfaces, are prescribed to correct the condition. ■

TEST FOR NEAR-POINT ACCOMMODATION

The elasticity of the lens decreases dramatically with age, resulting in difficulty in focusing for near or close vision. This condition is called **presbyopia**—literally, old vision. Lens elasticity can be tested by measuring the **near point of accommodation.** The near point of vision is about 7.5 cm (or 3 inches) at age 10, 9 cm (or 3.5 inches) at age 20, 17 cm (or 6.75 inches) at age 40, and 83 cm (or 33 inches) at the age of 60.

To determine your near point of accommodation, hold a common straight pin at arm's length in front of one eye. Slowly move the pin toward that eye until the pin image becomes distorted. Have your lab partner measure the distance from your eye to the pin at this point, and record the distance below. Repeat the procedure for the other eye.

Near point for right eye _____

Near point for left eye _____

TEST FOR VISUAL ACUITY
Visual acuity, or sharpness of vision, is generally tested with a Snellen eye chart, which consists of letters of various sizes printed on a white card. This test is based on the fact that letters of a certain size can be seen clearly by eyes with normal vision at a specific distance. The distance at which the normal, or emmetropic, eye can read a line of letters is printed at the end of that line.

1.   Have your partner stand 20 feet from the posted Snellen eye chart and cover one eye with a card or hand. As your partner reads each consecutive line aloud, check for accuracy. (If this individual wears glasses, take the test twice—first with glasses off and then with glasses on.)

2.   Record the number of the line with the smallest-sized letters read. If it is 20/20, the person's vision for that eye is normal. If it is 20/40 (or any ratio with a value less than one), he or she has less than the normal visual acuity. (Such an individual is myopic.) If the visual acuity is 20/15, vision is better than normal, because this person can read letters at 20 feet from the chart that are only discernible by the normal eye at 15 feet.

3.   Have your partner test and record your visual acuity.

Visual acuity right eye _____

Visual acuity left eye _____

TEST FOR ASTIGMATISM   The astigmatism chart (Figure 19.9) is designed to test for defects in the refracting surface of the lens and/or cornea.

View the chart first with one eye and then with the other, focusing on the center of the chart. If all the radiating lines appear equally dark and distinct, there is no distortion of your refracting surfaces. If some of the lines are blurred or appear less dark than others, at least some degree of astigmatism is present.

Is astigmatism present in your left eye? _____

Right eye? _____

Test for Color Blindness

Ishihara's color plates are designed to test for deficiencies in the cones or color photoreceptor cells. Studies suggest that there are three cone types, each containing a different photoreceptor pigment. One type primarily absorbs the red wavelengths of the visible light spectrum, another the blue wavelengths, and a third the green wavelengths. Nerve impulses reaching the brain from these different photoreceptor types are then interpreted (seen) as red, blue, and green, respectively.

The interpretation of the intermediate colors of the visible light spectrum is a result of overlapping input from more than one cone type.

1.   View the various color plates in bright light or sunlight while holding them about 30 inches away and at right angles to your line of vision. Report to your laboratory partner what you see in each plate. (Take no more than 3 seconds for each decision.)

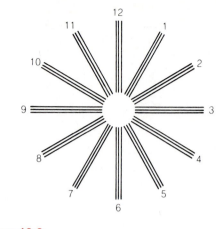

Figure 19.9

Astigmatism testing chart.

2.   Your partner is to write down your responses and then check their accuracy with the correct answers given at the front of the color plate book. Is there any indication that you have some degree of color

blindness? _____

If so, what type? _____

Repeat the procedure to test your partner's color vision.

Relative Positioning of Rods and Cones on the Retina

The test subject in a demonstration of the relative positioning of rods and cones should have shown no color vision problems during the test for color blindness. Students may work in pairs, or two students may perform the test for the entire class. White, red, blue and green paper disks and chalk will be needed for this demonstration.

1.   Position a test subject about 1 foot away from the blackboard.

2.   Make a white chalk circle on the board immediately in front of the subject's right eye. Have her close her left eye and stare fixedly at the circle with the right eye throughout the test.

3.   To map the extent of the rod field, begin to move a white paper disk into the field of vision from various sides of the visual field (beginning at least 2 feet away from the white chalk circle) and plot with white chalk dots the points at which the disk first becomes visible to the test subject.

4.   Repeat the procedure, using red, green, and blue paper disks and like-colored chalk to map the cone fields. Color (not object) identification is required.

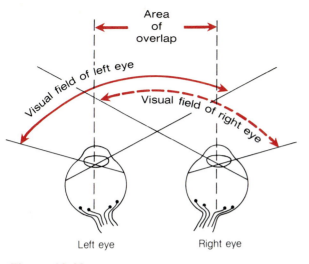

Figure 19.10

Overlapping of the visual fields.

5. After the test has been completed for all four disks, connect all dots of the same color. It should become apparent that each of the color fields has a different radius and that the rod and cone distribution on the retina is not uniform. (Normal fields have white outermost, followed by blue, red, and green as the fovea is approached.)

6. Record the rod and cone color field distribution observed in the laboratory review section using appropriately colored pencils.

Tests for Binocular Vision

Humans, cats, predatory birds, and most primates are endowed with **binocular,** or two-eyed, **vision.** Although both eyes look in approximately the same direction, they see slightly different views. Their visual fields, each about 170 degrees, overlap to a considerable extent; thus there is two-eyed vision at the overlap area (Figure 19.10).

In contrast, the eyes of many animals (rabbits, pigeons, and others) are more on the sides of their head. Such animals see in two different directions and thus have a panoramic field of view (**panoramic vision**).

Although both types of vision have their good points, we now know that binocular vision provides an accurate means of locating objects in space. This three-dimensional vision is called **stereopsis.** The slight differences between the views seen by the two eyes are fused by the higher centers of the visual cortex to give us depth perception. Because of the manner in which the visual cortex resolves these two different views into a single image, it is often referred to as the "cyclopean eye of the binocular animal."

1. To demonstrate that a slightly different view is seen by each eye, perform the following simple experiment: hold a pencil at arm's length directly in front of the left eye. Hold another pencil directly beneath it and move it about half the distance toward you after closing the right eye. As you move the lower pencil, make sure it remains in the *same plane* as the stationary pencil (so that the two pencils continually form a straight line). Then, without moving the pencils, close your left eye and open your right eye. Note that with only the left eye open the pencil stays in the same plane as the fixed pencil, but when viewed with the right eye, the moving pencil is displaced laterally away from the plane of the fixed pencil.

To demonstrate the importance of two-eyed binocular vision for depth perception, perform this second simple experiment.

2. Have your laboratory partner hold a test tube erect about arm's length in front of you. With both eyes open, quickly insert a pencil into the test tube. Remove the pencil, bring it back close to your body, close one eye, and quickly and without hesitation insert the pencil into the test tube. Repeat with the other eye closed.

Was it as easy to dunk the pencil with one eye closed

as with both eyes open? _____

Tests of Eye Reflexes

Both intrinsic (internal) and extrinsic (external) muscles are necessary for proper eye functioning. The intrinsic muscles, controlled by the autonomic nervous system, are those of the ciliary body (which alters the lens curvature in focusing) and the radial and circular muscles of the iris (which control pupillary size and thus regulate the amount of light entering the eye). The extrinsic muscles are the rectus and oblique muscles, which are attached to the eyeball exterior. These muscles control eye movement and make it possible to keep moving objects focused on the fovea centralis. They are also responsible for **convergence,** or medial eye movements, which is essential for near vision. When convergence occurs, both eyes are directed toward the near object viewed. The extrinsic eye muscles are controlled by the somatic nervous system.

Involuntary activity of both of these muscle types is brought about by reflex actions that can be observed in the following experiments.

PHOTOPUPILLARY REFLEX A sudden illumination of the retina by a bright light causes the pupil to contract reflexly in direct proportion to the light intensity. This protective response prevents damage to the delicate photoreceptor cells.

Obtain a laboratory lamp or penlight. Have your laboratory partner sit with eyes closed and hands over eyes. Turn on the light and position it so that it shines on the subject's right hand. After 1 minute, ask your partner to expose and open his right eye. Quickly observe the pupil of that eye. What happens to the

pupil? _____

_____

Shut off the light and ask your partner to expose and open the opposite eye. What are your observations?

_____

_____

ACCOMMODATION PUPILLARY REFLEX
Have your partner gaze for approximately 1 minute at a distant object in the lab—*not* toward the windows or another light source. Observe your partner's pupils, then hold some printed material 6 to 10 inches from his or her face.

How does pupil size change as your partner focuses

on the printed material? _____

_____

Explain the value of this reflex. _____

_____

_____

_____

CONVERGENCE REFLEX
Repeat the previous experiment, this time using a pen or pencil as the close object to be focused on. Note the position of your partner's eyeballs both while he or she is gazing at the distant and at the close object. Do they change position as the object of focus is changed?

_____

In what way? _____

_____

_____

_____

Explain the importance of the convergence reflex.

_____

_____

_____

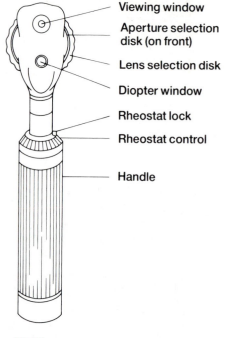

— Viewing window
— Aperture selection disk (on front)
— Lens selection disk
— Diopter window
— Rheostat lock
— Rheostat control
— Handle

Figure 19.11

Structure of an ophthalmoscope.

Ophthalmoscopic Examination of the Eye (Optional)

The opthalmoscope is an instrument used to examine the fundus, or eyeball interior, to determine visually the condition of the retina, optic disk, and internal blood vessels. Certain pathologic conditions such as diabetes, arteriosclerosis, and degenerative changes of the optic nerve and retina can be detected by such an examination. The ophthalmoscope consists of a set of lenses mounted on a rotating disk (the **lens selection disk**), a light source regulated by a **rheostat control,** and a mirror that reflects the light so that the eye interior can be illuminated (Figure 19.11).

The lens selection disk is positioned in a small slit in the mirror, and the examiner views the eye interior through this slit, appropriately called the *viewing window.* The focal length of each lens is indicated in diopters preceded by a + sign if the lens is convex and by a − sign if the lens is concave. When the zero (0) is seen in the *diopter window,* there is no lens in position in the slit. The depth of focus for viewing the eye interior is changed by changing the lens.

The light is turned on by depressing the red *rheostat lock* button and then rotating the rheostat control in the clockwise direction. The aperture selection disk on the front of the instrument allows the nature of the light beam to be altered. (Generally, a green light beam allows for clearest viewing of the blood vessels in the eye interior and is most comfortable for the subject.) Now that you are familiar with the ophthalmoscope, you are ready to conduct an eye examination.

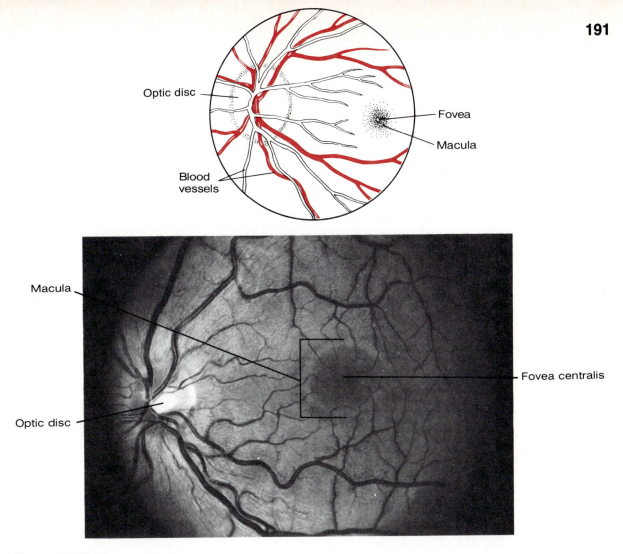

Figure 19.12

Posterior portion of the left retina. (Photograph taken with slit-lamp camera.)

1. Conduct the examination in a dimly lit or darkened room with the test subject comfortably seated and gazing straight ahead. To examine the right eye, sit face-to-face with the subject, and hold the instrument in your right hand. Use your right eye to view the eye interior. To view the left eye, use your left eye, and hold the instrument in your left hand. When the opthalmoscope is correctly set, the fundus should appear as shown in Figure 19.12.

2. Begin the examination with the 0 (no lens) in position. Hold the instrument so that the lens disk may be rotated with the index finger. Hold the ophthalmoscope about 6 inches from the subject's eye, and direct the light into the pupil at a slight angle—through the pupil edge rather than directly through its center. You will see a red circular area that is the illuminated eye interior.

3. Move in as close as possible to the subject's cornea as you continue to observe the area. Steady your instrument-holding hand on the subject's cheek if necessary. If both your eye and that of the subject are normal, the fundus can be viewed clearly without further adjustment of the ophthalmoscope. If the fundus cannot be focused, slowly rotate the lens disk counterclockwise until the fundus can be clearly seen. (Note: If a positive (convex) lens is required and your eyes are normal, the subject has hyperopia. If a negative (concave) lens is necessary to view the fundus and your eyes are normal, the subject is myopic.)

When the examination is proceeding correctly, the subject can often see images of retinal vessels in his own eye that appear rather like cracked glass. If you are unable to achieve a sharp focus or to see the optic disc, move medially or laterally and begin again.

4. Examine the optic disc for color, elevation, and sharpness of outline. Observe the blood vessels radiating from near its center. Locate the macula, which is lateral to the optic disc. It is a darker area in which blood vessels are absent, and the fovea appears to be a slightly lighter area in its center. The macula is most easily seen when the subject looks directly into the light of the ophthalmoscope.

Special Senses: Hearing and Equilibrium

OBJECTIVES

1. To identify the anatomic structures of the external, middle, and internal ear by appropriately labeling a diagram.

2. To describe the anatomy of the organ of hearing (organ of Corti in the cochlea) and explain its function in sound reception.

3. To describe the anatomy of the equilibrium organs of the inner ear (ampullae crystallaris and maculae), and to explain their relative function in the maintenance of equilibrium.

4. To define or explain *central deafness, conduction deafness,* and *nystagmus.*

5. To state the purpose of the Rinne and Romberg tests.

6. To explain how one is able to localize the source of sounds.

7. To describe the effects of acceleration on the semicircular canals.

8. To explain the role of vision in the maintenance of equilibrium.

MATERIALS

Three-dimensional dissectible ear model and/or chart of ear anatomy
Histologic slides of the cochlea of the ear
Compound microscope
Pocket watch that ticks
Tuning forks (range of frequencies)
Rubber mallet
Absorbent cotton
Otoscope (if available)
Alcohol swabs
12-inch ruler

ANATOMY OF THE EAR

Gross Anatomy

The ear is a complex structure containing sensory receptors for hearing and equilibrium. The ear is divided into three major areas: the **outer,** or **external ear,** the **middle ear,** and the **inner ear** (Figure 20.1). The outer and middle ear structures serve the needs of the sense of hearing *only,* while the inner ear structures function both in equilibrium and hearing reception.

Obtain a dissectible ear model and identify the structures described below. Refer to Figure 20.1 as you work.

The outer ear is composed primarily of the **pinna,** or **auricle,** and the **external auditory canal.** The pinna is the skin-covered cartilaginous structure encircling the auditory canal opening. In many animals, it collects and directs sound waves into the auditory canal. In humans this function of the pinna is largely lost.

The external auditory canal is a short, narrow (about 1 inch long by ¼ inch wide) chamber carved into the temporal bone. In its skin-lined walls are wax-secreting glands called **ceruminous glands.** The sound waves that enter the external auditory canal eventually encounter the **tympanic membrane,** or **eardrum,** which vibrates at exactly the same frequency as the sound wave(s) hitting it. The membranous eardrum separates the outer from the middle ear.

The middle ear is essentially a small chamber—the **tympanic cavity**—found within the temporal bone. The cavity is spanned by three small bones, the **ossicles*** (hammer, anvil, and stirrup), which articulate to form a lever system that transmits the vibratory motion of the eardrum to the fluids of the inner ear via the **oval window.**

Connecting the middle ear chamber with the nasopharynx is the **auditory,** or **eustachian, tube.** Normally this tube is flattened and closed, but swallowing or yawning can cause it to open temporarily

---

*The ossicles are often referred to by their Latin names, that is, **malleus, incus,** and **stapes,** respectively.

to equalize the pressure of the middle ear cavity with external air pressure. This is an important function, since the eardrum does not vibrate properly unless the pressure on both of its surfaces is the same.

Because the mucosal membranes of the middle ear cavity and the nasopharynx are continuous through the eustachian tube, **otitis media,** or inflammation of the middle ear, is a fairly common condition, especially among youngsters prone to sore throats. ■

The inner ear consists of a system of bony and rather tortuous chambers called the **osseous,** or **bony, labyrinth,** which is filled with an aqueous fluid called **perilymph** (Figure 20.2). Suspended in the perilymph is the **membranous labyrinth,** a system structured much like the osseous labyrinth. The interior of the membranous labyrinth is filled with a more viscous fluid called **endolymph.** The three subdivisions of the bony labyrinth are the **cochlea,** the **vestibule,** and the **semicircular canals,** with the vestibule situated between the cochlea and semicircular canals.

The snaillike cochlea (see Figures 20.2 and 20.3) contains the sensory receptors for hearing. The cochlear membranous labyrinth, the **cochlear duct,** is a soft wormlike tube about 1½ inches long. It winds through the full two and three-quarter turns of the cochlea and separates the cochlear cavity into upper and lower perilymph-containing chambers, the **scala vestibuli** and **scala tympani,** respectively. The scala vestibuli terminates at the oval window. The scala tympani is bounded by a membranous area called the **round window.** The cochlear duct, itself filled with endolymph, supports the **organ of Corti,** which contains the receptors for hearing—the sensory hair cells and nerve endings of the cochlear division of the vestibulocochlear nerve (VIII).

Otoscopic Examination of the Ear (Optional)

1.   Obtain an otoscope and an alcohol swab. To examine the eardrum of your partner's ear, grasp the ear pinna firmly and pull it up, back, and slightly laterally. (If your partner experiences pain or discomfort when the pinna is manipulated, an inflammation or infection of the external ear may be present. If this occurs, do not attempt to examine the ear canal.)

2.   Turn on the light switch of the otoscope and carefully insert the speculum of the otoscope into the external auditory canal in a downward and forward direction only far enough to permit examination of the shape and color and vascular network of the tympanic membrane or eardrum. The healthy tympanic membrane is pearly white.

3.   Note during the examination if there is any discharge or redness in the canal and identify earwax.

4.   After the examination, thoroughly clean the speculum with an alcohol swab before returning it to the supply area.

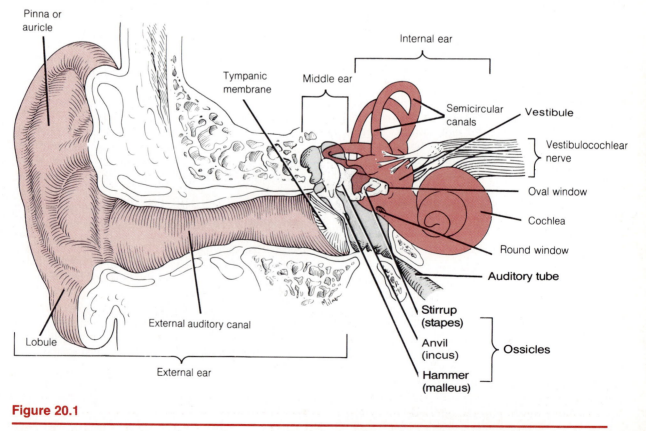

Figure 20.1

Anatomy of the ear.

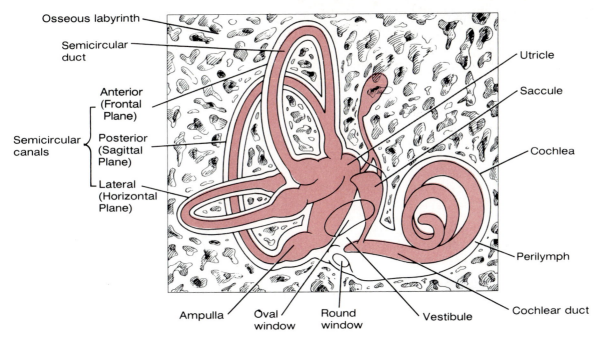

Figure 20.2

Inner ear: right membranous labyrinth shown within the bony labyrinth.

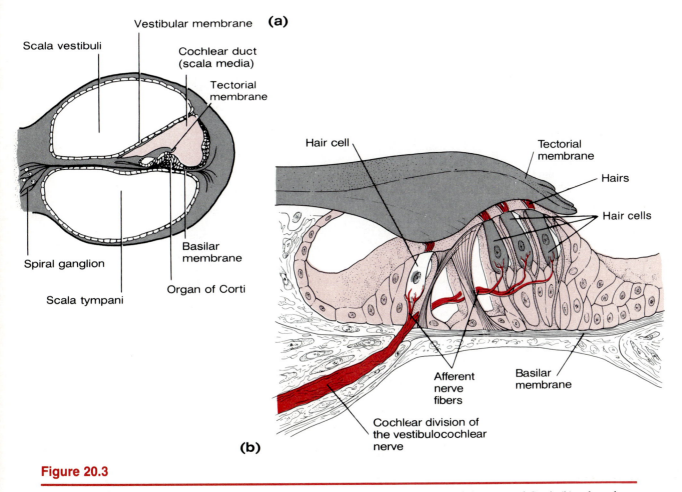

Figure 20.3

Organ of Corti: (a) cross section through one turn of the cochlea showing the position of the organ of Corti; (b) enlarged view of the organ of Corti.

Microscopic Anatomy of the Receptors

ORGAN OF CORTI AND THE MECHANISM OF HEARING

The anatomic details of the organ of Corti are shown in Figure 20.3. The hair (auditory receptor) cells rest on the **basilar membrane,** which forms the floor of the cochlear duct, and their cilia project into a gelatinous membrane, the **tectorial membrane,** which overlies them. The roof of the cochlear duct is called the **vestibular membrane.** The endolymph-filled chamber of the cochlear duct is called the **scala media.**

 Obtain a compound microscope and a prepared microscope slide of the cochlea and identify the areas noted in Figure 20.3(a). Draw "your" version of the organ of Corti in the space below.

The mechanism of hearing begins as sound waves pass through the external auditory canal and through the middle ear into the inner ear, where the vibration eventually reaches the organ of Corti, which contains the receptors for hearing. Many theories have attempted to explain how the organ of Corti actually responds to sound.

The popular "traveling wave" hypothesis of Von Békésy suggests that vibrations at the oval window initiate traveling waves that cause maximal displacements of the basilar membrane where they peak and stimulate the hair cells of the organ of Corti in that region. Since the area at which the traveling waves peak is a high-pressure area, the vestibular membrane is compressed at this point and, in turn, compresses the endolymph and the basilar membrane of the cochlear duct. The resulting pressure on the perilymph in the scala tympani causes the membrane of the round window to bulge outward into the middle ear chamber, thus acting as a relief valve for the compressional wave (Figure 20.4). Von Békésy found that high-frequency waves (high-pitched sounds) peaked close to the oval window and that low-frequency waves (low-pitched sounds) peaked farther up the basilar membrane near the apex of the cochlea. Although the mechanism of sound reception by the organ of Corti is incompletely understood, we do know that hair cells on the basilar membrane are uniquely stimulated by sounds of various frequencies and amplitude and that once stimulated they depolarize and begin the chain of nervous impulses to the auditory centers of the temporal lobe cortex. This series of events results in the phenomenon we call hearing.

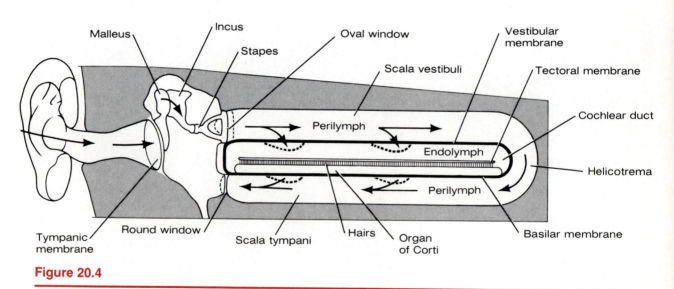

Figure 20.4

Fluid movement in the cochlea following stirrup thrust on the oval window. Compressional wave created causes the round window to bulge into the middle ear.

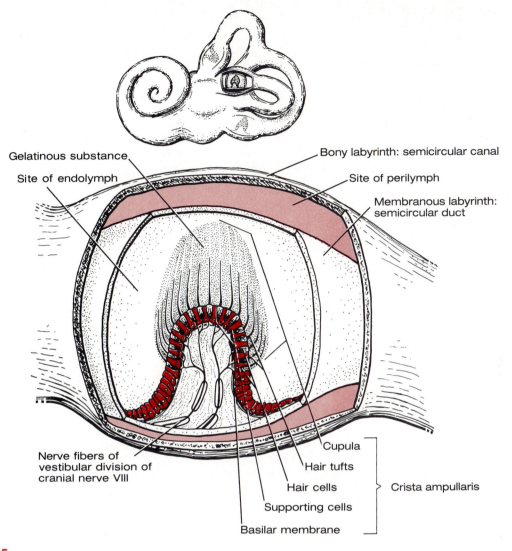

Gelatinous substance

Site of endolymph

Bony labyrinth: semicircular canal

Site of perilymph

Membranous labyrinth: semicircular duct

Nerve fibers of vestibular division of cranial nerve VIII

Cupula

Hair tufts

Hair cells

Crista ampullaris

Supporting cells

Basilar membrane

Figure 20.5

Crista ampullaris in the semicircular canal.

EQUILIBRIUM APPARATUS AND MECHANISMS OF EQUILIBRIUM   The equilibrium apparatus of the inner ear is in the vestibular and semicircular canal portions of the bony labyrinth. Their chambers are filled with perilymph, in which membranous labyrinth structures are suspended. The vestibule contains the saclike **utricle** and **saccule,** and the semicircular canals contain membranous **semicircular ducts.** Like the cochlear duct, these membranes are filled with endolymph and contain receptor cells that are activated by the disturbance of their cilia.

The semicircular canals are centrally involved in the **mechanism of dynamic equilibrium.** They are about ½ inch in circumference and are oriented in three planes—horizontal, frontal, and sagittal. At the base of each membranous canal is an enlarged region, the **ampulla,** which communicates with the utricle of the vestibule. Within each ampulla is a receptor

region called a **crista ampullaris,** which consists of a tuft of hair cells covered with a gelatinous cap, or **cupula** (Figure 20.5). When your head position changes in an angular direction, as when twirling on the dance floor or when taking a rough boat ride, the endolymph in the canal lags behind and moves in the opposite direction, pushing the cupula—like a swinging door—in a direction opposite to that of the angular motion. This movement initiates the action potential of the hair cells. These impulses are then transmitted up the vestibular division of the eighth cranial nerve to the brain. Likewise, when the angular motion stops suddenly, the inertia of the endolymph causes it to continue to move, bending the cupula in the direction of its movement and again initiating electrical impulses in the hair cells. (This phenomenon accounts for the reversed motion sensation you feel when you stop suddenly after twirling.) If you begin to move at a constant rate of motion,

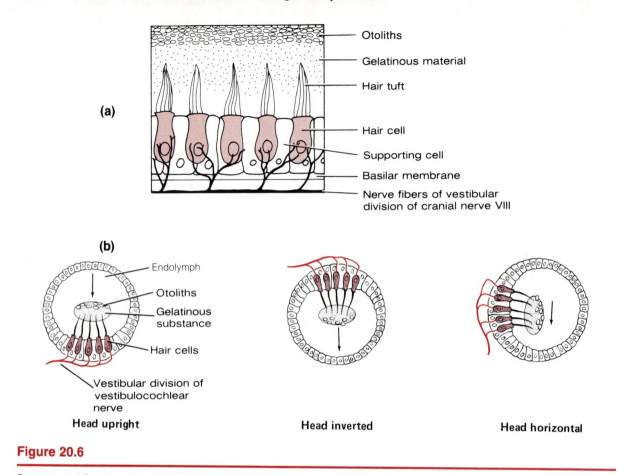

(a)

- Otoliths
- Gelatinous material
- Hair tuft
- Hair cell
- Supporting cell
- Basilar membrane
- Nerve fibers of vestibular division of cranial nerve VIII

(b)

- Endolymph
- Otoliths
- Gelatinous substance
- Hair cells
- Vestibular division of vestibulocochlear nerve

Head upright   **Head inverted**   **Head horizontal**

Figure 20.6

Structure and function of static equilibrium receptors (maculae): (a) diagrammatic view of a portion of a macula; (b) stimulation of the maculae by movement of otoliths in the gelatinous material that creates a pull on the hair cells. Arrows indicate direction of gravitational pull.

the cupula gradually returns to its original position. The hair cells, no longer bent, send no new signals, and you lose the sensation of spinning. Thus the response of these dynamic equilibrium receptors is a reaction to *changes* in angular motion rather than to motion itself.

The vestibule contains receptors (**maculae**) that are essential to the **mechanism of static equilibrium.** The maculae respond to gravitational pull, thus providing information on which way is up or down, and to linear or straightforward changes in speed. They are located on the walls of the saccule and utricle. A gelatinous material, containing small grains of calcium carbonate (**otoliths**), overrides the hair cells in each macula. As the head moves, the otoliths move in response to variations in gravitational pull and deflect different hair cells, thus triggering nerve impulses along the vestibular nerve (Figure 20.6).

Although the receptors of the semicircular canals and the vestibule are responsible for dynamic and static equilibrium, respectively, they rarely act individually. Complex interaction of many of the receptors is often the rule. Also, the information these equilibrium, or balance, senses provide is signifi-

cantly augmented by the proprioceptors and sight, as some of the following laboratory experiments demonstrate.

LABORATORY TESTS

Hearing Tests

Perform the following hearing tests in a quiet area.

ACUITY TEST   Have your lab partner pack one ear with cotton and sit quietly with eyes closed. Obtain the pocket watch and hold it very close to his or her unpacked ear. Then slowly move it away from the ear until your partner signals that the ticking is no longer audible. Record the distance in inches at which ticking is inaudible.

Right ear _____ Left ear _____

Is the threshold of audibility sharp or indefinite?

_____

SOUND LOCALIZATION Ask your partner to close both eyes. Hold the pocket watch at an audible distance (about 6 inches) from his or her ear, and move it to various locations (front, back, sides, and above his or her head). Have your partner locate the position by pointing in each instance. Can the sound be

localized equally well at all positions? _____

If not, at what position(s) was the sound less easily

located? _____

_____

The ability to localize the source of a sound depends on two factors—the difference in the loudness of the sound reaching each ear and the time of arrival of the sound at each ear. How does this information

help to explain your findings? _____

_____

_____

FREQUENCY RANGE OF HEARING Obtain three tuning forks: one with a low frequency (75 to 100 cps), one with a frequency of approximately 1000 cps, and one with a frequency of 4000 to 5000 cps. Strike the lowest-frequency fork with a rubber mallet, and hold it close to your partner's ear. Repeat with the other two forks.

Which fork was heard most clearly and comfortably?

_____ cps.

Which was heard least well? _____ cps.

WEBER TEST TO DETERMINE CONDUCTIVE AND PERCEPTIVE DEAFNESS Strike a tuning fork with the rubber mallet, and place the handle of the tuning fork medially on your forehead. Is the tone equally loud in both ears, or is it louder in one ear?

_____

If it is equally loud in both ears, you have equal hearing or loss of hearing in both ears. If perceptive (sensorineural) deafness is present in one ear, the tone will be heard in the unaffected ear but not in the ear with sensorineural deafness. If conduction deafness is present, the sound will be heard more strongly in the ear in which there is a hearing loss. Conduction deafness can be simulated by plugging one ear with cotton to interfere with the conduction of sound to the inner ear.

RINNE TEST FOR COMPARING BONE- AND AIR-CONDUCTION HEARING

1. Strike the tuning fork, and place its handle on your partner's mastoid process.
2. When your partner indicates that the sound is no longer audible, hold the still-vibrating prongs close to his auditory canal. If your partner hears the fork again when it is moved to that position (by air conduction), hearing is not impaired, and the test result is to be recorded as positive (+). (Record below.)
3. Repeat the test, but this time test air conduction hearing first.
4. After the tone is no longer heard by air conduction, hold the handle of the tuning fork on the bony mastoid process. If the subject hears the tone again by bone conduction after hearing by air conduction is lost, there is some conductive deafness and the result is recorded as negative (−).
5. Repeat the sequence for the opposite ear.

   Right ear _____ Left ear _____

   Does the subject hear better by bone or by air

   conduction? _____

Equilibrium Experiments

The function of the semicircular canals and vestibule are not routinely tested in the laboratory, but the following simple tests should serve to illustrate normal equilibrium apparatus functioning.

BALANCE TEST Have your partner walk a straight line, placing one foot directly in front of the other.

Is he or she able to walk without undue wobbling

from side to side? _____

Did he or she experience any dizziness? _____

The ability to walk with balance and without dizziness, unless subject to rotational forces, indicates normal function of the equilibrium apparatus.

Was nystagmus* present? _____

---

*__Nystagmus__ is the involuntary rolling of the eyes in any direction or the trailing of the eyes slowly in one direction, followed by their rapid movement in the opposite direction. It is normal after rotation; abnormal otherwise. The direction of nystagmus is that of its quick phase on acceleration.

BARANY TEST (INDUCTION OF NYSTAGMUS AND VERTIGO*)   This experiment evaluates the semicircular canals and should be conducted as a group effort to protect test subject(s) from possible injury. The following precautionary notes should be read before beginning:

● The subject(s) chosen should not be easily inclined to dizziness on rotational or turning movements.
● Rotation should be stopped immediately if the subject complains of feeling nauseous.
● Because the subject(s) will experience vertigo and loss of balance as a result of the rotation, several classmates should be prepared to catch, hold, or support the subject(s) as necessary until the symptoms pass.

1.   Instruct the subject to sit on a rotating chair or stool, and to hold on to the arms or seat of the chair, feet on stool rungs. The subject's head should be tilted forward approximately 30 degrees (almost touching the chest). The horizontal semicircular canal will be stimulated when the head is in this position.

2.   Four classmates should position themselves so that the subject is surrounded on all sides. The classmate posterior to the subject will rotate the chair.

3.   Rotate the chair to the subject's right approximately 10 revolutions in 10 seconds, and then suddenly stop the rotation.

4.   Note the direction of the subject's resultant nystagmus; and ask him or her to describe the feelings of movement, indicating speed and direction sensation. Record below.

_____

_____

_____

_____

If the semicircular canals are operating normally, the subject will experience a sensation that the stool is still rotating immediately after it has stopped and *will* demonstrate nystagmus.

    When the subject is rotated to the right, the cupula will be bent to the left, causing nystagmus during rotation in which the eyes initially move slowly to the left and then quickly to the right. Nystagmus will continue until the cupula has returned to its initial position. Then, when rotation is stopped abruptly, the cupula will be bent to the right, producing nystagmus with its slow phase to the right and its rapid

phase to the left. In many subjects, this will be accompanied by a feeling of vertigo and a tendency to fall to the right.

ROMBERG TEST   The Romberg test determines the integrity of the dorsal white column of the spinal cord, which transmits impulses to the brain from the proprioceptors involved with posture.

1.   Have your partner stand with his or her back to the blackboard.

2.   Draw one line parallel to each side of your partner's body. He or she should stand erect, with eyes open and staring straight ahead for 2 minutes while you observe any movements. Did you see any gross

swaying movements? _____

3.   Repeat the test. This time the subject's eyes should be closed. Note and record the degree of side-to-side

movement. _____

_____

4.   Repeat the test with the subject's eyes first open and then closed. This time, however, the subject should be positioned with his or her left shoulder toward, but not touching, the board so that you may observe and record the degree of front-to-back sway-

ing. _____

Do you think the equilibrium apparatus of the inner ear was operating equally well in all these tests?

_____

The proprioceptors? _____

Why was the observed degree of swaying greater

when the eyes were closed? _____

_____

_____

What conclusions can you draw regarding the factors necessary for maintaining body equilibrium and bal-

ance? _____

_____

_____

_____

_____

***Vertigo** is a sensation of dizziness and rotational movement when such movement is not occurring or has ceased.

ROLE OF VISION IN MAINTAINING EQUILIB-
RIUM   To further demonstrate the role of vision in
maintaining equilibrium, perform the following
experiment. (Ask your lab partner to record observations and act as a "spotter.") Stand erect, with your
eyes open. Raise your left foot approximately 1 foot
off the floor, and hold it there for 1 minute.

Record the observations: _____

_____

_____

Raise for 1 or 2 minutes; and then repeat the experiment with your other foot raised, but with your eyes

closed. Record the observations: _____

_____

_____

_____

_____

_____

Special Senses: Taste and Olfaction

OBJECTIVES

1. To describe the structure and function of the taste receptors.

2. To describe the location and cellular composition of the olfactory epithelium.

3. To name the four basic types of taste sensation and list the chemical substances that elicit these sensations.

4. To point out on a diagram of the tongue the predominant location of the basic types of taste receptors (salty, sweet, sour, bitter).

5. To explain the interdependence between the senses of smell and taste.

6. To name two factors other than olfaction that influence taste appreciation of foods.

7. To define *olfactory adaptation*.

MATERIALS

Prepared histologic slides: the tongue showing taste buds; the nasal epithelium (longitudinal section)

Compound microscope

Paper towels

Small mirror

Granulated sugar

Cotton-tipped swabs; absorbent cotton

Paper cups; paper plates

Prepared vials of 10% NaCl, 0.1% quinine or Epsom salt solution, 5% sucrose solution, and 1% acetic acid; beaker containing 10% bleach solution

Prepared dropper bottles of oil of cloves, oil of peppermint, and oil of wintergreen or corresponding flavors found in the condiment section of a supermarket

Equal-size food cubes of cheese, apple, raw potato, dried prunes, banana, raw carrot, and hard-cooked egg white; equality of sample size is important

Chipped ice

The receptors for taste and olfaction are classified as **chemoreceptors** because they respond to chemicals or volatile substances in solution. Although four relatively specific types of taste receptors have been identified, the olfactory receptors are considered sensitive to a much wider range of chemical sensations. The sense of smell is the least understood of the special senses.

LOCALIZATION AND ANATOMY OF TASTE BUDS

The **taste buds,** specific receptors for the sense of taste, are widely but not uniformly distributed in the oral cavity. Most are located on the dorsal surface of the tongue (as described next). A few are found on the soft palate, epiglottis, and inner surface of the cheeks.

The dorsal tongue surface is covered with small projections, or **papillae,** of three major types: sharp filiform papillae and the rounded fungiform and circumvallate papillae. The taste buds are located primarily on the sides of the circumvallate papillae (arranged in a V-formation on the posterior surface of the tongue) and on the more numerous fungiform papillae, which look rather like minute mushrooms and are widely distributed on the tongue. (See Figure 21.1.)

Use a mirror to examine your tongue. Can you pick out the various papillae types? _____

If so, which? _____

_____

Each taste bud consists largely of a globular arrangement of two types of modified epithelial cells: the **gustatory,** or **taste, cells** which are the actual receptor cells, and the **supporting cells.** Several nerve fibers enter each taste bud and supply sensory nerve endings to each of the receptor cells. The microvilli of the receptor cells penetrate the epithelial surface through an opening called the **taste pore.** When these microvilli, called **gustatory hairs,** contact specific chemicals in the solution, the receptor cells depolarize. The afferent fibers from the taste buds to the sensory cortex in the postcentral gyrus of the brain are carried in three cranial nerves: the facial nerve (VII) serves the anterior two-thirds of the tongue;

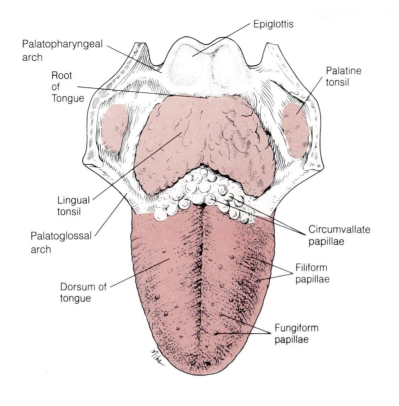

Figure 21.1

Dorsal surface of the tongue, showing major structures.

the glossopharyngeal nerve (IX) serves the posterior third of the tongue; and the vagus nerve (X) carries a few fibers from the pharyngeal region.

 Obtain a microscope and a prepared slide of a tongue cross section. Use Figure 21.2 as a guide to aid you in locating the taste buds on the tongue papillae. Make a detailed drawing of one taste bud in the space provided to the right. Label the taste pore and sensory hairs if observed.

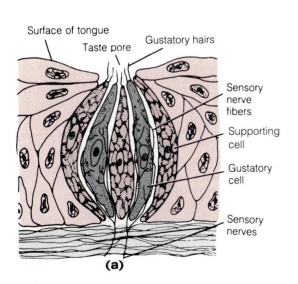

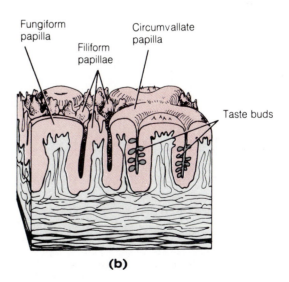

Figure 21.2

Taste bud anatomy and localization: (a) diagrammatic view of a taste bud; (b) enlarged diagrammatic view of the tongue papillae, showing positioning of taste buds.

LOCALIZATION AND ANATOMY OF THE OLFACTORY RECEPTORS

The **olfactory epithelium** (organ of smell) occupies an area of about 2.5 cm in the roof of each nasal cavity. Since the air entering the human nasal cavity must make a hairpin turn to enter the respiratory passages below, the nasal epithelium is in a rather poor position for performing its function. This is why sniffing, which brings more air into contact with the receptors, intensifies the sense of smell.

The specialized receptor cells in the olfactory epithelium are surrounded by supporting cells, non-sensory, epithelial cells. The receptor cells are believed to be bipolar neurons whose olfactory hairs (actually cilia) extend outward from the epithelium. Emerging from their basal ends are axonal nerve fibers that penetrate the cribriform plate of the ethmoid bone and proceed as the olfactory nerves to synapse in the olfactory bulbs lying on either side of the crista galli of the ethmoid bone. Impulses from neurons of the olfactory bulbs are then conveyed to the olfactory portion of the cortex (uncus).

Obtain a longitudinal section of olfactory epithelium. Examine it closely, comparing it to Figure 21.3.

LABORATORY EXPERIMENTS

Stimulation of Taste Buds

 Dry the dorsal surface of your tongue with a paper towel, and place a few sugar crystals on it. Do *not* close your mouth. Time how long it takes to taste the sugar.

_____ sec

Why couldn't you taste the sugar immediately?

_____

_____

Plotting Taste Bud Distribution

There are four basic taste sensations, which correspond to the stimulation of four major types of taste buds. Although all taste buds are believed to respond in some degree to all four classes of chemical stimuli, each type responds optimally to only one. This characteristic makes it possible to map the tongue to show the relative density of each type of taste bud.

The sweet receptors respond to a number of seemingly unrelated compounds such as sugars

(fructose, sucrose, glucose), saccharine, and some amino acids. Some believe the common factor is the hydroxyl (OH^-) group. Sour receptors respond to hydrogen ions (H^+) or the acidity of the solution, bitter receptors to alkaloids, and salty receptors to metallic ions in solution.

1. Prepare to make a taste sensation map of your lab partner's tongue by obtaining the following: cotton-tipped swabs, one vial each of NaCl, quinine or Epsom salt solution, sucrose solution, acetic acid, paper cups, and a flask of distilled or tap water.

2. Before each test, the subject should rinse his or her mouth thoroughly with water and lightly dry his or her tongue with a paper towel.

3. Moisten a swab with 5% sucrose solution and touch it to the center, back, tip, and sides of the dorsal surface of the subject's tongue.

4. Map, with an O on the tongue outline below, the location of the sweet receptors.

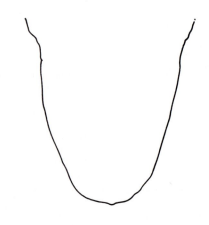

5. Repeat the procedure with quinine (or Epsom salt solution) to map the location of the bitter receptors (use the symbol B), with NaCl to map the salt receptors (symbol +), and with acetic acid to map the sour receptors (symbol −). *Use a fresh swab for each test and dispose of the swabs in the bleach—containing beaker immediately after use.*

What area of the dorsum of the tongue seems to lack

taste receptors? _____

How closely does your localization of the different taste receptors coincide with the information in your

textbook? _____

_____

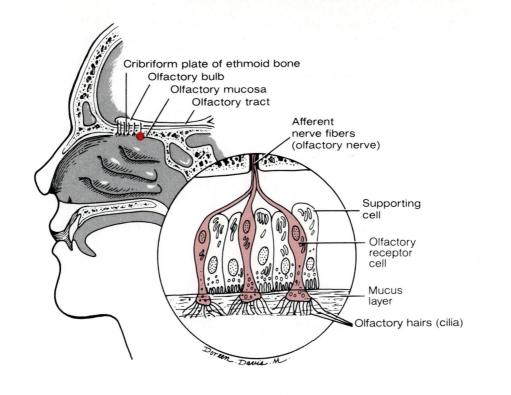

Figure 21.3

Diagrammatic representation of the cellular composition of olfactory epithelium.

Combined Effects of Smell, Texture, and Temperature on Taste

1. Ask the subject to sit with eyes closed and to stuff her nostrils with absorbent cotton.

2. Obtain samples of the food items listed in the chart below.

3. For each test, place a cube of food in the subject's mouth and ask her to identify the food by using the following sequence of activities:

- First, by manipulating the food with the tongue
- Second, by chewing the food
- If a positive identification is not made with these techniques along with the taste sense, ask the subject to remove the cotton and continue chewing with the nostrils open to determine if a positive identification can be made.

Method of Identification

Food	Texture only	Chewing with nostrils packed	Chewing with nostrils open	Identification not made
Cheese	_____	_____	_____	_____
Apple	_____	_____	_____	_____
Raw potato	_____	_____	_____	_____
Banana	_____	_____	_____	_____
Dried prunes	_____	_____	_____	_____
Raw carrot	_____	_____	_____	_____
Hard-cooked egg white	_____	_____	_____	_____

At no time should the subject be allowed to see the foods being tested. Record the results on the chart by checking the appropriate column. (Use an out-of-sequence order of food testing.)

Was the sense of smell equally important in all cases?

_____

Where did it seem to be important and why? _____

_____

_____

Effect of Olfactory Stimulation on Taste

There is no question that what is commonly referred to as the sense of taste depends heavily on stimulation of the olfactory receptors, particularly in the case of strongly odoriferous substances. The following experiment should illustrate this fact.

1. Obtain vials of oil of wintergreen, peppermint, and cloves and some fresh cotton-tipped swabs. Ask the subject to sit, so that he cannot see which vial is being used, and to dry his tongue and close his nostrils.
2. Apply a drop of one of the oils to his tongue.

    Can he distinguish the flavor? _____

3. Have the subject open his nostrils and record the

    change in sensation he reports. _____

4. Have the subject rinse his mouth well and dry his tongue.
5. Prepare two swabs, each with one of the two remaining oils.
6. Hold one swab under the subject's open nostrils, while touching the second swab to his tongue.

    Record the reported sensations. _____

    _____

    _____

    _____

7. Appropriately dispose of the swabs in the bleach—containing beaker before continuing.

Which sense, taste or smell, appears to be more important in the proper identification of a strongly

flavored volatile substance? _____

In addition to the effect that olfaction and textural factors play in determining our taste sensations, the temperature of foods may also determine if the food is appreciated or even tasted. To illustrate this, have your partner hold some chipped ice on her tongue for approximately a minute and then close her eyes. Immediately place any of the foods previously identified in her mouth and ask for an identification.

Olfactory Adaptation

Obtain some absorbent cotton and two of the following oils (oil of wintergreen, peppermint, or cloves). Press one nostril shut or stuff it with cotton. Hold the bottle of oil under the open nostril and exhale through the mouth. Record the time required for the odor to disappear (for olfactory adaptation to occur).

_____sec

Repeat the procedure with the other nostril.

_____sec

Immediately test another oil with the nostril that has just experienced olfactory adaptation. What are the

results? _____

_____

_____

What conclusions can you draw? _____

_____

_____

_____

_____

Anatomy and Basic Function of the Endocrine Glands

OBJECTIVES

1. To identify and name the major endocrine glands and tissues of the body when provided with an appropriate diagram.

2. To list the hormones produced by the endocrine glands and discuss the general function of each.

3. To indicate the means by which hormones contribute to body homeostasis by giving appropriate examples of hormonal actions.

4. To cite the mechanism by which the endocrine glands are stimulated to release their hormonal products.

5. To describe the physiologic relationship between the hypothalamus and the pituitary gland.

6. To describe a major pathologic consequence of hypersecretion and hyposecretion of each hormone considered.

7. To correctly identify the histologic structure of the anterior and posterior pituitary, thyroid, parathyroid, adrenal cortex and medulla, and pancreas by microscopic inspection or when presented with an appropriate photomicrograph or diagram. (Optional)

8. To name and point out the specialized hormone-secreting cells in the above tissues as studied in the laboratory. (Optional)

MATERIALS

Human torso model
Anatomic chart of the human endocrine system
Compound microscopes
Colored pencils
Histologic slides of the anterior–posterior pituitary (differential staining), thyroid gland, parathyroid glands, adrenal gland, pancreas tissue, ovary and testis tissue
Dissection animal, dissection tray, and instruments (if desired by the instructor)

The endocrine system is the second major controlling system of the body. Acting with the nervous system, it helps to coordinate and integrate the activity of the body's cells. However, the nervous system employs electrochemical impulses to bring about rapid control, while the more slowly acting endocrine system employs chemical "messengers," or **hormones,** which are released into the blood to be transported throughout the body.

The term *hormone* comes from a Greek word meaning to arouse. The body's hormones, which are steroids or amino-acid based, arouse the body's tissues and cells by stimulating changes in their metabolic activity. These changes lead to growth and development and to the physiologic homeostasis of many body systems. Hormones affect the body cells primarily by altering (increasing or decreasing) a metabolic process rather than by initiating a new one. Although all hormones are blood-borne, a given hormone affects only the biochemical activity of a specific organ or organs. Organs that respond to a particular hormone are referred to as the *target organs* of that hormone. The ability of the target tissue to respond seems to depend on the ability of the hormone to bind with specific receptors (proteins), occurring on the cell membranes or within the cells.

Although the function of some hormone-producing glands (the anterior pituitary, thyroid, adrenals, parathyroids) is purely endocrine, the function of others (the pancreas and gonads) is mixed—both endocrine and exocrine. Both types of glands are derived from epithelium, but the endocrine, or ductless glands release their product (always hormonal) directly into the blood. The exocrine glands release their products at the body's surface or outside an epithelial membrane via ducts. In addition, there are varied numbers of hormone-producing cells within the intestine, stomach, kidney, and placenta, organs whose functions are primarily nonendocrine.

You will use anatomic charts and models to localize the endocrine organs in this exercise and wait to identify the endocrine glands of the laboratory animals along with their other major organ systems in subsequent units. Partial dissection would cause the laboratory animals to become desiccated. However, if your instructor wishes you to perform an animal dissection to identify the endocrine organs, Figure 22.1 may be used as a guide for this purpose. Your instructor will provide methods of making incisions that minimize desiccation and damage to other organs.

GROSS ANATOMY AND BASIC FUNCTION OF THE ENDOCRINE GLANDS

 As the endocrine organs are described, *locate and identify them by name* on Figure 22.2. When you have completed the descriptive material, also locate the organs on the anatomic charts or torso.

Pituitary Gland (Hypophysis)

The pituitary gland, or hypophysis, is located in the concavity of the sella turcica of the sphenoid bone. It consists largely of two functional areas, the **adenohypophysis,** or **anterior pituitary,** and the **neurohypophysis,** or **posterior pituitary,** and is attached to the hypothalamus by a stalk called the **infundibulum.**

ADENOHYPOPHYSEAL HORMONES   The adenohypophyseal hormones consist of the following tropic hormones or hormone groups: The **gonadotropins—follicle–stimulating hormone (FSH)** and **luteinizing hormone (LH)**—regulate the hormonal activity of the gonads (ovaries and testes). **Adrenocorticotropic hormone (ACTH)** regulates the endocrine activity of the adrenal cortex portion of the adrenal gland. **Thyrotropic hormone (TSH)** influences the growth and activity of the thyroid gland. These adenohypophyseal hormones are all **tropic** hormones. In each case, a tropic hormone released by the anterior pituitary stimulates its target organ, which is also an endocrine gland, to secrete its hormones. Target organ hormones then exert their effects on other body organs and tissues.

Three other major hormones produced by the anterior pituitary are not directly involved in the regulation of other endocrine glands of the body. **Somatotropin (STH),** or **growth hormone (GH),** is a general metabolic hormone that plays an important role in determining body size. It is essential for normal retention of body protein and affects many tissues of the body. Its major effects, however, are exerted on the growth of muscle and the long bones of the body. **Prolactin (PRL)** promotes and maintains lactation by the mammary glands after childbirth. Its function in males is unknown. **Melanocyte-stimulating hormone (MSH)** does not appear to be of major significance in humans, but it darkens the skin of reptiles and amphibians and may similarly affect the melanocytes that produce pigment in human skin.

The anterior pituitary controls the activity of so many other endocrine glands that it has often been called the *master endocrine gland.* Its removal dramatically interferes with body metabolism: the gonads and adrenal and thyroid glands atrophy, and changes resulting from subsequent inadequate hormone production become obvious. However, the anterior pituitary is not autonomous in its control because the release of the anterior pituitary hormones is controlled by neurosecretions, releasing or inhibiting hormones, produced by the hypothalamus. These hypothalamic hormones are liberated into the hypophyseal portal system, which serves the circulatory needs of the anterior pituitary (Figure 22.3).

NEUROHYPOPHYSEAL HORMONES   The neurohypophysis, or posterior pituitary, is not an endocrine gland in a strict sense, because it does not synthesize the hormones it releases. (This relationship is also indicated in Figure 22.3.) Instead, it acts as a storage area for two hormones transported to it from the hypothalamus. The first of these hormones is **oxytocin,** which stmulates powerful uterine contractions during birth and coitus and also causes milk ejection in the lactating mother. The second, **antidiuretic hormone (ADH),** causes the kidney tubule cells to resorb water from the urinary filtrate, thereby reducing urine production and conserving body water. It also plays a role in increasing blood pressure because of its vasoconstrictor effect on the arterioles.

Hyposecretion of ADH results in dehydration from excessive urine output, a condition called **diabetes insipidus.** Individuals with this condition experience an insatiable thirst. ∎

Thyroid Gland

The thyroid gland is composed of two lobes joined by a central mass, or isthmus. It is located in the throat, just inferior to the larynx. It produces two major hormones, thyroid hormone (TH) and calcitonin.

Because the primary function of **thyroid hormone** (actually two physiologically active substances known as T_3 and T_4) is to control the rate of body metabolism and cellular oxidation, it affects every cell in the body. In addition, it is an important regulator of tissue growth and development, especially in the reproductive and nervous systems.

Hyposecretion of thyroxine leads to a condition of mental and physical sluggishness, which is called **myxedema** in the adult. ∎

Calcitonin decreases blood calcium levels by causing calcium to be deposited in the bones. It acts antagonistically to parathyroid hormone, the hormonal product of the parathyroid glands.

● Try to palpate your thyroid gland by placing your fingers against your windpipe. As you swallow, the thyroid gland will move up and down on the sides and front of the windpipe.

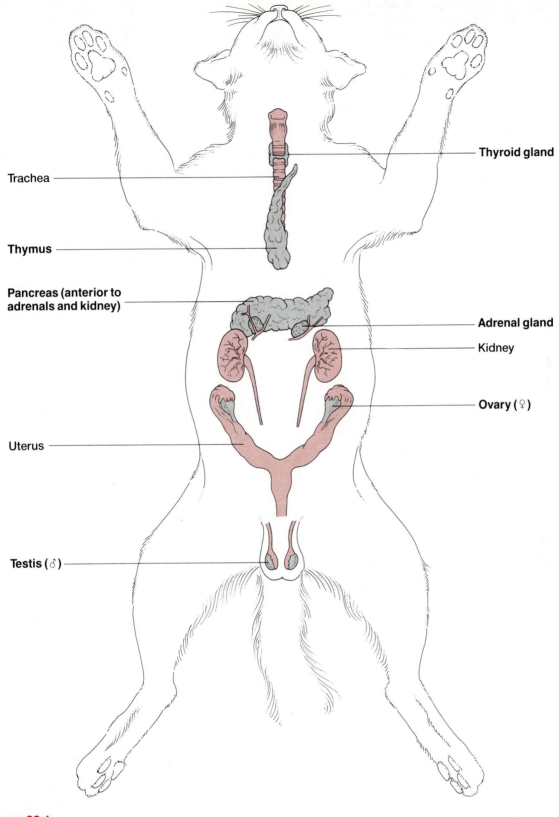

Figure 22.1

Endocrine organs in the cat.

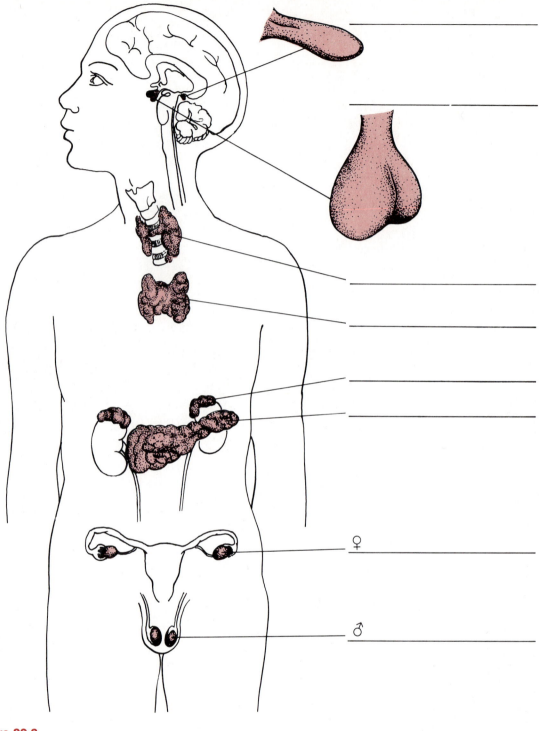

Figure 22.2

Human endocrine organs.

Parathyroid Glands

The parathyroid glands are found on the posterior surface of the thyroid gland. Typically, there are two small oval glands on each lobe. They secrete **parathyroid hormone (PTH),** the most important regulator of calcium–phosphate ion homeostasis of the blood. When blood calcium levels decrease below a certain critical level, the parathyroids release PTH, which causes bone to release calcium from the matrix and causes the kidney to increase calcium reabsorption and decrease reabsorption of phosphate from the filtrate. If blood calcium levels fall too low, **tetany** results and may be fatal.

Adrenal Glands

The two bean-shaped adrenal, or suprarenal, glands are located atop or close to the kidneys. Anatomically, the adrenal medulla develops from neural crest tissue, and the **adrenal medulla** is directly controlled by sympathetic nervous system neurons. The cells of the medulla respond to this stimulation by releasing **epinephrine** (80%) or **norepinephrine** (20%), which act in conjunction with the sympathetic nervous system to elicit the "flight or fight" response to stressors.

The **adrenal cortex** produces three major groups of steroid hormones, collectively called the **corticosteroids**—the mineralocorticoids, the glucocorticoids, and the gonadocorticoids. The **mineralocorticoids,** chiefly **aldosterone,** regulate water and electrolyte balance in the extracellular fluids primarily by regulating sodium ion reabsorption by kidney tubules. The **glucocorticoids** (cortisone, hydrocortisone, and corticosterone) enable the body to resist long-term stressors, primarily by increasing blood glucose levels. Because of their anti-inflammatory properties at pharmacologic levels, they are often administered to decrease tissue edema and counteract vascular dilation. The **gonadocorticoids,** or **sex hormones,** produced by the adrenal cortex are chiefly androgens (male sex hormones), but some estrogens (female sex hormones) are also formed. The gonadocorticoids are produced throughout life in relatively insignificant amounts; however, hypersecretion of these hormones produces abnormal hairiness **(hirsutism),** and masculinization occurs.

Pancreas

The pancreas, which functions as both an endocrine and exocrine gland, produces digestive enzymes as well as **insulin** and **glucagon,** important hormones concerned with the regulation of blood sugar levels. The pancreas is found close to the stomach.

Elevated blood glucose levels stimulate release of insulin, which decreases blood sugar levels, primarily by accelerating the transport of glucose into the body cells, where it is oxidized for energy or converted to glycogen or fat for storage.

Hyposecretion of insulin leads to **diabetes mellitus,** which is characterized by the inability of body cells to utilize glucose and the subsequent loss of glucose in the urine. Alterations of protein and fat metabolism also occur, but these are probably secondary to derangements in carbohydrate metabolism. ■

Glucagon acts antagonistically to insulin. Its release is stimulated by low blood glucose levels, and its action is basically hyperglycemic. Its primary target organ is the liver. It stimulates the liver to break down its glycogen stores to glucose and subsequently to release the glucose to the blood.

The Gonads

The female gonads, or ovaries, are paired, almond-sized organs located in the pelvic cavity. In addition

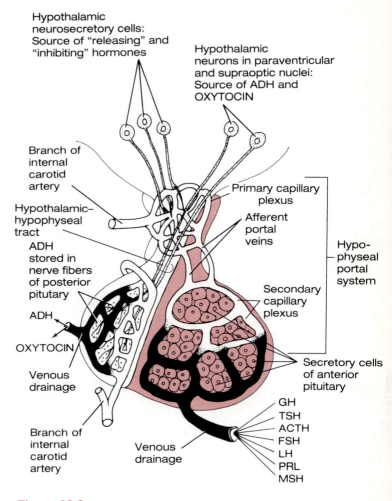

Figure 22.3

Neural and vascular relationships between the hypothalamus and the anterior and posterior lobes of the pituitary. (Note: PRL = prolactin.)

to producing the female sex cells (ova), the ovaries produce two steroid hormone groups, the **estrogens** and **progesterone.** The endocrine and exocrine functions of the ovaries do not begin until the onset of puberty, when the anterior pituitary gonadotropic hormones prod the ovary into action that produces rhythmic ovarian cycles in which ova develop and hormonal levels rise and fall. The estrogens are responsible for the development of the secondary sex characteristics of the female at puberty (primarily maturation of the reproductive organs and development of the breasts) and act with progesterone to bring about cyclic changes of the uterine lining that occur during the menstrual cycle. The estrogens also help maintain pregnancy and prepare the mammary glands for lactation.

Progesterone, as already noted, acts with estrogen to bring about the menstrual cycle. During pregnancy it maintains the uterine musculature in a quiescent state and helps to prepare the breast tissue for lactation.

The paired oval testes of the male are suspended in a pouchlike sac, the scrotum, outside the pelvic cavity. In addition to the male sex cells, sperm, the testes also produce the male sex hormone, **testosterone.** Testosterone promotes the maturation of the reproductive system accessory structures, brings about the development of the secondary sex characteristics, and is responsible for the male sexual drive, or libido. Both the endocrine and exocrine functions of the testes begin at puberty under the influence of the anterior pituitary gonadotropins.

Two glands not mentioned earlier as major endocrine glands should also be briefly considered here, the thymus and the pineal gland.

Thymus

The thymus is a bilobed gland situated in the upper thorax, posterior to the sternum and overlying the heart and lungs. Conspicuous in the infant, it begins to atrophy at puberty, and by adulthood it is relatively inconspicuous. Researchers now believe that the thymus produces a hormone called **thymosin** and that during early life the thymus acts as an incubator for the maturation and specialization of a unique population of white blood cells called T-lymphocytes. The T-lymphocytes are responsible for the cellular immunity aspect of body defense; that is, rejection of foreign grafts, tumors, or virus-infected cells.

Pineal Body

The pineal body, or **epiphysis cerebri,** is a small cone-shaped gland located in the roof of the third ventricle of the brain. Like the thymus gland, it has a limited functional life and begins to atrophy at an early age—presumably around the age of 7. The major secretion of the pineal body is **melatonin.**

The endocrine role of the pineal body in humans is still controversial (early philosophers regarded it as the seat of the soul), but melatonin appears to exert some inhibitory effect on the reproductive system (especially on the ovaries) that prevents precocious sexual maturation. Serotonin, a neurotransmitter, is produced by many regions of brain tissue and is involved in normal neural transmission of impulses. In the pineal body, serotonin is known to be a chemical precursor of melatonin.

The endocrine functions of the intestine, stomach, and kidney will be considered in conjunction with their respective organ systems.

Once you are satisfied that you can name and locate the endocrine organs and have labeled Figure 22.2, use an appropriate reference to define the following pathologic conditions (which have not been described here) resulting from hypersecretion or hyposecretion of the various hormones.

acromegaly _____

_____

_____

_____

_____

Addison's disease _____

_____

_____

_____

_____

cretinism _____

_____

_____

Cushing's syndrome _____

_____

_____

_____

_____

diabetes insipidus _____

_____

_____

pituitary dwarfism _____

_____

_____

_____

eunuchism _____

_____

_____

_____

exophthalmic goiter _____

_____

_____

_____

gigantism _____

_____

_____

_____

hypercalcemia _____

_____

_____

_____

Simmond's disease _____

_____

_____

_____

MICROSCOPIC ANATOMY OF SELECTED ENDOCRINE GLANDS (OPTIONAL)

 To prepare for the histologic study of the endocrine glands, obtain a microscope and one each of the slides listed in the list of materials. We will study only organs in which it is possible to identify the endocrine-producing cells. As applicable, compare your observations with the tissue photomicrographs in color plates G, H, and I and inside the back cover.

Thyroid Gland

1. Scan the thyroid under low power, noting the spherical sacs (follicles), containing a pink-stained material (colloid). Stored thyroxine is attached to the protein colloidal material in the follicles as **thyroglobulin** and is released gradually to the blood.
2. Observe the tissue under high power. Note that the walls of the follicles are formed by simple cuboidal epithelial cells that synthesize the follicular products. The **parafollicular, or C, cells** you see between the follicles are responsible for calcitonin production.
3. Sketch two or three follicles in the space below. Label the colloid, cuboidal epithelial cells (follicular cells), and parafollicular cells.

When the thyroid gland is actively secreting, the follicles appear small, and the colloidal material has a ruffled border. When the thyroid is hypoactive or inactive, the follicles are large and plump and the follicular epithelium appears to be squamous-like. What is the physiologic state of the tissue you have

been viewing? _____

_____

Parathyroid Glands

1. Observe the parathyroid tissue under low power to localize its two major cell types, the **chief cells** and **oxyphil cells,** which are arranged in thick branching cords. The chief cells, which are thought to bear the major responsibility for the synthesis of PTH, are small and abundant. The function of the scattered, much larger oxyphil cells is unknown.
2. Sketch a small portion of the parathyroid tissue in the space provided. Label the chief cell, oxyphil cells, and the connective tissue matrix.

Pancreas

1. Observe pancreas tissue under low power to identify the roughly circular **islets of Langerhans,** the endocrine portions of the pancreas. The islets are scattered amid the more numerous acinar cells and stain differentially (usually lighter), which makes their identification possible.

2. Focus on an islet and examine its cells under high power. Note that the islet cells are densely packed and have no definite arrangement. In contrast, the cuboidal acinar cells are arranged around secretory ducts. Unless special stains are used, it will not be possible to distinguish the **alpha cells,** which produce glucagon, from the **beta cells,** which synthesize insulin. With these specific stains, the beta cells are larger and stain blue, and the alpha cells are smaller and appear to contain pink granules. What is the product of

   the acinar cells? _____

   _____

3. Draw a section of the pancreas in the space provided. Label the islets and the acinar cells. Differentiate between the alpha and beta cells if possible.

Pituitary Gland

1. Observe the general structure of the pituitary gland under low power to differentiate between the glandular anterior pituitary and the neural posterior pituitary.

2. Using the h.p. lens, focus on the nests of cells of the anterior pituitary. It is possible to identify the specialized cell types that secrete the specific hormones when differential stains are used. Locate the pink-staining **acidophil cells,** which produce growth hormone and prolactin, and the **basophil cells,** whose deep-blue granules are responsible for the production of the other four anterior pituitary hormones (TSH, ACTH, FSH,

and LH). **Chromophobes,** the third cellular population, do not take up the stain and appear rather dull and colorless. The role of the chromophobes is controversial, but they apparently are not directly involved in hormone production.

3. Switch your focus to the posterior pituitary. Observe the nerve fibers (axons of hypophyseal neurons) that compose most of this portion of the pituitary. Also note the **pituicytes,** glial cells which are randomly distributed among the nerve fibers.

   What two hormones are stored here? _____

   _____

   What is their source? _____

   _____

4. Make sketches of the anterior pituitary and posterior pituitary tissue. Use appropriately colored pencils to indicate the acidophils, basophils, and chromophobes. Label these cells as well as the nerve fibers and pituicytes of the posterior pituitary.

Anterior pituitary   Posterior pituitary

Adrenal Gland

1. Hold the slide of the adrenal gland up to the light to distinguish outer cortical and inner medulla areas. Then scan the cortex under low power to distinguish the differences in cell appearance and arrangement in the three cortical areas. Identify the following cortical areas:

   - Connective tissue capsule of the adrenal gland.
   - The outermost **zona glomerulosa,** where most mineralocorticoid production occurs and where the tightly packed cells are arranged in spherical clusters.
   - The deeper intermediate **zona fasciculata,** which produces glucocorticoids. This is the thickest part of the cortex. Its cells are arranged in parallel cords.

● The innermost cortical zone, the **zona reticularis,** abutting the medulla, which produces sex hormones and some glucocorticoids. The cells here stain deeply and form interconnections with each other.

2. Switch focus to view the lightly stained cells of the adrenal medulla under high power. Note their clumped arrangement and that they are relatively large ovoid-shaped cells.

What hormonal products are produced by the

medulla? _____

and _____

3. Draw a representative area of each of the adrenal regions, indicating in your sketch the differences in relative cell size and arrangement.

2. Examine the vesicular follicle under high power, identifying the follicular cells that produce estrogen, the antrum (fluid-filled cavity), and developing ovum (if present). The ovum will be the largest cell in the follicle.
3. Draw a vesicular follicle below, labeling the antrum, follicle cells, and developing ovum.
4. Switch to low power, and scan the slide to find a **corpus luteum,** a large amorphous-looking area that produces progesterone.

Zona glomerulosa   Zona fasciculata

Testis

1. Examine a section of a testis under low power. Identify the seminiferous tubules, which produce sperm and the **interstitial cells,** which produce testosterone. The interstitial cells are scattered between the seminiferous tubules in the connective tissue matrix.
2. Draw a representative area of the testis in the space provided. Label the seminiferous tubules and area of the interstitial cells.

Zona reticularis   Adrenal medulla

Ovary

Because you will consider the ovary in greater histologic detail when you study the reproductive system, the objective in this laboratory exercise is just to identify the endocrine-producing parts of the ovary.

1. Scan an ovary slide under low power, and look for a **vesicular (Graafian) follicle,** a circular arrangement of cells enclosing a central cavity. This structure synthesizes estrogens.

Blood

OBJECTIVES

1. To name the two major components of blood and state their average percentages in whole blood.

2. To cite the composition of plasma and state the functional importance of many of its constituents.

3. To define *formed elements* and list the cell types composing them, cite their relative percentages, and describe their major functions.

4. To identify red blood cells, basophils, eosinophils, monocytes, lymphocytes, and neutrophils when provided with a microscopic preparation or appropriate diagram.

5. To conduct the following blood test determinations in the laboratory, and to state their norms and the importance of each.

   hematocrit
   hemoglobin determination
   clotting time
   sedimentation rate
   differential white blood cell count
   total white blood cell count
   total red blood cell count
   ABO and Rh blood typing

6. To discuss the reason for transfusion reactions resulting from the administration of mismatched blood.

7. To define *anemia, polycythemia, leukopenia,* and *leukocytosis* and to cite a possible reason for each condition.

MATERIALS

Compound microscope
Immersion oil
Clean microscope slides

Sterile lancets
Alcohol swabs (wipes)
Wright's stain in dropper bottle
Distilled water in dropper bottle
Three-dimensional models of blood cells (if available)
Plasma (obtained from an animal hospital or prepared by centrifuging animal blood)
Wide-range pH paper
Test tubes
Test tube racks
Disposable gloves

Because many blood tests are to be conducted in this exercise, it seems advisable to set up a number of appropriately labeled supply areas for the various tests. These are designated here.

General supply area:
Heparinized blood (dog blood obtained from an animal hospital if desired by the instructor)
Sterile blood lancets
Alcohol swabs
Absorbent cotton balls
Millimeter ruler
Heparinized capillary tubes
Test tubes
Test tube racks
Pipette cleaning solutions—(1) 10% household bleach solution, (2) distilled water, (3) 70% ethyl alcohol, (4) acetone
Beaker containing 10% household bleach solution for slide and glassware disposal
Disposable autoclave bag
Spray bottles containing 10% bleach solution

Blood cell count supply area:
Hemacytometer
RBC and WBC dilution fluids (Hayem's solution for RBCs; 1% acetic acid for WBCs)*
RBC and WBC dilution pipettes (Thoma pipettes) and tubing*
Mechanical hand counters

Hematocrit supply area:
Heparinized capillary tubes
Microhematocrit centrifuge and reading gauge (if the reading gauge is not available, millimeter ruler may be used)
Seal-ease (Clay Adams Co.) or modeling clay

Hemoglobin determination supply area:
Tallquist hemoglobin scales and test paper
Sahli hemoglobin determination kit (or another type of hemoglobinometer such as Spencer or American Optical)
1% HCl
Slender glass stirring rod

Note to the Instructor: See directions for handling of soiled glassware and disposable items on page 15.

*Alternatively Unopette apparatus obtainable from Becton-Dickinson may be used. Test #5851 (RBC) and #5855 (WBC).

Sedimentation rate supply area:
Landau Sed-rate pipettes with tubing and rack
Wide-mouthed bottle of 5% sodium citrate
Mechanical suction device

Coagulation time supply area:
Capillary tubes (nonheparinized)
Triangular file

Blood typing supply area:
Blood typing sera (anti-A, anti-B, and anti-Rh [D])
Rh typing box
Wax marker
Toothpicks
Clean microscope slides

In this exercise you will study the plasma and formed elements of blood and conduct various hematological tests. These tests are extremely useful diagnostic tools for the physician because blood composition (number and types of blood cells, and chemical composition) reflects the status of many body functions and malfunctions.

ALERT: The decision to use animal blood for testing or to have students test their own blood will be made by the instructor in accordance with the educational purpose of the student group. For example, for students in the nursing or laboratory technician curricula, learning how to safely handle human blood or other human wastes is essential. If blood samples are provided and they are *human* blood samples, gloves should be worn while conducting the blood tests. If human blood is being tested, yours or that obtained from a clinical agency, precautions provided in the text for disposal of human waste **must be observed.** All soiled glassware is to be immersed in household bleach solution immediately after use, and disposable items (lancets, cotton balls, alcohol swabs, etc.) are to be placed in a disposable autoclave bag so that they can be sterilized before disposal.

COMPOSITION OF BLOOD

The blood circulating to and from the body cells within the vessels of the vascular system is a rather viscous substance that varies from bright scarlet to a dull brick red, depending on the amount of oxygen it is carrying. The circulatory system of the average adult contains about 5.5 liters of blood.

Blood is classified as a type of connective tissue, since it is composed of a nonliving fluid matrix (the **plasma**), in which the living cells (**formed elements**) are suspended. The fibers typical of a connective tissue matrix become visible in blood only when clotting occurs. They then appear as fibrin threads, which form the structural basis for clot formation.

Over 100 different substances are dissolved or suspended in the plasma, which is over 90% water. These include nutrients, gases, hormones, various wastes and metabolites, many types of functional proteins, and mineral salts. The composition of plasma varies continuously as cells remove or add substances to the blood.

Three types of formed elements are present in blood. The most numerous are the **erythrocytes,** or red blood cells (RBCs), which are literally sacs of hemoglobin molecules that transport the bulk of the oxygen carried in the blood (and a small percentage of the carbon dioxide). **Leukocytes,** or white blood cells (WBCs), are part of the body's nonspecific defenses and the immune system, and the **thrombocytes,** or platelets, function in hemostasis (blood clot formation). Formed elements normally constitute 45% of whole blood, plasma the remaining 55%.

Physical Characteristics of Plasma

Go to the supply area and carefully pour a few milliliters of plasma into a test tube. Also obtain wide-range pH paper and then return to your laboratory bench to make the following simple observations.

pH OF PLASMA   Test the pH of the plasma with wide-range pH paper. Record the pH observed.

_____

COLOR AND CLARITY OF PLASMA   Hold the test tube up to a source of natural light and note its color and degree of transparency. Is it clear, translucent, or opaque?

Color _____

Degree of transparency _____

CONSISTENCY   Dip your finger and thumb into the plasma and then press them firmly together for a few seconds. Gently pull them apart. How would you describe the consistency of plasma? Slippery, watery, sticky, or granular?

Record your observations. _____

_____

Formed Elements of Blood

Conduct your observations of blood cells on a slide prepared from your own blood.

1.   Obtain two glass slides, dropper bottles of Wright's stain and distilled

water, two or three lancets, cotton balls, and alcohol swabs. Bring this equipment to the laboratory bench.

2.   Clean the slides thoroughly and dry. Open the alcohol swab packet and scrub your third or fourth finger with the swab. (Because the pricked finger may be a little sore later, it is better to prepare a finger on the hand used less often.) Circumduct your hand (swing it around) for 10 to 15 seconds. This will dry the alcohol and cause your fingers to become engorged with blood. Then, open the lancet packet and grasp the lancet by its blunt end. Quickly jab the pointed end into the prepared finger to produce a free flow of blood. It is *not* a good idea to squeeze or "milk" the finger, as this forces out tissue fluid as well as blood. If the blood is not flowing freely, another puncture should be made. _Under no circumstances is a lancet to be used for more than one puncture._ Dispose of the lancets in the disposable autoclave bag *immediately* after use.

3.   Wipe away the first drop of blood with a cotton ball, and allow another large drop of blood to form. Touch the blood to one of the cleaned slides approximately ½ inch from the end. Then quickly (to prevent clotting) use the second slide to form a blood smear as shown in Figure 23.1. The blood smear when properly prepared is uniformly thin. If it appears streaked, the blood probably began to clot or coagulate before the smear was made, and another slide should be prepared.

4.   Dry the slide by waving it in the air. When it is completely dry, it will look dull. Place it on a paper towel, and flood it with Wright's stain. Count the number of drops of stain used. Allow the stain to remain on the slide for 3 to 4 minutes and then flood the slide with an equal number of drops of distilled water. Allow the water and Wright's stain mixture to remain on the slide for 4 or 5 minutes or until a metallic green film or scum is apparent on the fluid surface. Blow on the slide gently every minute or so to keep the water and stain mixed during this interval.

5.   Rinse the slide with a stream of distilled water. Then flood it with distilled water, and allow it to lie flat until the slide becomes translucent and takes on a pink cast. Then stand the slide on its long edge on the paper towel, and allow it to dry completely.

6.   Once the slide is dry, you can begin your observations. Obtain a microscope and scan the slide under low power to find the area where the blood smear is the thinnest. Use the oil immersion lens to identify the formed elements. Read the following descriptions of cell types, and find each one on Figure 23.2. Then observe the slide carefully to identify each cell type.

ERYTHROCYTES   Erythrocytes, or red blood cells, which range in size from 5 to 10 μm in diameter

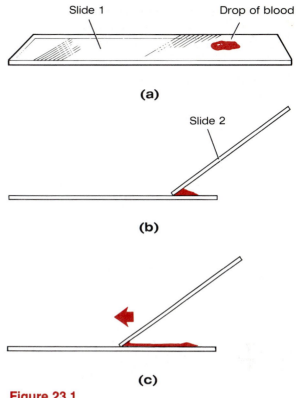

Figure 23.1

Procedure for making a blood smear: (a) place drop of blood on slide 1 approximately 1/2 inch from one end; (b) hold slide 2 at a 30° to 40° angle to slide 1 (it should touch the drop of blood); allow blood to spread along entire bottom edge of angled slide; (c) smoothly advance slide 2 to end of slide 1 (blood should run out before reaching the end of slide 1).

(averaging 7.5 μm), are cells whose color varies from brick red to pale pink, depending on the effectiveness of the stain. They have a distinctive biconcave disk shape and appear paler in the center than at the edge.

Red blood cells differ from the other blood cells because they are anucleate when mature and circulating in the blood. As a result, they are unable to reproduce and have a limited life span of 100 to 120 days, after which they begin to fragment and are destroyed in the spleen and other reticuloendothelial tissues of the body.

In various anemias, the red blood cells may appear pale (an indication of decreased hemoglobin content) or may be nucleated (an indication that the bone marrow is turning out cells prematurely).

As you observe the slide, note that the red blood cells are by far the most numerous blood cells seen in the field. Their number averages 4.5 million to 5.0 million cells per cubic millimeter of blood (for women and men, respectively).

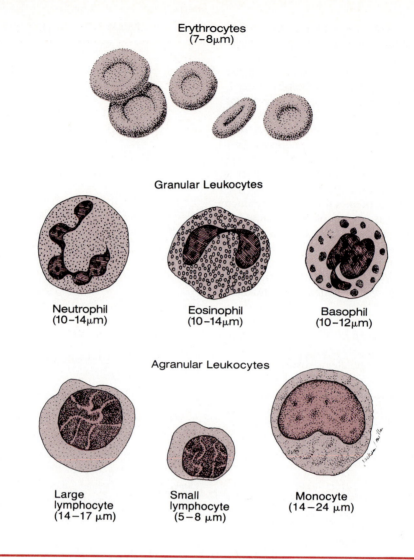

Erythrocytes
(7–8μm)

Granular Leukocytes

Neutrophil
(10–14μm)

Eosinophil
(10–14μm)

Basophil
(10–12μm)

Agranular Leukocytes

Large
lymphocyte
(14–17 μm)

Small
lymphocyte
(5–8 μm)

Monocyte
(14–24 μm)

Figure 23.2

Diagrammatic representation of blood cells (relative size relationships indicated). Also see Color Plate F.

LEUKOCYTES   Leukocytes, or white blood cells, are nucleated cells that are formed in the bone marrow from the same stem cell (hemocytoblast) as red blood cells. They are much less numerous than the red blood cells, averaging from 4000 to 11,000 cells per cubic millimeter. Basically, white blood cells are protective, pathogen-destroying cells that are transported to all parts of the body in the blood or lymph. Important to their protective function is their ability to move in and out of blood vessels, a process called **diapedesis,** and wander through body tissues by **ameboid motion** to reach sites of inflammation or tissue destruction. They are classified into two major groups, depending on whether or not they contain conspicuous granules in their cytoplasm.

Granulocytes comprise the first group. The granules in their cytoplasm stain differentially with Wright's stain, and they have peculiarly lobed nuclei, which often consist of expanded nuclear regions connected by thin strands of nucleoplasm. Additional information about the three types of granulocytes follows:

Neutrophil: The most abundant of the white blood cells (40% to 70% of the leukocyte population); nucleus consists of 3 to 7 lobes and the pale lilac cytoplasm contains fine cytoplasmic granules, which are generally indistinguishable; function as active phagocytes whose number increases exponentially during acute infections.

Eosinophil: Represent 1% to 4% of the leukocyte population; nucleus is generally figure 8 or bilobed in shape; contain large cytoplasmic granules that stain red-orange with Wright's stain. Precise function is unknown, but they increase in number during allergy and parasite infections and may selectively phagocytize antigen–antibody complexes.

Basophil: Least abundant leukocyte type representing less than 1% of the population; large U- or S-shaped nucleus with two or more indentations. Cytoplasm contains coarse sparse granules that stain deep purple with Wright's stain. The granules contain histamine which is discharged on exposure to antigens and helps to mediate the inflammatory response.

The second group, **agranulocytes,** or **agranular leukocytes,** contain no observable cytoplasmic granules. Although found in the bloodstream, they are much more abundant in lymphoid tissues and lymph. Their nuclei tend to be closer to the norm, that is, spherical, oval, or kidney-shaped. Specific characteristics of the two types of agranulocytes are listed below.

Lymphocyte: The smallest of the leukocytes, approximately the size of a red blood cell. The dark blue–purple staining nucleus is generally spherical or slightly indented. Sparse cytoplasm appears as a thin blue rim around the nucleus. Concerned with immunologic responses in body; one population, the B-lymphocytes, oversees the production of antibodies that are released to blood; the second population, T-lymphocytes, plays surveillance role and destroys grafts, tumors, and virus-infected cells and activates B-lymphocytes. Represent 20% to 45% of the WBC population.

Monocyte: The largest of the leukocytes; approximately twice the size of red blood cells. Represent 4% to 8% of the leukocyte population. Dark blue nucleus is generally kidney-shaped; abundant cytoplasm stains grey-blue. Function as active phagocytes (the "long-term cleanup team"), which increase dramatically in number during chronic infections such as tuberculosis.

THROMBOCYTES Thrombocytes, or platelets, are cell fragments of large multinucleate cells (**megakaryocytes**) formed in the bone marrow. They appear as darkly staining, irregular bodies interspersed among the other blood cells. The normal platelet count in blood is 250,000 to 500,000 /mm^3. Platelets are instrumental in the clotting process that occurs in plasma when blood vessels are ruptured.

After you have identified these cell types on your slide, observe three-dimensional models of blood cells if these are available. _Do not dispose of your slide,_ as it will be used later for the differential white blood cell count.

HEMATOLOGIC TESTS

When one enters a hospital as a patient, several hematologic studies are routinely done to determine general level of health as well as the presence of pathologic conditions. You will be conducting the most common of these studies in this exercise.

Materials such as cotton balls, lancets, and alcohol swabs are used in nearly all of the following diagnostic tests. These supplies are at the general supply area and should be properly disposed of immediately after use—place soiled glassware in the bleach-containing beaker and disposable items in the autoclave bag. Other necessary supplies and equipment are at specific supply areas marked according to the test with which they are used. Since nearly all of the tests require a finger stab to obtain blood, it might be wise to quickly read through the various tests to determine in which instances more than one preparation can be done from the same finger stab. For example, the hematocrit capillary tubes and sedimentation rate samples might be prepared at the same time the chamber is being prepared for the total blood cell counts. A little preplanning will save you the discomfort of a multiply punctured finger.

An alternative to using blood obtained from the finger stab technique is using heparinized blood samples supplied by your instructor. The purpose of using heparinized tubes is to prevent the blood from clotting. Thus blood collected and stored in such tubes will be suitable for all tests except coagulation time testing.

Differential White Blood Cell Count

To make a differential white blood cell count, 100 WBCs are counted and classified according to type. Such a count is routine in a physical examination and in sickness, since any abnormality or significant elevation in the normal percentages of the specific types of WBCs may indicate the possible source of pathology. Use the slide prepared for the identification of the blood cells (pp. 218–219) for the count.

1. Begin at the edge of the smear and move the slide in a systematic manner on the microscope stage—either up and down or from side to side as indicated in Figure 23.3.
2. Record each type of white blood cell you observe by making a count on the chart on page 222 (for example, ⫫⫫ // = 7 cells) until you have observed and recorded a total of 100 WBCs. Using the following equation, compute the percentage of each:

$$\text{Percent (\%)} = \frac{\#\ \text{observed}}{\text{total}\ \#\ \text{counted}\,(100)} \times 100$$

Record the percentages on the data sheet at the end of this exercise.

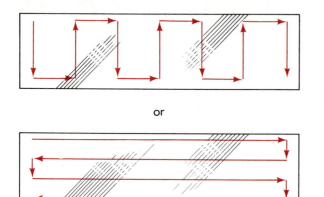

or

Figure 23.3

Alternative methods of moving the slide for a differential WBC count.

Cell type	Number observed
Neutrophils	
Eosinophils	
Basophils	
Lymphocytes	
Monocytes	

How does your differential white blood cell count correlate with the percentages given for each type

on pages 220–221? _____

_____

Total White and Red Blood Cell Counts

To conduct a total or direct WBC or RBC count, you will dilute a known volume of blood with a fluid that prevents blood coagulation and place the mixture into a counting chamber of known volume—a hemacytometer. The cells are then counted by microscopic inspection, and corrections for dilution and chamber volume are made to obtain the result in cells per cubic millimeter. The white and red blood cell diluents differ. They are isotonic for the cell type to be counted and cause dissolution of the cell types not being counted. In other words, if the WBC diluent is used, the RBCs are hemolyzed and thus do not interfere with the WBC counting process. Thus, the diluents must not be intermixed or used haphazardly. Figure 23.4 depicts the apparatus and procedure used for the counts. (Note that this technique is rather outdated, since most clinical agencies now have computerized equipment for performing blood counts.)

TOTAL WHITE BLOOD CELL COUNT   Since white blood cells are an important part of the body's defense system, it is essential to note any abnormalities in them. **Leukocytosis,** an abnormally high WBC count, may indicate bacterial or viral infection, metabolic disease, hemorrhage, or poisoning by drugs or chemicals. A decrease in the white cell number below 5000/mm^3 (**leukopenia**) may indicate typhoid fever, measles, infectious hepatitis or cirrhosis, tuberculosis, or excessive antibiotic or x-ray therapy. A person with leukopenia lacks the usual protective mechanisms.

1.   Obtain a hemacytometer and cover glass, cotton balls, lancets, alcohol swabs, and a hand counter. Also obtain _either_ the Unopette WBC testing apparatus _or_ WBC diluent (1% acetic acid), WBC diluting pipette (white bead in bulb) and tubing.

2.   Cleanse your finger thoroughly, and obtain a second drop of blood as described on page 219.

3.   If using the Unopette, follow the directions provided with the test kit. Otherwise proceed as follows to mix the blood sample with the diluent. Attach the tubing (with attached mouthpiece) to the pipette, place the tip of the pipette in the blood drop, and fill the WBC pipette exactly to the 0.5 ml mark by drawing air through the pipette via the mouthpiece. _Hold the pipette horizontally during the filling process._ Wipe the tip of the pipette with a cotton ball and then immerse it into WBC diluent and draw it up into the pipette until the mixture is exactly at the 11.0 mark. Rotate the pipette as the fluid is being aspirated to ensure proper mixing. The dilution of the blood is now 1:20.

4.   Still holding the pipette in the horizontal position, remove the tubing from the pipette with a gentle rocking motion. Place your thumb at one end and the second finger at the opposite end of the pipette. Mix the pipette contents for 2 to 3 minutes, moving the pipette in a back and forth arclike movement approximately parallel with the bench top.

5.   Place a cover slip on the hemacytometer, and prepare to charge it. Discard the first 2 to 3 drops—which is only diluent—from the pipette or Unopette onto a paper towel. When your finger is on the distal end of the pipette, the mixture will remain in the pipette. When you remove your finger, the mixture will be discharged. Place 1 drop of the mixture at each end of the cover slip at the junction of the hemacytometer and cover slip. _Do this quickly;_ otherwise the hemacytometer will overfill, and the mixture will flow under the cover slip by capillary action. You now have two charged chambers. If the chambers are properly charged, no fluid will appear in the moat.

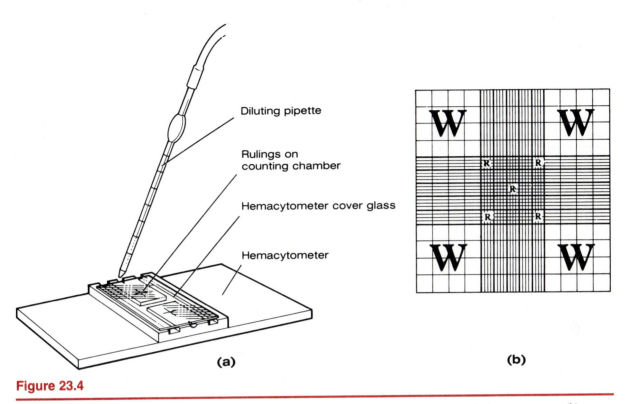

Figure 23.4

Apparatus and preparation for blood cell counts: (a) procedure for introducing diluted blood to the hemacytometer; (b) areas of the hemacytometer used for red and white blood cell counts.

6.   Place the hemacytometer on the microscope stage and allow it to remain undisturbed for 2 to 3 minutes to allow the cells to become evenly distributed.

7.   Bring the grid lines into focus under low power (10× objective lens) and check for uniformity of white blood cell distribution. If it is uneven, switch to the other chamber for observation.

8.   Move the slide into position so that one of the W areas can be seen (see Figure 23.4), and proceed to count all WBCs observed in the four corner W areas. Use a hand counter to facilitate the counting process. *To prevent overcounts of cells at boundary lines, count only those that touch the left and upper boundary lines of the W area but not those touching the right and lower boundaries.* Record the number of WBCs counted and multiply this number by 50 to obtain the number of WBCs per cubic millimeter.*

No. WBCs: _____ WBC/mm^3: _____

*The factor of 50 is obtained by multiplying the volume dilution factor (20) by the volume correction factor (2.5): 20 × 2.5 = 50. The 2.5 volume correction factor is obtained in the following manner: Each W area on the grid is exactly 1 mm^2 × 0.1 mm deep; therefore, the volume of each W section is 0.1 mm^3. Since WBCs in four W areas are counted (a total of 0.4 mm^3), this volume must be multiplied by 2.5 (correction factor) to obtain the number of cells in 1 mm^3.

Record this information on the data sheet on page 229. How does your result compare with the norms

for a white blood cell count? _____

9.   Obtain four test tubes; put bleach in the first, distilled water in the second, alcohol in the third, and acetone in the fourth. Clean the pipette by drawing up the bleach, distilled water, alcohol, and acetone in succession. *If you use mouth suction, watch the fluid column rise to prevent aspiration of the cleaning solutions.* Discharge the contents each time before drawing up the next solution. If animal blood was used, return the cleaned pipette to the supply area; otherwise place it in the beaker of household bleach. Clean the hemacytometer and coverslip in preparation for the red blood cell count. Do not dispose of the test tubes containing the cleaning solutions because they will be used twice again.

TOTAL RED BLOOD CELL COUNT   The red blood cell count, like the white blood cell count, determines the total number of this cell type per unit volume of blood. Since RBCs are absolutely necessary for oxygen transport, the doctor should investigate any excessive change in their number immediately. An increase in the number of RBCs (**polycythemia**) may result from bone marrow cancer or from living at high altitudes where less oxygen is available. A decrease in the number of RBCs results in anemia. (The term *anemia* simply indicates a decreased oxygen-carrying capacity of blood that may result from a decrease in RBC number or size or a decreased hemoglobin content of the RBCs.) A decrease in RBCs may result suddenly from hemorrhage or more gradually from conditions that increase RBC destruction or decrease RBC production. ■

1.   Obtain a RBC diluting pipette (red bead in bulb), RBC diluent (Hayem's solution) or Unopette apparatus for performing a RBC count. Also obtain other supplies as required for the WBC count. Note that the RBC pipette has a much larger bulb. Follow the same procedure for obtaining blood and charging the chamber as for the WBC count, except use Hayem's solution diluent, and draw it up to the 101 mark after 0.5 ml of blood have been aspirated. This provides a dilution of 1:200. If a Unopette system is used, follow the directions provided with the test kit.

2.   Allow the hemacytometer to rest on the microscope stage as before. Using the h.p. lens, count all RBCs in the R-marked areas (see Figure 23.4). Again, at the boundary lines, count only those cells touching the left and upper lines.

3.   Record the number of cells counted, and multiply that number by 10,000 to compute the RBC/mm^3. (Note: the dilution factor (200) × volume correction factor (50) = 10,000).

No. RBCs: _____ RBCs/mm^3: _____

How does your result compare with normal RBC

values? _____

4.   Clean the glass pipette as described in Step 9 in the instructions for performing the WBC count (p. 223). Place the tubing mouthpiece and all used glassware (including the hemacytometer) in the beaker containing bleach. Put disposable items, including the Unopette apparatus, in the disposable autoclave bag.

Hematocrit

The hematocrit, or packed cell volume (PCV), is routinely determined when anemia is suspected. Centrifuging whole blood causes the formed elements to spin to the bottom of the tube, with plasma forming the top layer. Since the blood cell population is primarily RBCs, the PCV is generally considered equivalent to the RBC volume, and this is the only value reported. However, the relative percentage of WBCs can be differentiated, and both WBC and plasma volume will be reported here. Normal PCV values for the male and female, respectively, are 47.0 ± 7 and 42.0 ± 5.

The PCV is determined by the micromethod, so only a drop of blood is needed. If possible, all members of the class should prepare their capillary tubes at the same time so the centrifuge can be properly balanced and run only once.

1.   Obtain two heparinized capillary tubes, Seal-ease or modeling clay, a lancet, alcohol swabs, and some cotton balls.

2.   Cleanse the finger, and allow the blood to flow freely. Wipe away the first few drops and, holding the red-line-marked end of the capillary tube to the blood drop, allow the tube to fill at least three-fourths full by capillary action (Figure 23.5). If the blood is not flowing freely, the end of the capillary tube will not be completely submerged in the blood during filling, air will enter, and you will have to prepare another sample.

3.   Plug the blood-containing end by pressing it into the Seal-ease or clay. Prepare a second tube in the same manner.

4.   Place the prepared tubes opposite one another in the radial grooves of the centrifuge with the sealed ends abutting the rubber gasket at the centrifuge periphery. This loading procedure balances the centrifuge and prevents blood from spraying everywhere by centrifugal force. Make a note of the numbers of the grooves your tubes are in. When all the tubes have been loaded, make sure the centrifuge is properly balanced, and secure the centrifuge cover. Turn the centrifuge on, and set the timer for 4 or 5 minutes.

5.   Determine the percentage of RBCs, WBCs, and plasma by using the microhematocrit reader. The RBCs are the bottom layer, the plasma is the top layer, and the WBCs are the buff-colored layer between the two. If the reader is not available, use a millimeter ruler to measure the length of the filled capillary tube occupied by each element, and compute its percentage by using the following formula:

$$\frac{\text{Height of the column composed by the element (mm)}}{\text{Height of the original column of whole blood (mm)}} \times 100$$

Record your calculations below and on the data sheet.

% RBC _____ % WBC _____ % plasma _____

Usually WBCs constitute 1% of the total blood volume. How do your blood values compare to this figure and to the normal percentages for RBCs and plasma? (See p. 218.)

_____

As a rule, the PCV is considered a more accurate test for determining the RBC composition of the blood than the total RBC count. A hematocrit within the

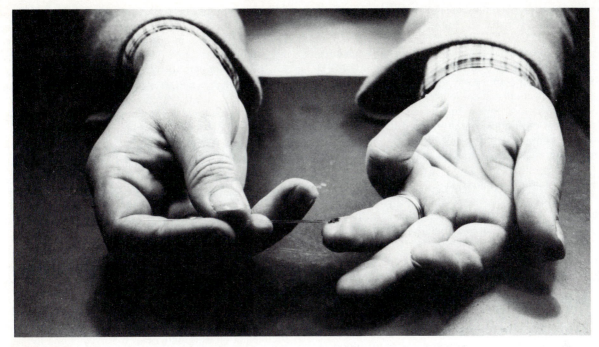

Figure 23.5

Filling a capillary tube with blood. (Photo courtesy of Ann Allworth.)

normal range is generally indicative of a normal RBC number, whereas an abnormally high or low PCV is cause for concern.

Hemoglobin Concentration Determination

As noted earlier, a person can be anemic even with a normal RBC count. Since hemoglobin is the RBC protein responsible for oxygen transport, perhaps the most accurate way of measuring the oxygen-carrying capacity of the blood is to determine its hemoglobin content. Oxygen, which combines reversibly with the heme (iron-containing portion) of the hemoglobin molecule, is picked up by the blood cells in the lungs and unloaded in the tissues. Thus, the more hemoglobin molecules the RBCs contain, the more oxygen they will be able to transport. Normal blood contains 12 to 16 g hemoglobin per 100 ml blood. Hemoglobin content in men is slightly higher (13 to 18 g) than in women (12 to 16 g).

Several techniques have been developed to estimate the hemoglobin content of blood, ranging from the old, rather inaccurate Tallquist method to expensive colorimeters, which are precisely calibrated and yield highly accurate results. A procedure using the Sahli-Adams hemoglobinometer is used in the hemoglobin determination here. (If another type of hemoglobinometer is available, it may be used in lieu of the Sahli equipment. Simply follow the instructions provided with the instrument, and record your results noting the equipment used.)

Obtain a Sahli hemoglobin determination kit, lancets, alcohol swabs, cotton balls, slender glass stirring rod, dropper bottle of 1% HCl, and a dropper bottle of distilled water. Read through the directions for the procedure before beginning.

In the Sahli-Adams method, hemoglobin is converted to a brown acid (hematin) by mixing the blood with hydrochloric acid (HCl). This technique provides a fairly accurate estimation of hemoglobin content because the hemoglobin is released from the blood cells.

1. Fill the graduated tube to the 2-g mark with 1% HCl. Produce a free flow of blood from a finger stab, and draw the blood into the pipette to the 20 mm^3 marking.

2. Wipe the tip of the pipette with a cotton ball, and blow the blood from the pipette into the acid solution in the tube.

3. Draw the mixture up into the pipette, and blow it back into the tube gently two or three times to ensure proper mixing.

4. Place the tube in the hemoglobinometer and turn it so that the graduated scale will not interfere with your color observations later.

5. Allow the mixture to stand for 10 minutes and then compare the color of your test sample with the color standard. Be sure you have a good light source

behind the color standard before attempting to match the colors.

6.   If the test sample is darker than the standard, add distilled water, a drop at a time, mixing with a glass stirring rod after each addition, until the colors of the test sample and the standard are identical. When the colors appear identical to you, read the two scales on the test tube to record the percentage of hemoglobin and the grams of hemoglobin per 100 ml blood. Record here and on the data sheet.

% hemoglobin concentration _____

g/100 ml _____

Are your values within the normal range? _____

7.   Clean the pipette with the cleaning solutions before returning it and the other equipment to the proper containers at the supply area.

Generally speaking, the relationship between the PCV and grams of hemoglobin per 100 ml blood is 3:1. How do your values compare?

_____

Record this value (obtained from your data) on the data sheet on page 229.

Blood Indices and Corpuscular (or Blood) Constants

Most health agencies perform tests to determine absolute values of red blood cell count, hematocrit, and hemoglobin concentration, and also report the interrelationships in terms of ratios or indices. To be meaningful the indices must be calculated on very accurate RBC counts and hematocrit and hemoglobin determinations. When this is so, the indices aid in the diagnosis of the various types of anemias, allowing them to be classified as macrocytic, normocytic, or microcytic, relative to the size of the RBCs, and hypochromic, normochromic, or hyperchromic, according to the amount of hemoglobin contained in the RBCs. These indices will not be computed here but are defined as follows:

Mean corpuscular volume (**MCV**): the volume of the average RBC in a given sample of blood. The normal volume range is $80-94$ μ^3. Higher values indicate enlarged (macrocytic) RBCs, whereas lower values indicate smaller than normal (microcytic) RBCs.

Mean corpuscular hemoglobin (**MCH**): the amount of hemoglobin (by weight) in the average RBC in a given sample of blood. Normal value is $27-32$ $\mu\mu g$.

Mean corpuscular hemoglobin concentration (**MCHC**): the weight to volume ratio of hemoglobin in the average RBC in a given sample of blood. Normal value is 33%–38%. The MCHC relates the blood samples to norms of hemoglobin content relative to RBC size. RBCs that have lower than normal MCHC values are pale (hypochromic) and considered to be hemoglobin-deficient.

Sedimentation Rate

The speed at which red blood cells settle to the bottom of a vertical tube when allowed to stand is called the sedimentation rate. The normal rate for adults is 0 to 6 mm/hr (averaging 3 mm/hr) and for children 0 to 8 mm/hr (averaging 4 mm/hr). Sedimentation of RBCs apparently proceeds in three stages: rouleaux formation, rapid settling, and final packing. Rouleaux formation (alignment of RBCs like a stack of pennies) does not occur with abnormally shaped red blood cells (as in sickle cell anemia); therefore, the sedimentation rate is decreased. The size and number of RBCs affect the packing phase. In anemia the sedimentation rate increases; in polycythemia the rate decreases. The sedimentation rate is greater than normal during menses and pregnancy, and very high sedimentation rates may be indicative of infectious conditions or tissue destruction occurring somewhere in the body. Although this test is nonspecific, it alerts the diagnostician to the need for further tests to pinpoint the site of pathology. The Landau micromethod, which uses just one drop of blood, is used here.

1.   Obtain lancets, cotton balls, alcohol swabs, and Landau Sed-rate pipette and tubing, the Landau rack, a wide-mouthed bottle of 5% sodium citrate, a mechanical suction device, and a millimeter ruler.

2.   Use the mechanical suction device to draw up the sodium citrate to the first (most distal) marking encircling the pipette. Prepare the finger for puncture, and produce a free flow of blood.

3.   Wipe off the first drop, and then draw the blood into the pipette until the mixture reaches the second encircling line. Keep the pipette tip immersed in the blood to avoid air bubbles.

4.   Thoroughly mix the blood with the citrate (an anticoagulant) by drawing the mixture into the bulb and then forcing it back down into the lumen. Repeat this mixing procedure six times, and then adjust the top level of the mixture as close to the zero marking as possible. *If any air bubbles are introduced during the mixing process, discard the sample and begin again.*

5.   Seal the tip of the pipette by holding it tightly against the tip of your index finger, and then carefully remove the suction device from the upper end of the pipette.

6.   Stand the pipette in an exactly vertical position on the Sed-rack with its lower end resting on the base of the rack. Record the time, and allow it to stand for exactly 1 hour.

time _____

7.   After 1 hour, measure the number of millimeters of visible clear plasma (which indicates the amount of settling of the RBCs), and record this figure here and on the data sheet.

sedimentation rate _____ mm/hr

8.   Clean and dispense with the pipette as instructed for previous tests.

Coagulation Time

Blood clotting, or coagulation, is a protective device that minimizes blood loss when blood vessels are ruptured. This process requires the interaction of many substances normally present in the plasma (clotting factors, or procoagulants) as well as some released by platelets and injured tissues. Basically hemostasis proceeds as follows: The injured tissues and platelets release **thromboplastin** and PF$_3$ respectively, which trigger the clotting mechanism, or cascade. Thromboplastin and PF$_3$ interact with other blood protein clotting factors and calcium ions to convert **prothrombin** (present in the plasma) to **thrombin.** Thrombin then acts enzymatically to polymerize the soluble **fibrinogen** proteins (present in plasma) into insoluble **fibrin,** which forms a meshwork of strands that traps the RBCs and forms the basis of the clot (Figure 23.6). Normally, blood removed from the body clots within 3 to 6 minutes.

1.   Obtain a *nonheparinized* capillary tube, a lancet, cotton balls, a triangular file, and alcohol swabs.

2.   Clean and prick the finger to produce a free flow of blood.

3.   Place one end of the capillary tube in the blood drop, and hold the opposite end at a lower level to collect the sample.

4.   Lay the capillary tube on a paper towel and record the time. _____

5.   At 30-sec intervals, make a small nick on the tube close to one end with the triangular file, and then carefully break the tube. Slowly separate the ends to see if a gellike thread of fibrin spans the gap.

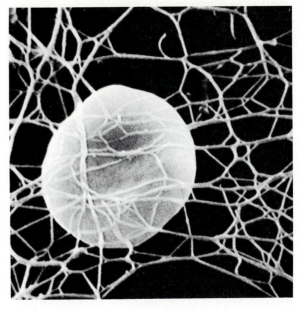

Figure 23.6

Photomicrograph of an RBC trapped in a fibrin mesh.

When this occurs, record the time. _____

Record on the data sheet the time for coagulation to occur. Are your results within the normal time range?

_____

6.   Dispose of the capillary tube and used supplies in the disposable autoclave bag.

Blood Typing

Blood typing is a system of blood classification based on the presence of specific proteins (combined with polysaccharides) on the outer surface of the RBC plasma membrane. Such proteins are called **antigens,** or **agglutinogens,** and are genetically determined. In many cases, these antigens are accompanied by other proteins found in the plasma, **antibodies** or **agglutinins,** which react with RBCs bearing different antigens, causing them to become clumped, agglutinated, and eventually hemolyzed. It is because of this phenomenon that a person's blood must be carefully typed before a whole blood or packed cell transfusion.

Several blood typing systems exist, based on the various possible antigens, but the factors usually typed for are the antigens of the ABO and Rh blood groups that are most commonly involved in transfusion reactions. Other blood factors, such as Kell, Lewis, M, and N, are not routinely typed for unless the individual is expected to require multiple transfusions. The basis of the ABO typing is shown on page 228.

ABO blood type	Antigens present on RBC membranes	Antibodies present in plasma	% of U.S. population White	Black	Asian
A	A	Anti-B	41	27	28
B	B	Anti-A	9	20	27
AB	A and B	None	3	4	5
O	Neither	Anti-A and anti-B	47	49	40

Individuals whose red blood cells carry the Rh antigen are considered to be Rh positive (approximately 85% of the U.S. population); those lacking the antigen are Rh negative. Unlike ABO blood groups neither the blood of the Rh-positive nor Rh-negative individuals carries preformed anti-Rh antibodies. This is understandable in the case of the Rh-positive individual. However, Rh-negative persons who receive transfusions of Rh-positive blood become sensitized by the Rh antigens of the donor RBCs and their systems begin to produce anti-Rh antibodies. On subsequent exposures to Rh-positive blood, typical transfusion reactions occur, resulting in the clumping and hemolysis of the donor blood cells.

1. Obtain two clean microscope slides, a wax marking pencil, anti-A, anti-B, and anti-Rh typing sera, toothpicks, lancets, alcohol swabs, and the Rh typing box.

2. Divide slide 1 into two equal halves with the wax marking pencil. Label the lower left hand corner "anti-A" and the lower right hand corner "anti-B." Mark the bottom of slide 2 "anti-Rh."

3. Place one drop of anti-A serum on the *left* side of slide 1. Place one drop of anti-B serum on the *right* side of slide 1. Place one drop of anti-Rh serum in the center of slide 2.

4. Cleanse your finger with an alcohol swab, pierce the finger with a lancet, and wipe away the first drop of blood. Obtain 3 drops of freely flowing blood, placing one drop on each side of slide 1 and a drop on slide 2.

5. Quickly mix each blood–antiserum sample with a *fresh* toothpick.

6. Place slide 2 on the Rh typing box and rock gently back and forth. (A slightly higher temperature is required for precise Rh typing than for ABO typing.)

7. After 2 minutes, observe all three blood samples for evidence of clumping. The agglutination that occurs in the positive test for the Rh factor is very fine and difficult to perceive; thus if there is any question, observe the slide under the microscope. Record your observations below:

	Observed (+)	Not observed (−)
Presence of clumping with anti-A		
Presence of clumping with anti-B		
Presence of clumping with anti-Rh		

8. Interpret your results in light of the following information: Slide 1: If clumping occurs on both sides, your ABO blood group is AB. If clumping occurs only in the blood sample mixed with anti-A serum, you are ABO type A. If clumping occurs only in the blood sample mixed with anti-B serum, you are type B. If clumping was not observed with either serum, you are type O. Slide 2: If clumping was observed, you are Rh positive. If not, you are Rh negative.

9. Record your blood type on the data sheet.

10. Put the used slides in the beaker containing bleach; put disposable supplies in the autoclave bag. Before leaving the laboratory, use a paper towel saturated with bleach solution to wash down your laboratory bench.

Hematologic Test Data Sheet

Differential WBC count:

_____% granulocytes _____% agranulocytes

_____% neutrophils _____% lymphocytes

_____% eosinophils _____% monocytes

_____% basophils

Total WBC count _____ WBCs/mm^3

Total RBC count _____ RBCs/mm^3

Hematocrit (PCV):

_____ RBC % of blood volume

_____ WBC % of blood volume ⎫ not

_____ Plasma % of blood volume ⎬ generally
reported

Hemoglobin content:

_____ %

_____ g/100 ml blood

Ratio (PCV/grams Hb per 100 ml blood: _____

Sedimentation rate _____ mm/hr

Coagulation time _____

Blood typing:

ABO group _____ Rh factor _____

Anatomy of the Heart

<table>
<tr><td>

OBJECTIVES

1. To describe the location of the heart.

2. To name and locate the major anatomical areas and structures of the heart when provided with an appropriate model, diagram, or dissected sheep heart, and to explain the function of each.

3. To trace the pathway of blood through the heart.

4. To explain why the heart is called a double pump, and to compare the pulmonary and systemic circuits.

5. To explain the operation of the atrioventricular and semilunar valves.

6. To name the functional blood supply of the heart.

7. To describe the histologic structure of cardiac muscle, and to note the importance of its intercalated disks and spiraling arrangement in the heart.

</td><td>

MATERIALS

Torso model or laboratory chart showing heart anatomy

Red and blue pencils

Preserved sheep heart, pericardial sacs intact (if possible)

Dissecting pan and instruments

Pointed glass rods for probes

Three-dimensional model of cardiac muscle

Heart model (three-dimensional)

X ray of the human thorax for observation of the position of the heart *in situ;* X ray viewing box

Compound microscope

Histological slides of cardiac muscle (longitudinal section)

</td></tr>
</table>

When they hear the term *circulatory system,* most people immediately think of the heart. People have long recognized the vital importance of the heart, and for centuries have alluded to it in song and poetry.

From a more scientific standpoint, it is important to understand why the heart and circulatory system play such a vital role in human physiology. The major function of the circulatory system is transportation. Using blood as the transport vehicle, the system carries oxygen, digested foods, cell wastes, electrolytes, and many other substances vital to the body's homeostasis to and from the body cells. The system's propulsive force is the contracting heart, which can be compared to a muscular pump equipped with one-way valves. As the heart contracts, it forces blood into a closed system of large and small plumbing tubes (blood vessels) within which the blood is confined and circulated. This exercise deals with the structure of the heart or circulatory pump. The anatomy of the blood vessels is considered separately in Exercise 25.

GROSS ANATOMY OF THE HUMAN HEART

The heart is a cone-shaped organ approximately the size of a fist and is located within the mediastinum of the thorax. It is flanked laterally by the lungs, posteriorly by the vertebral column, and anteriorly by the sternum. Its more pointed **apex** extends slightly to the left and rests on the diaphragm, approximately at the level of the fifth intercostal space. Its broader **base,** from which the great vessels emerge, lies beneath the second rib. *In situ,* the right ventricle of the heart forms most of its anterior surface.

If an X ray of a human thorax is available, verify the relationships described above.

Figure 24.1 shows two views of the heart—an external anterior view and a frontal section. As anatomical areas are described in the text, consult the figure. When you have pinpointed all the structures, observe the human heart model, and reidentify the same structures without reference to the figure.

The heart is enclosed within a double-walled fibroserous sac called the pericardium. The thin **visceral pericardium,** or **epicardium,** which is closely applied to the heart muscle, reflects downward at the base of the heart to form its companion serosa, the loosely applied **parietal pericardium,** which is attached at the apex to the diaphragm. Serous fluid produced by these serosae allows the heart to beat in a relatively frictionless environment. The parietal serous layer, in turn, lines the loosely-fitting superficial **fibrous pericardium** composed of dense connective tissue.

Inflammation of the pericardium, **pericarditis,** causes painful adhesions between the pericardial layers. These adhesions interfere with heart movements. ■

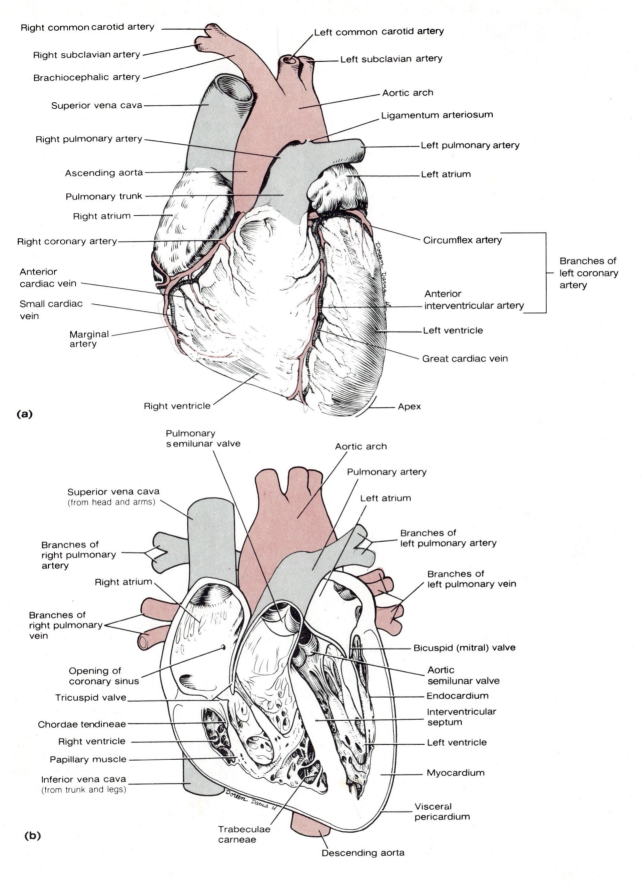

(a)

Right common carotid artery
Right subclavian artery
Brachiocephalic artery
Superior vena cava
Right pulmonary artery
Ascending aorta
Pulmonary trunk
Right atrium
Right coronary artery
Anterior cardiac vein
Small cardiac vein
Marginal artery
Right ventricle

Left common carotid artery
Left subclavian artery
Aortic arch
Ligamentum arteriosum
Left pulmonary artery
Left atrium
Circumflex artery
Branches of left coronary artery
Anterior interventricular artery
Left ventricle
Great cardiac vein
Apex

(b)

Pulmonary semilunar valve
Superior vena cava (from head and arms)
Branches of right pulmonary artery
Right atrium
Branches of right pulmonary vein
Opening of coronary sinus
Tricuspid valve
Chordae tendineae
Right ventricle
Papillary muscle
Inferior vena cava (from trunk and legs)
Trabeculae carneae

Aortic arch
Pulmonary artery
Left atrium
Branches of left pulmonary artery
Branches of left pulmonary vein
Bicuspid (mitral) valve
Aortic semilunar valve
Endocardium
Interventricular septum
Left ventricle
Myocardium
Visceral pericardium
Descending aorta

Figure 24.1

Anatomy of the human heart: (a) external anterior view; (b) frontal section.

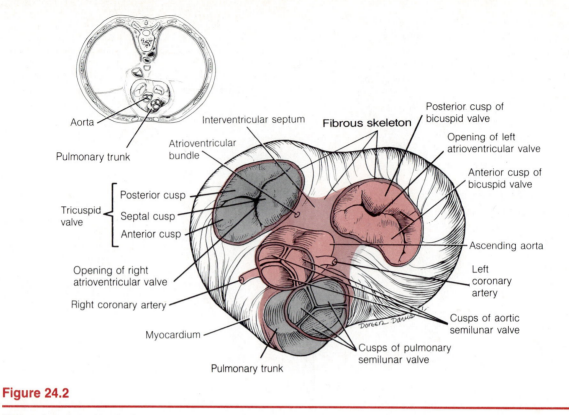

Aorta

Pulmonary trunk

Interventricular septum

Atrioventricular bundle

Fibrous skeleton

Posterior cusp of bicuspid valve

Opening of left atrioventricular valve

Anterior cusp of bicuspid valve

Posterior cusp

Septal cusp

Anterior cusp

Tricuspid valve

Opening of right atrioventricular valve

Right coronary artery

Myocardium

Pulmonary trunk

Ascending aorta

Left coronary artery

Cusps of aortic semilunar valve

Cusps of pulmonary semilunar valve

Figure 24.2

Heart valves (superior view).

The walls of the heart are composed primarily of cardiac muscle—the **myocardium**—which is reinforced internally by a dense fibrous connective tissue network. This network, the fibrous skeleton of the heart, is more elaborate and thicker in certain areas, for example, around the valves and at the base of the great vessels leaving the heart (see Figure 24.2).

Heart Chambers

The heart is divided into four chambers: two **atria** and two **ventricles,** each lined with a thin serous lining called the **endocardium.** The septum that divides the heart longitudinally is referred to as the **interatrial** or **interventricular septum,** depending on which chambers it partitions. The superiorly located atria are receiving chambers and are relatively ineffective as pumps. Blood flows into the atria under low pressure from the veins of the body. The right atrium receives relatively oxygen-poor blood from the body via the **superior** and **inferior vena cavae.** Four **pulmonary veins** deliver oxygen-rich blood from the lungs to the left atrium.

The inferior thick-walled ventricles, which form the bulk of the heart, are the discharging chambers. They force blood out of the heart into the large arteries that emerge from its base. The right ventricle pumps blood into the **pulmonary trunk,** which carries the blood to the lungs to be oxygenated. The left ventricle discharges blood into the **aorta,** from

which all systemic arteries of the body diverge to supply the body tissues. Discussions of the heart's pumping action usually refer to ventricular activity.

Pulmonary and Systemic Circulations

The heart functions as a double pump. The right side serves as the **pulmonary circulation** pump, shunting the carbon dioxide–rich blood entering its chambers to the lungs to unload carbon dioxide and pick up oxygen, and then back to the left side of the heart. The function of this circuit is strictly to provide for gas exchange. The second circuit, which carries oxygen-rich blood from the left heart through the body tissues and back to the right heart is called the **systemic circulation.** It supplies the functional blood supply to all body tissues.

 Trace the pathway of blood through the heart by adding arrows to the frontal section diagram (see Figure 24.1b). Use red arrows for the oxygen-rich blood and blue arrows for the less oxygen-rich blood.

Heart Valves

Four valves enforce a one-way blood flow through the heart chambers. The **atrioventricular valves,** located between the atrial and ventricular chambers on each side, prevent backflow into the atria when the ventricles begin to contract. The left atrioventricular valve, also called the **mitral** or **bicuspid valve,** consists of two cusps, or flaps, of endocardium. The right atrioventricular valve, the **tricuspid valve,** has three cusps (see Figure 24.2). Tiny white collagenic cords called the **chordae tendineae** (literally, tendinous cords) anchor the cusps to the ventricular walls. The chordae tendineae originate from small bundles of cardiac muscle, **papillary muscles,** that project from the myocardial wall. When blood is flowing passively into the atria and then into the ventricles during **diastole** (the period of ventricular relaxation), the atrioventricular valve flaps hang passively into the ventricular chambers. When the ventricles begin to contract (**systole**) and blood in their chambers is compressed, the intraventricular blood pressure begins to rise, causing the valve flaps to be reflected superiorly, which closes the AV valves. The chordae tendineae anchor the flaps in a closed position preventing backflow into the atria during ventricular contraction. If unanchored, the flaps would blow upward into the atria rather like an umbrella being turned inside out by a strong wind.

The second set of valves, the **semilunar valves,** each composed of three pocketlike cusps, guards the bases of the two large arteries leaving the ventricular chambers. These are referred to as the **pulmonary** and **aortic semilunar valves.** The valve cusps are forced open and flatten against the walls of artery as the ventricles discharge their blood into the large arteries during systole. However, when the ventricles relax, blood begins to flow backward toward the heart and the cusps fill with blood, closing the semilunar valves and preventing arterial blood from reentering the heart.

Cardiac Circulation

Even though the heart chambers are almost continually bathed with blood, this contained blood does not nourish the myocardium. The functional blood supply of the heart is provided by the right and left **coronary arteries,** which issue from the base of the aorta just above the aortic semilunar valve and encircle the heart in the **atrioventricular groove** at the junction of the atria and ventricles. They then ramify over the heart's surface, the right coronary artery supplying the posterior surface of the ventricles and the lateral aspect of the right side of the heart, largely through its posterior interventricular and marginal artery branches. The left coronary artery supplies the anterior ventricular walls and the laterodorsal part of the left side of the heart via its two major branches, the *anterior interventricular artery* and the *circumflex artery.* The coronary arteries and their branches are compressed during systole and fill when the heart is relaxed. The myocardium is drained by the **cardiac veins,** most of which empty into the **coronary sinus,** which in turn empties into the right atrium.

DISSECTION OF THE SHEEP HEART

Dissection of the sheep heart is valuable because of the similarity of its size and structure to the human heart. Also, a dissection experience allows you to view structures in a way not possible with models and diagrams. Refer to Figure 24.3 as you proceed with the dissection.

1.   Obtain a preserved sheep heart, a dissection tray, and dissecting instruments. Rinse the sheep heart in cold water to remove excessive preservatives and to flush out any trapped blood clots. Now you are ready to make your observations.

2.   Observe the texture of the pericardium. Also, note its point of attachment to the heart. Where is it

attached? _____

_____

3.   If the pericardial sac is still intact, slit open the parietal pericardium and cut it from its attachments. Observe the visceral pericardium (epicardium). Using a sharp scalpel, carefully pull a little of this serous membrane away from the myocardium. How does its position, thickness, and apposition to the heart differ

from those of the parietal pericardium? _____

_____

_____

_____

4.   Examine the external surface of the heart. Notice the accumulation of adipose tissue, which in many cases marks the separation of the chambers and the location of the coronary arteries that nourish the myocardium. Carefully scrape away some of the fat with a scalpel to expose the coronary blood vessels.

5.   Identify the base and apex of the heart, and then identify the two wrinkled **auricles,** earlike flaps of tissue projecting from the atrial chambers. The balance of the heart muscle is ventricular tissue. To identify the left ventricle, compress the ventricular chambers on each side of the longitudinal fissures carrying the coronary blood vessels. The side that feels thicker and more solid is the left ventricle. The right ventricle feels much thinner and somewhat flabby on compression. This difference reflects the greater demand placed on the left ventricle, which must pump blood through the much longer systemic circulation,

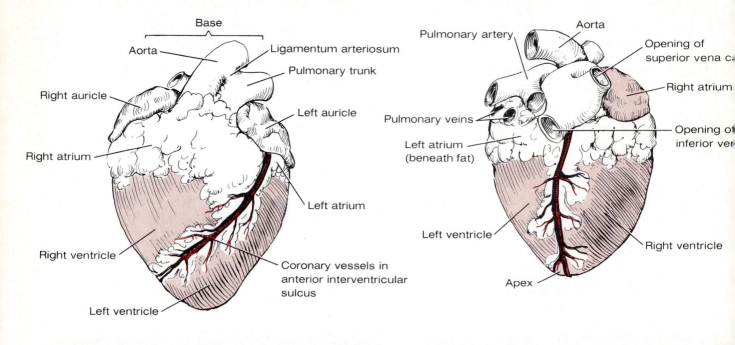

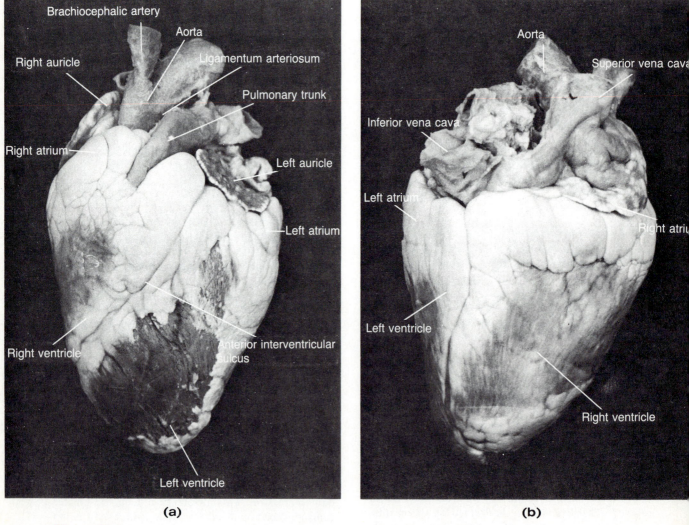

Figure 24.3

Anatomy of the sheep heart: (a) anterior view; (b) posterior view. (Photographs courtesy of Ann Allworth.)

a pathway with much higher resistance than the pulmonary circulation served by the right ventricle. Hold the heart in its anatomic position (Figure 24.3(a)), with the anterior surface uppermost. In this position the left ventricle composes the entire apex and the left side of the heart.

6. Identify the pulmonary trunk and the aorta extending from the superior aspect of the heart. The pulmonary trunk is the most anterior, and you may see its division into the right and left pulmonary arteries if it has not been cut too closely to the heart. The aorta, which is thicker walled and branches almost immediately, is located just beneath the pulmonary trunk. The first branch of the sheep aorta, the **brachiocephalic artery,** is identifiable unless the aorta has been cut immediately as it leaves the heart. The brachiocephalic artery splits to form the right carotid and subclavian arteries, which supply the right side of the head and right forelimb, respectively. Carefully clear away some of the fat between the pulmonary trunk and the aorta to expose the **ligamentum arteriosum,** a cordlike remnant of the **ductus arteriosus.** (In the fetus the ductus arteriosus allows blood to pass directly from the pulmonary trunk to the aorta, thus bypassing the nonfunctional fetal lungs.)

7. Cut through the wall of the aorta until you see the aortic semilunar valve. Identify the two openings to the coronary arteries just above the valve. Place a probe into one of these holes to see if you can follow the course of a coronary artery between the right and left ventricles.

8. Turn the heart to view its posterior surface. The heart will appear as shown in Figure 24.3(b). The right and left ventricles appear equal-sized in this view. Identify the four thin-walled pulmonary veins entering the left atrium. (It may or may not be possible to locate the pulmonary veins from this vantage point, depending on how they were cut as the heart was removed.) Identify the superior and inferior vena cavae entering the right atrium. Compare the approximate diameter of the superior vena cava with the diameter of the aorta.

Which is larger? _____

Which has thicker walls? _____

Why do you suppose these differences exist? _____

_____

_____

9. Insert a probe into the superior vena cava and use scissors to cut through its wall so that you can view the interior of the right atrium. Do not extend your cut entirely through the right atrium or into the ventricle. Observe the right atrioventricular valve.

How many flaps does it have? _____

Pour some water into the right atrium and allow it to flow into the ventricle. Slowly and gently squeeze the right ventricle to watch the closing action of this valve. (If you squeeze too vigorously, you'll get a face full of water!) Drain the water from the heart before continuing.

10. Return to the pulmonary trunk and cut through its anterior wall until you can see the pulmonary semilunar valve. Pour some water into the base of the pulmonary trunk to observe the closing action of this valve. How does its action differ from that of

the atrioventricular valve? _____

_____

_____

After observing valve action, drain the heart once again. Return to the superior vena cava, and continue the cut made in its wall through the right atrium and right atrioventricular valve into the right ventricle. Parallel the anterior border of the interventricular septum until you "round the corner" to the dorsal aspect of the heart (Figure 24.4).

11. Reflect the cut edges of the superior vena cava, right atrium, and right ventricle to obtain the view seen in Figure 24.4. Observe the comblike ridges of muscle throughout most of the right atrium. This is called **pectinate muscle** (*pectin* means comb). Identify, on the ventral atrial wall, the large opening of the inferior vena cava and follow it to its external opening with a probe. Notice that the atrial walls in the vicinity of the vena cava are smooth and lack the roughened appearance (pectinate musculature) of the other regions of the atrial walls. Just below the inferior vena caval opening, identify the opening of the **coronary sinus,** which returns venous blood of the coronary circulation to the heart.

12. Identify the papillary muscles in the right ventricle, and follow their attached chordae tendineae to the flaps of the tricuspid valve. Notice the pitted and ridged appearance (**trabeculae carneae**) of the inner ventricular muscle.

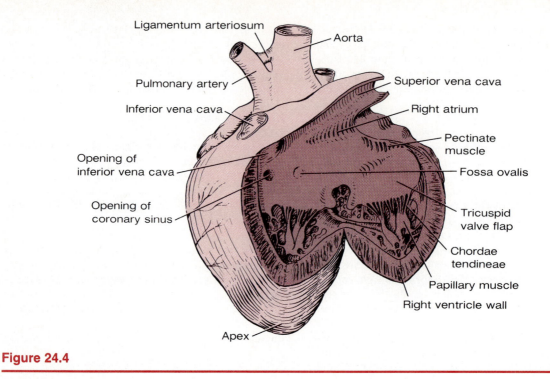

Figure 24.4

Right side of the sheep heart opened and reflected to reveal internal structures (diagrammatic view).

13. Make a longitudinal incision through the aorta and continue it into the left ventricle. Notice how much thicker the myocardium of the left ventricle is as compared to that of the right ventricle. Compare the *shape* of the left ventricular cavity to the shape

of the right ventricular cavity. _____

_____

_____

_____

Are the papillary muscles and chordae tendineae observed in the right ventricle also present in the left

ventricle? _____ Count the number of cusps in the left atrioventricular valve. How does this compare with the number seen in the right

atrioventricular valve? _____

_____

_____

_____

How do the sheep valves compare with their human

counterparts? _____

_____

14. Continue your incision from the left ventricle superiorly into the left atrium. Reflect the cut edges of the atrial wall, and attempt to locate the entry points of the pulmonary veins into the left atrium. Follow them to the heart exterior with a probe. Note how thin-walled these vessels are. Locate an oval depression, the **fossa ovalis,** in the interatrial septum. This depression marks the site of an opening in the fetal heart, the **foramen ovale,** which allows blood to pass from the right to the left atrium, thus bypassing the fetal lungs.

15. Properly dispose of the organic debris, and clean the dissecting tray and instruments.

MICROSCOPIC ANATOMY OF CARDIAC MUSCLE

Cardiac muscle is found in only one place—the heart. The heart acts as a vascular pump, propelling blood to all tissues of the body; cardiac muscle is thus very important to life. Cardiac muscle is involuntary, thus ensuring a constant blood supply.

The cardiac cells, only sparingly invested in connective tissue, are arranged in spiral or figure-8 shaped bundles (Figure 24.5). When the heart contracts, its internal chambers become smaller (or are temporarily obliterated), forcing the blood into the large arteries leaving the heart.

1.   Observe the three-dimensional model of cardiac muscle, noting its branching cells and the areas where the cells interdigitate, the **intercalated discs.** These two structural features provide a continuity to cardiac muscle not seen in other muscle tissues and allow close coordination of heart activity.

2.   Note the similarities and differences between cardiac muscle and skeletal muscle.

3.   Obtain and observe a longitudinal section of cardiac muscle under high power, and draw a small section of the tissue in the space provided below. Label the nucleus, striations, intercalated disks, and sarcolemma. Compare your observations to Figure 24.6.

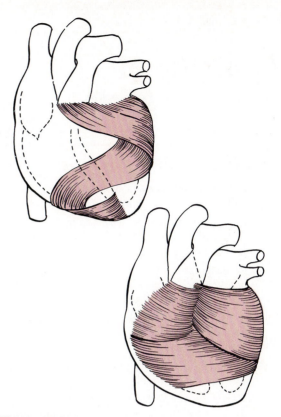

Figure 24.5

Longitudinal view of the heart chambers showing the spiral arrangement of the cardiac muscle fibers.

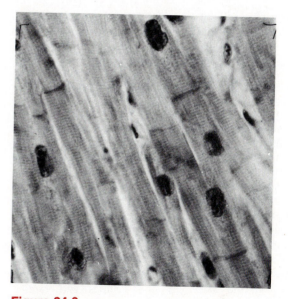

Figure 24.6

Photomicrograph of cardiac muscle (160X).

Anatomy of Blood Vessels

OBJECTIVES

1. To describe the tunics of arterial and venous walls and state the function of each layer.

2. To correlate differences observed in artery, vein, and capillary structures with the functions these vessels perform.

3. To recognize a cross-sectional view of an artery and vein when provided with a microscopic view or appropriate diagram.

4. To list and/or identify the major arteries arising from the aorta, and to indicate the body region supplied by each.

5. To list and/or identify the major veins draining into the superior and inferior vena cavae, and to indicate the body regions drained.

6. To point out and/or discuss the unique features of special circulations in the body:

   - The pattern of blood flow in the hepatic portal system and its importance
   - The vascular supply of the brain and the importance of the circle of Willis
   - Components of the pulmonary circulation
   - Structures unique to the fetal circulation and the importance of each

7. To point out anatomic differences between the vascular system of the human and the laboratory dissection specimen.

MATERIALS

Anatomical charts of human arteries and veins (or a three-dimensional model of the human circulatory system)

Anatomical charts of the following specialized circulations: pulmonary circulation, hepatic portal circulation, arterial supply and circle of Willis of the brain (or a brain model showing this circulation), fetal circulation

Compound microscope

Prepared microscope slides showing cross sections of an artery and vein

Dissection animals

Dissecting pans and instruments

Blood circulates within the blood vessels, which constitute a closed transport system. As the heart contracts, blood is propelled into the large arteries leaving the heart. It moves into successively smaller arteries and then to the arterioles, which feed the capillary beds in the tissues. Capillary beds are drained by the venules, which in turn empty into veins that ultimately converge on the great veins entering the heart. Thus arteries, carrying blood away from the heart, and veins, which drain the tissues and return blood to the heart, function simply as conducting vessels or conduits. Only the tiny capillaries that connect the arterioles and venules and ramify throughout the tissues directly serve the needs of the body's cells. It is through the capillary walls that exchanges between tissue cells and blood occur.

Respiratory gases, nutrients, and wastes move along diffusion gradients; thus, oxygen and nutrients diffuse from the blood to the tissue cells, and carbon dioxide and metabolic wastes move from the cells to the blood.

In this exercise you will examine the microscopic structure of blood vessels and will identify the major arteries and veins of the systemic circulation and other special circulations.

MICROSCOPIC STRUCTURE OF THE BLOOD VESSELS

Except for the microscopic capillaries, the walls of blood vessels are constructed of three coats, or tunics (Figure 25.1). The **tunica intima,** or **interna,** which lines the lumen of a vessel, is a single thin layer of endothelium (squamous cells underlain by a scant basal lamina) that is continuous with the endocardium of the heart. Its cells fit closely together, forming an extremely smooth blood vessel lining that helps to decrease resistance to blood flow.

The **tunica media** is the more bulky middle coat and is composed primarily of smooth muscle and elastic tissue. The smooth muscle, under the control

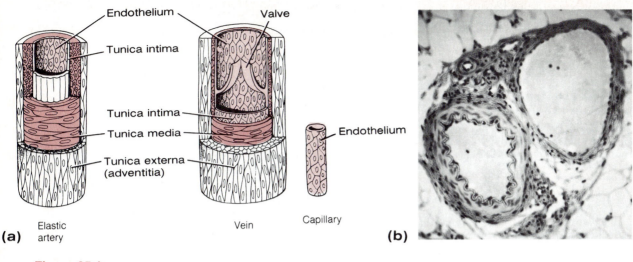

(a)   Elastic artery   Vein   Capillary   **(b)**

Figure 25.1

Structure of arteries, veins, and capillaries. Photomicrograph shows a small artery (left) and vein (right), cross-sectional view (160X).

of the sympathetic nervous system, plays an active role in reducing or increasing the diameter of the vessel, which in turn increases or decreases the peripheral resistance and blood pressure.

The **tunica externa,** or **adventitia,** the outermost tunic, is composed of areolar or fibrous connective tissue. Its function is basically supportive and protective.

In general, the walls of arteries are much thicker than those of veins. The tunica media in particular tends to be considerably heavier and contains substantially more smooth muscle and elastic tissue. This anatomic difference reflects a functional difference in the two types of vessels. Arteries, which are closer to the pumping action of the heart, must be able to expand as an increased volume of blood is propelled into them and then recoil passively as the blood flows off into the circulation during diastole. Their walls must be sufficiently strong and resilient to withstand such pressure fluctuations. Since these larger arteries have such large amounts of elastic tissue in their media, they are often referred to as *elastic arteries*. Smaller arteries, further along in the circulatory pathway, are exposed to less extreme pressure fluctuations. They have somewhat less elastic tissue but still have substantial amounts of smooth muscle in their media. For this reason, they are called *muscular arteries*.

In contrast, the veins, which are far removed from the heart in the circulatory pathway, are not subjected to such pressure fluctuations and are essentially low-pressure vessels. Thus, veins may be thinner-walled without jeopardy. However, the low-pressure condition itself requires structural modifications to ensure that venous return equals cardiac output; thus, the lumens of veins tend to be substantially larger than those of corresponding arteries.

Because blood returning to the heart often flows against gravity, there are other aids to venous return.

Valves in the larger veins function to prevent backflow of blood in much the same manner as the semilunar valves of the heart. Skeletal muscle activity also promotes venous return; as the skeletal muscles surrounding the veins contract and relax, the blood is "milked" through the veins toward the heart. (Anyone who has had to stand relatively still for an extended time will be happy to show you their swollen ankles, caused by blood pooling in their feet during the period of muscle inactivity!) Finally, pressure changes that occur in the thorax during breathing also facilitate the return of blood to the heart.

To demonstrate the efficiency of the venous valves in preventing backflow of blood, perform the following simple experiment. Allow one hand to hang by your side until the blood vessels on the dorsal aspect become distended. Place two fingertips against one of the distended veins and, pressing firmly, move the superior finger proximally along the vein and then release this finger. The vein will remain flattened and collapsed despite gravity. Then remove the distal fingertip and observe the rapid filling of the vein.

The transparent walls of the tiny capillaries are only one cell layer thick, consisting of just the endothelium or tunica intima. Because of this exceptional thinness, exchanges are easily made between the blood and tissue cells.

1.   Obtain a cross-sectional view preparation of blood vessels and a microscope.

2.   Scan the section to identify a thick-walled artery. Very often, but not always, its lumen will appear scalloped due to the constriction of its walls by the elastic tissue of the media.

3.   Identify a vein. Its lumen may appear elongated or irregularly shaped and collapsed, and its walls will be considerably thinner. Note the difference in the relative amount of elastic fibers in the media of the two vessels. Also, note the thinness of the intima layer, which is composed of flat squamous-type cells.

4.   Make a drawing of your observations of the two vessel types below, and label the tunics. Try to indicate the proper size relationships relative to wall thickness and the tunic widths.

Artery

Vein

MAJOR SYSTEMIC ARTERIES OF THE BODY

The aorta is the largest artery of the body. Extending upward as the ascending aorta from the left ventricle, it arches posteriorly and to the left (aortic arch) and courses downward as the descending aorta through the thoracic cavity. It penetrates the diaphragm to enter the abdominal cavity just anterior to the vertebral column.

Figure 25.2 depicts the course of the aorta and its major branches. As you locate the arteries on the figure, be aware of ways in which you can make your memorization task easier—in many cases the name of the artery reflects the body region traversed (axillary, subclavian, brachial, popliteal), the organ served (renal, hepatic), or the bone followed (tibial, femoral, radial, ulnar). Once you have identified these arteries on the figure, attempt to locate and name them (without a reference) on a large anatomic chart or a three-dimensional model of the circulatory system vessels. (All arteries described here are shown in the figure, but some that are not described are also named and illustrated. Ask your instructor which arteries you are required to identify.)

Ascending Aorta

The only branches of the ascending aorta are the **right** and the **left coronary arteries,** which supply the myocardium. The coronary arteries are described in Exercise 24 in conjunction with heart anatomy.

Aortic Arch

The **brachiocephalic** ("arm-head") **artery** is the first branch of the aortic arch. It persists briefly before dividing into the right **common carotid artery** and the right **subclavian artery.** The common carotid divides to form the **internal carotid artery,** which serves the brain, and the **external carotid artery,** which supplies the extracranial tissues of the neck and head. The subclavian artery gives off three branches to the head and neck, the most important being the **vertebral artery,** which runs up the posterior neck to supply a portion of the brain. In the axillary region, the subclavian artery becomes the **axillary artery** and then the **brachial artery** as it enters the arm. At the elbow, the brachial artery divides into the **radial** and **ulnar arteries,** which follow the same-named bones to supply the forearm and hand.

The **left common carotid artery,** the second branch of the aortic arch, supplies the left side of the head and neck in the same manner the right common carotid serves the right side. The third branch is the **left subclavian artery,** which supplies the left upper extremity and subdivides as described for the right subclavian artery.

Descending Aorta

Not shown in Figure 25.2 are the 9 or 10 pairs of *intercostal arteries* that supply the muscles of the thoracic wall and are small branches of the descending aorta, as are the *phrenic arteries,* which supply the diaphragm. Other more major branches of the descending aorta supply the abdominal region. The **celiac trunk** is an unpaired artery that subdivides into three branches: the **left gastric** artery supplying the stomach, the **splenic artery** supplying the spleen, and the **common hepatic artery,** which provides the functional blood supply of the liver. The largest branch of the descending aorta, the **superior mesenteric artery,** supplies most of the small intestine and the first half of the large intestine. The small paired **suprarenal arteries** emerge at approximately the same level as the superior mesenteric artery and run laterally to supply the adrenal glands. (These are not shown in Figure 25.2.) The paired **renal arteries** supply the kidneys, and the **gonadal arteries,** arising from the ventral surface of the aorta slightly below the renal arteries, run inferiorly to serve the gonads. They are called **ovarian arteries** in the female and **testicular** (or **internal spermatic**) **arteries** in the male. Since these vessels must travel through the inguinal canal to supply the testes in the male, they are considerably longer in the male than in the female. The small unpaired artery supplying the second half of the large intestine is the **inferior mesenteric artery.**

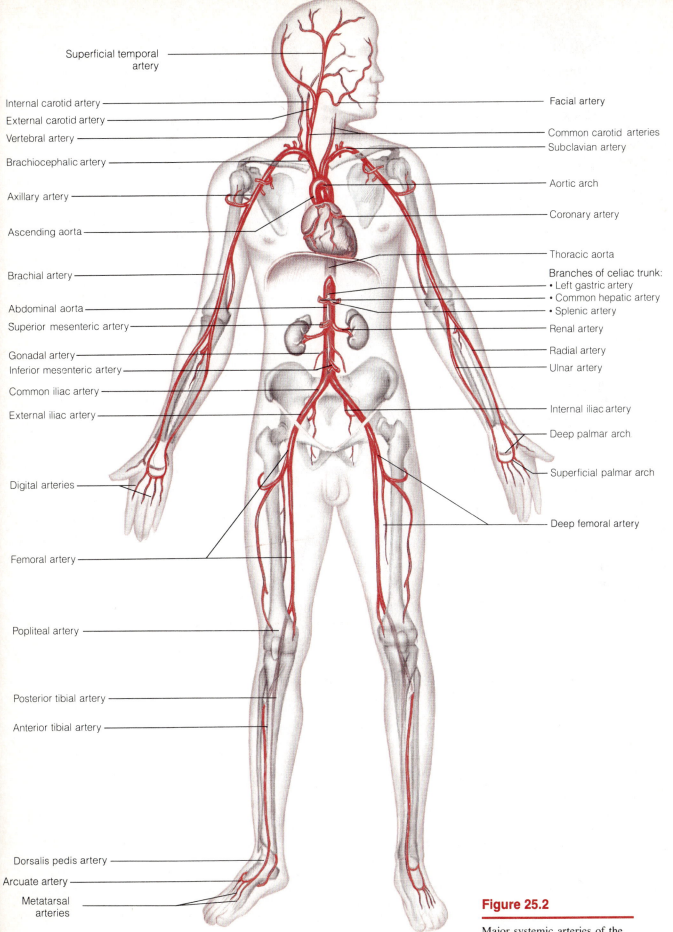

Superficial temporal artery

Internal carotid artery

External carotid artery

Vertebral artery

Brachiocephalic artery

Axillary artery

Ascending aorta

Brachial artery

Abdominal aorta

Superior mesenteric artery

Gonadal artery

Inferior mesenteric artery

Common iliac artery

External iliac artery

Digital arteries

Femoral artery

Popliteal artery

Posterior tibial artery

Anterior tibial artery

Dorsalis pedis artery

Arcuate artery

Metatarsal arteries

Facial artery

Common carotid arteries

Subclavian artery

Aortic arch

Coronary artery

Thoracic aorta

Branches of celiac trunk:
• Left gastric artery
• Common hepatic artery
• Splenic artery

Renal artery

Radial artery

Ulnar artery

Internal iliac artery

Deep palmar arch

Superficial palmar arch

Deep femoral artery

Figure 25.2

Major systemic arteries of the body.

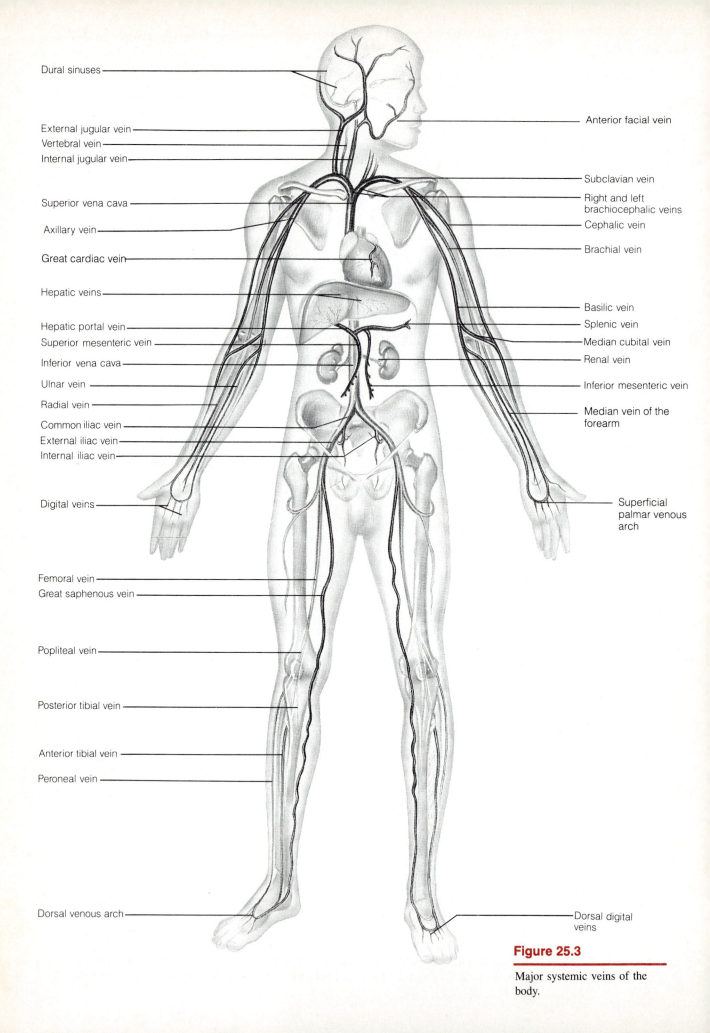

Dural sinuses

External jugular vein
Vertebral vein
Internal jugular vein

Superior vena cava

Axillary vein

Great cardiac vein

Hepatic veins

Hepatic portal vein
Superior mesenteric vein

Inferior vena cava

Ulnar vein

Radial vein

Common iliac vein

External iliac vein
Internal iliac vein

Digital veins

Femoral vein
Great saphenous vein

Popliteal vein

Posterior tibial vein

Anterior tibial vein

Peroneal vein

Dorsal venous arch

Anterior facial vein

Subclavian vein

Right and left
brachiocephalic veins

Cephalic vein

Brachial vein

Basilic vein
Splenic vein
Median cubital vein

Renal vein

Inferior mesenteric vein

Median vein of the
forearm

Superficial
palmar venous
arch

Dorsal digital
veins

Figure 25.3

Major systemic veins of the
body.

In the pelvic region, the descending aorta divides into the two large **common iliac arteries.** Each of these vessels extends for about 2 inches into the pelvis before it divides into the **internal iliac artery,** which supplies the pelvic organs (bladder, rectum, and some reproductive structures) and the **external iliac artery,** which continues into the thigh, where its name changes to the **femoral artery.** A branch of the femoral artery, the **deep femoral artery,** supplies the posterior thigh region. In the knee region, the femoral artery briefly becomes the **popliteal artery;** its subdivisions—the **anterior** and **posterior tibial arteries**—supply the leg, ankle, and foot. The posterior tibial gives off one main branch, the **peroneal artery** (not shown), which supplies the lateral calf (peroneal muscles). The anterior tibial artery terminates at the **dorsalis pedis** artery, which supplies the dorsum of the foot and continues on as the **arcuate artery.** The dorsalis pedis is often palpated in patients with circulation problems of the leg to determine the circulatory efficiency to the limb as a whole.

MAJOR SYSTEMIC VEINS OF THE BODY

Arteries are generally located in deep, well-protected body areas. However, many veins follow a more superficial course and are often easily visualized and palpated on the body surface (Figure 25.3). Most deep veins parallel the course of the major arteries; thus in many cases the naming of the veins and arteries is identical except for the designation of the vessels as veins. Whereas the major systemic arteries branch off the aorta, the veins tend to converge on the vena cavae, which enter the right atrium of the heart. Veins draining the head and upper extremities empty into the **superior vena cava,** and those draining the lower body empty into the **inferior vena cava.**

Veins Draining into the Superior Vena Cava

Veins draining into the superior vena cava are named from the superior vena cava distally; **but remember that the flow of blood is in the opposite direction.**

The **right** and **left brachiocephalic veins** drain the head, neck, and upper extremities and unite to form the superior vena cava. (Note that although there is only one brachiocephalic artery, there are two brachiocephalic veins.) Branches of the brachiocephalic veins include the **internal jugular veins,** large veins that drain the dural sinuses of the brain; the **vertebral veins,** which drain the posterior aspect of the head; and the **subclavian veins,** which receive the venous blood from the upper extremity. The **external jugular vein** joins the subclavian vein near its origin to return the venous drainage of the extracranial tissues of the head and neck. As the subclavian vein traverses the axilla, it becomes the **axillary vein** and then the **brachial vein** as it courses along the posterior aspect of

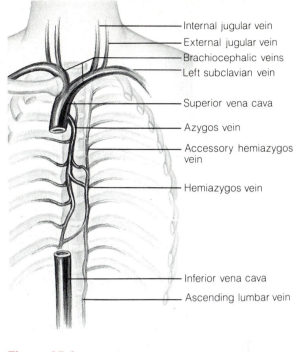

Internal jugular vein
External jugular vein
Brachiocephalic veins
Left subclavian vein
Superior vena cava
Azygos vein
Accessory hemiazygos vein
Hemiazygos vein
Inferior vena cava
Ascending lumbar vein

Figure 25.4

The azygos system.

the humerus. The brachial vein is formed by the union of the deep **radial** and **ulnar veins** of the forearm. The superficially located venous drainage of the arm includes the **cephalic vein,** which courses along the lateral aspect of the arm and empties into the axillary vein; the **basilic vein,** found on the medial aspect of the arm and entering the brachial vein; and the **median cubital vein,** which runs between the cephalic and basilic veins in the anterior aspect of the elbow (this vein is often the site of choice for removing blood for testing purposes).

The **azygos vein,** part of the *azygos system* that drains the intercostal muscles of the thorax and provides an accessory venous system to drain the abdominal wall, enters the dorsal aspect of the superior vena cava immediately before it enters the right atrium. The azygos system, depicted in Figure 25.4, also includes the **hemiazygos** and **accessory hemiazygos veins,** which together drain the left aspect of the thorax and empty into the azygos vein. The azygos vein drains the right aspect of the thorax.

Veins Draining into the Inferior Vena Cava

The inferior vena cava, a much longer vessel than the superior vena cava, returns blood to the heart from all body regions below the diaphragm (see Figure 25.3). It begins in the lower abdominal region with the union of the paired **common iliac veins,** which drain venous blood from the legs and pelvis. Each common iliac vein in turn is formed by the union of the **internal iliac vein,** draining the pelvis, and the

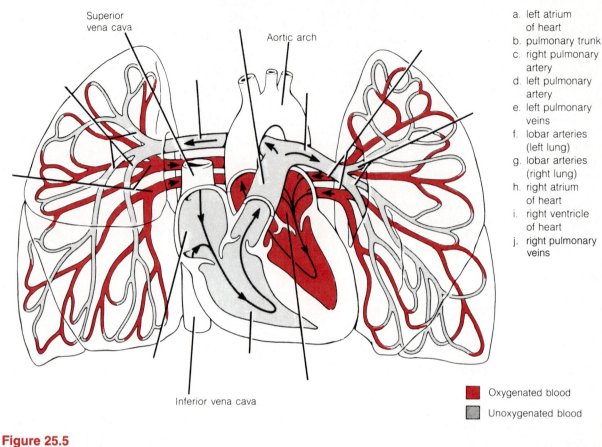

a. left atrium
   of heart
b. pulmonary trunk
c. right pulmonary
   artery
d. left pulmonary
   artery
e. left pulmonary
   veins
f. lobar arteries
   (left lung)
g. lobar arteries
   (right lung)
h. right atrium
   of heart
i. right ventricle
   of heart
j. right pulmonary
   veins

Superior vena cava

Aortic arch

Inferior vena cava

■ Oxygenated blood

■ Unoxygenated blood

Figure 25.5

The pulmonary circulation.

external iliac vein, which receives venous blood from the leg. Veins of the leg include the **anterior** and **posterior tibial veins,** which serve the calf and foot. The posterior tibial vein becomes the **popliteal vein** in the knee region and the **femoral vein** in the thigh. The femoral vein empties into the external iliac vein in the inguinal region. The **great saphenous vein,** a superficial vein, is the longest vein of the body. Beginning in the foot with the **dorsal venous arch,** it extends up the medial side of the leg, knee, and thigh to empty into the femoral vein. Moving superiorly into the abdominal cavity, the inferior vena cava receives blood from the **right gonadal vein** (testicular or spermatic vein in the male; ovarian vein in the female), which drains the right gonad. (The left testicular or ovarian vein drains into the left renal vein.) The **right** and **left renal veins** drain the kidneys, and the **right** and **left hepatic veins** drain the liver. The unpaired veins draining the digestive tract organs empty into a special vessel, the **hepatic portal vein,** which carries this blood through the liver before it enters the systemic venous system. (The hepatic portal system is discussed separately on page 246.)

Identify the important arteries and veins on the large anatomic chart or model without referring to the figures.

SPECIAL CIRCULATIONS

Pulmonary Circulation

The pulmonary circulation (discussed previously in relation to heart anatomy on page 232) differs in many ways from the systemic circulation, since it does not serve the metabolic needs of the body tissues with which it is associated (in this case, lung tissue). It functions instead to bring the blood into close contact with the alveoli of the lungs to permit gaseous exchanges that rid the blood of excess carbon dioxide and replenish its supply of vital oxygen. The arteries of the pulmonary circulation are structurally much like veins; thus they create a low pressure bed in the lungs. (If the arterial pressure in the

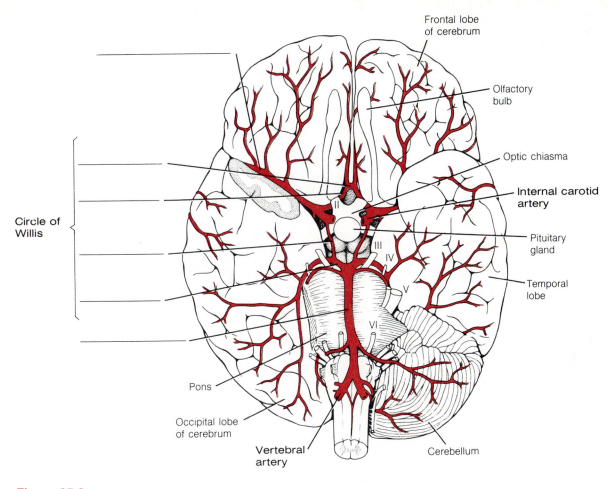

Frontal lobe
of cerebrum

Olfactory
bulb

Optic chiasma

Internal carotid
artery

Pituitary
gland

Temporal
lobe

Circle of
Willis

Pons

Occipital lobe
of cerebrum

Vertebral
artery

Cerebellum

Figure 25.6

Arterial supply of the brain. (The cerebellum is not shown on the left side of the diagram.)

systemic circulation is 120/80, the pressure in the pulmonary artery is likely to be approximately 70/10.) The functional blood supply of the lungs is provided by the **bronchial arteries,** which diverge from the thoracic portion of the descending aorta.

Pulmonary circulation begins with the large **pulmonary trunk,** which leaves the right ventricle and divides into the **right** and **left pulmonary arteries** about 2 inches above its origin (Figure 25.5). The right and left pulmonary arteries plunge into the lungs, where they subdivide into the **lobar arteries** (three on the right and two on the left), which accompany the main bronchi into the lobes of the lungs. Within the lungs, the lobar arteries branch extensively to form the arterioles, which finally terminate in the capillary networks surrounding the alveolar sacs of the lungs. Diffusion of the respiratory gases occurs across the walls of the alveoli and **pulmonary capillaries.** The pulmonary capillary beds are drained by venules, which converge to form sequentially larger veins and finally the four **pulmonary veins** (two leaving each lung), which return the blood to the left atrium of the heart.

Using the terms provided in Figure 25.5, *label* all structures provided with leader lines.

Arterial Supply of the Brain and the Circle of Willis

A continuous blood supply to the brain is crucial, since deprivation for even a few minutes causes irreparable damage to the delicate brain tissue. The brain is supplied by two pairs of arteries arising from the region of the aortic arch—the **internal carotid arteries** and the **vertebral arteries.** Figure 25.6 is a diagram of the brain's arterial supply. The internal carotid and vertebral arteries are labeled. As you read the description of the blood supply below, <u>complete the labeling of this diagram.</u>

The internal carotid arteries, branches of the common carotid arteries, follow a deep course through the neck and along the pharynx, entering the skull through the carotid canals of the temporal bone. Within the cranium, each divides into the **anterior** and **middle cerebral** arteries, which supply the bulk of the cerebrum. The internal carotid arteries also contrib-

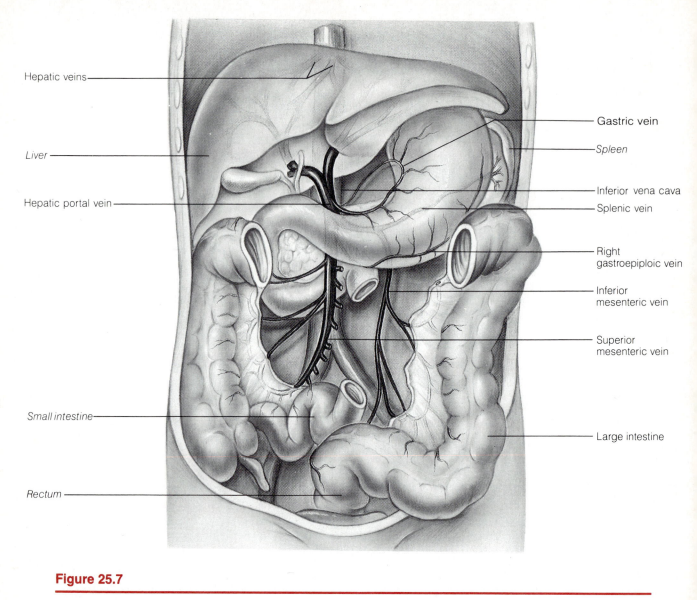

Hepatic veins

Liver

Hepatic portal vein

Small intestine

Rectum

Gastric vein

Spleen

Inferior vena cava

Splenic vein

Right gastroepiploic vein

Inferior mesenteric vein

Superior mesenteric vein

Large intestine

Figure 25.7

Hepatic portal circulation of the human.

ute to the formation of the **circle of Willis,** an arterial anastomosis at the base of the brain surrounding the pituitary gland and the optic chiasma, by forming a **posterior communicating artery** on each side. The circle is completed by the **anterior communicating artery,** a short shunt connecting the right and left anterior cerebral arteries.

The paired vertebral arteries diverge from the subclavian arteries and pass superiorly through the foramina of the transverse process of the cervical vertebrae to enter the skull through the foramen magnum. Within the skull, the vertebral arteries unite to form a single **basilar artery,** which continues superiorly along the ventral aspect of the brain stem, giving off branches to the pons, cerebellum, and inner ear. At the base of the cerebrum, the artery divides to form the **posterior cerebral arteries,** which supply portions of the temporal and occipital lobes of the cerebrum and also become part of the circle of Willis

by joining with the posterior communicating arteries.

The uniting of the blood supply of the internal carotid arteries and the vertebral arteries via the circle of Willis is a protective device that theoretically provides an alternate set of pathways for blood to reach the brain tissue in the case of arterial occlusion or impaired blood flow anywhere in the system. (In actuality, the communicating arteries are tiny, and in many cases the communicating system is defective.)

Hepatic Portal Circulation

Blood vessels of the hepatic portal circulation drain the digestive viscera, spleen, and pancreas and deliver this blood to the liver for processing via the **hepatic portal vein.** If a meal has recently been eaten, the hepatic portal blood will contain a high concentration of nutrient substances. Since the liver is a key body organ involved in maintenance of proper sugar,

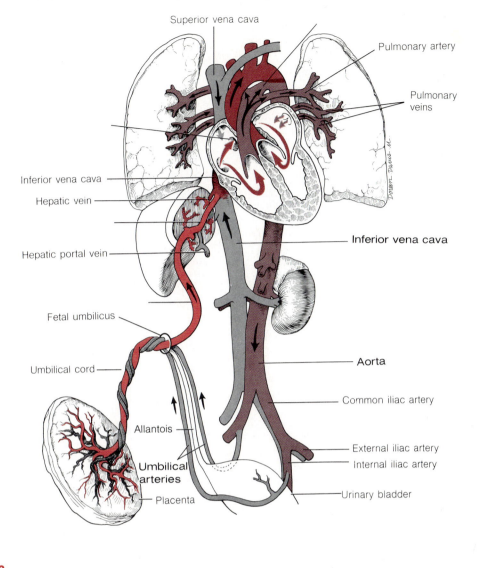

Figure 25.8

The fetal circulation.

fatty acid, and amino acid concentrations in the blood, this system ensures that these substances pass through the liver before entering the systemic circulation. As blood percolates through the liver sinusoids, some of the nutrients are removed to be stored or processed in various ways for release to the general circulation. The liver in turn is drained by the hepatic veins that enter the inferior vena cava.

The **inferior mesenteric vein,** draining the transverse and terminal portions of the large intestine, drains into the **splenic vein,** which drains the spleen, pancreas, and greater curvature of the stomach. The splenic vein and the **superior mesenteric vein,** which receives blood from the small intestine and the ascending colon, join to form the hepatic portal vein. The **gastric vein,** which drains the lesser curvature of the stomach, drains directly into the hepatic portal vein.

Identify the vessels named above on Figure 25.7.

Fetal Circulation

In a developing fetus, the lungs and digestive system are not yet functional, and all nutrient, excretory, and gaseous exchanges must occur through the placenta (see Figure 25.8). Thus nutrients and oxygen move across placental barriers from the mother's blood into fetal blood, and carbon dioxide and other metabolic wastes move from the fetal blood supply to the mother's blood.

Fetal blood travels through the umbilical cord which contains three blood vessels: two smaller **umbilical arteries** and one large **umbilical vein.** The umbilical vein carries blood rich in nutrients and oxygen to the fetus; the umbilical arteries carry carbon dioxide and waste-laden blood from the fetus to the placenta. The umbilical arteries meet the umbilical vein at the umbilicus (navel, or belly button) and wrap around the vein within the cord en route to their placental attachments. Newly oxygenated blood flows in the umbilical vein superiorly toward the fetal heart. En route, some of this blood perfuses the liver, but a larger proportion is ducted through the relatively nonfunctional liver via a vessel called the **ductus venosus** to enter the inferior vena cava, which carries the blood to the right atrium of the heart.

Since fetal lungs are nonfunctional and collapsed, two shunting mechanisms ensure that the lungs are almost entirely bypassed. Much of the blood entering the right atrium is shunted into the left atrium through a flaplike opening in the interatrial septum— the **foramen ovale.** The left ventricle then pumps the blood out the aorta to the systemic circulation. Blood that does enter the right ventricle and is pumped out of the pulmonary trunk encounters a second shunt, the **ductus arteriosus,** a short vessel connecting the pulmonary trunk and the aorta. Because the collapsed lungs present an extremely high-resistance pathway, blood more readily enters the systemic circulation through the ductus arteriosus.

The aorta carries blood to the tissues of the body; this blood ultimately finds its way back to the placenta via the umbilical arteries. The only fetal vessel that carries highly oxygenated blood is the umbilical vein; all other vessels contain varying degrees of oxygenated and deoxygenated blood.

At birth, or shortly after, the foramen ovale closes and becomes the **fossa ovalis,** and the ductus arteriosus collapses and is converted to the fibrous **ligamentum arteriosum.** Lack of blood flow through the umbilical vessels leads to their eventual obliteration, and the circulatory pattern becomes that of the adult. Remnants of the umbilical arteries exist as the **medial umbilical ligaments** on the inner surface of the anterior abdominal wall, of the umbilical vein as the **ligamentum teres** of the liver, and of the ductus venosus as a fibrous band called the **ligamentum venosus** on the inferior surface of the liver.

The pathway of fetal blood flow is indicated with arrows on Figure 25.8. Appropriately label all specialized fetal circulatory structures provided with leader lines.

DISSECTION OF THE BLOOD VESSELS OF THE CAT

Opening the Ventral Body Cavity

1.   Using scissors, make a longitudinal incision through the ventral body wall beginning just superiorly to the midline of the pubic bone. Continue the cut anteriorly to the rib cage.

2.   Angle the scissors slightly (½ inch) to the right or left of the sternum, and continue the cut through the rib cartilages, just lateral to the body midline, to the base of the throat.

3.   Make two lateral cuts on either side of the ventral body surface, anterior and posterior to the diaphragm. Leave the diaphragm intact. Spread the thoracic walls laterally to expose the thoracic organs.

Preliminary Organ Identification

A helpful prelude to identifying and tracing the blood supply of the various organs of the cat is a preliminary identification of ventral body cavity organs shown in Figure 25.9. Since you will study the organ systems contained in the ventral cavity in later units, the objective here is simply to identify the most important organs. Using Figure 25.9 as a guide, identify the following body cavity organs:

THORACIC CAVITY ORGANS

Heart: in the mediastinum enclosed by the pericardium
Lungs: flanking the heart
Thymus: superior to and partially covering the heart

ABDOMINAL CAVITY ORGANS

Liver: posterior to the diaphragm

Lift the large, drapelike, fat-infiltrated greater omentum covering the abdominal organs to expose the following:

Stomach: dorsally located and to the left side of the liver
Spleen: a flattened, brown organ curving around the lateral aspect of the stomach
Small intestine: continuing posteriorly from the stomach
Large intestine: taking a U-shaped course around the small intestine and terminating in the rectum

Blood Vessels of the Body Cavity and Lower Extremities

1.   Carefully clear away any thymus tissue or fat obscuring the heart and the large vessels associated with the heart. Before identifying the blood vessels, try to locate the phrenic nerve (from the cervical plexus), which innervates the diaphragm. The phrenic nerves are ventral to the root of the lung on each side, passing to the diaphragm. Also attempt to locate the vagus nerve (cranial nerve X) passing laterally along the trachea and dorsal to the root of the lung.

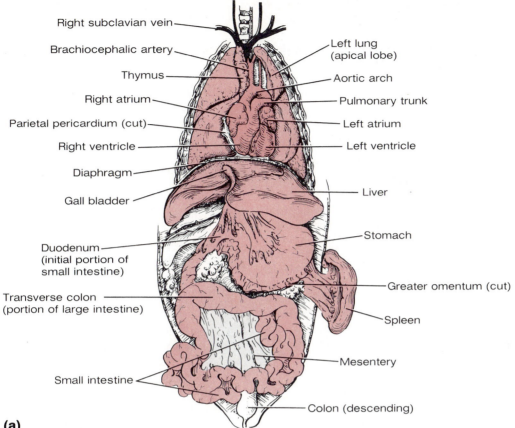

Right subclavian vein

Brachiocephalic artery

Thymus

Right atrium

Parietal pericardium (cut)

Right ventricle

Diaphragm

Gall bladder

Duodenum
(initial portion of
small intestine)

Transverse colon
(portion of large intestine)

Small intestine

Left lung
(apical lobe)

Aortic arch

Pulmonary trunk

Left atrium

Left ventricle

Liver

Stomach

Greater omentum (cut)

Spleen

Mesentery

Colon (descending)

(a)

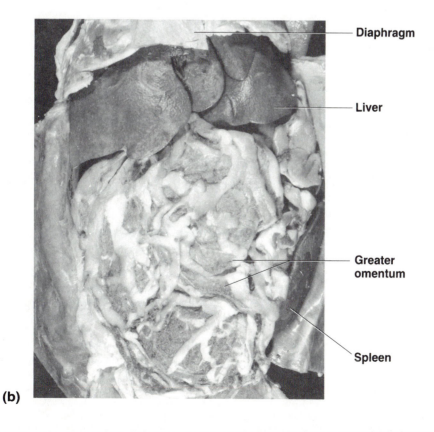

Diaphragm

Liver

Greater
omentum

Spleen

(b)

Figure 25.9

Ventral body cavity organs of the cat: (a) diagrammatic view, greater omentum removed; (b) photograph of abdominal cavity with greater omentum intact and in its normal position.

2.   Slit the parietal pericardium and reflect it superiorly; then cut it away from its heart attachments. Review the structures of the heart. Note its pointed inferior end (apex) and its broader superior portion. Identify the two atria, which appear darker than the inferior ventricles. Identify the coronary arteries in the sulcus on the ventral surface of the heart; these should be injected with red latex. (As an aid to blood vessel identification, the arteries of laboratory dissection specimens are injected with red latex; the veins are injected with blue latex. Exceptions to this will be noted as they are encountered.)

3.   Identify the two large vena cavae—the superior and inferior vena cavae—entering the right atrium. The superior vena cava is the largest dark-colored vessel entering the base of the heart. (These vessels are called the **precava** and **postcava,** respectively, in the cat.) The caval veins drain the same relative body areas as in humans. Also identify the pulmonary trunk (usually injected with blue latex) extending anteriorly from the right ventricle and the right and left pulmonary arteries. Trace the pulmonary arteries until they enter the lungs. Locate the pulmonary veins entering the left atrium and the ascending aorta arising from the left ventricle and running dorsally to the precava and to the left of the body midline.

VENOUS SUPPLY   Refer to Figure 25.10 and to Color Plate E as you study the venous system of the cat. Keep in mind that the vessels are named for the region drained, not for the point of union with other veins. (As you continue with the dissection, note that not all vessels shown on Figure 25.10 are discussed.)

1.   Re-identify the postcaval vein, and trace it to its passage through the diaphragm. Note as you follow its course that the intercostal veins drain into a much smaller vein lying dorsal to the postcava, the azygos vein, which drains most of the thoracic wall and empties into the precaval vein. Trace the azygos vein to its union with the precava just anterior to its point of entry into the right atrium.

2.   Attempt to identify the hepatic veins entering the postcava from the liver. These may be seen if some of the anterior liver tissue is scraped away where the postcava enters the liver.

3.   Displace the intestines to the left side of the body cavity, and proceed posteriorly to identify the following veins in order. All of these veins empty into the postcava and drain the organs served by the same-named arteries. In the cat, variations in the connections of the veins to be localized are common, and in some cases the postcaval vein may be double below the level of the renal veins. If you observe deviations, call them to the attention of your instructor.

Adrenolumbar veins: from the adrenal glands and body wall
Renal veins: from the kidneys (It is common to find two renal veins on the right side.)
Genital veins (testicular or **ovarian veins):** the left vein of this venous pair enters the left renal vein anteriorly
Iliolumbar veins: drain muscles of the back
Common iliac veins: form the postcava by their union

The common iliac veins are formed in turn by the union of the internal iliac and external iliac veins. The more medial internal iliac veins receive branches from the pelvic organs and gluteal region whereas the external iliac vein receives venous drainage from the lower extremity. As the external iliac vein enters the thigh by running beneath the inguinal ligament, it receives the deep femoral vein, which supplies the thigh and the external genital region and then becomes the femoral vein, which receives blood from the thigh, leg, and foot. Follow the femoral vein down the thigh to identify the great saphenous vein, a superficial vein that courses up the inner aspect of the calf and across the inferior portion of the gracilis muscle (accompanied by the great saphenous artery and nerve) to enter the femoral vein. The femoral vein is formed by the union of this vein and the popliteal vein. The popliteal vein can also be located deep in the thigh beneath the semimembranous and semitendinous muscles in the popliteal space accompanying the popliteal artery. Trace the popliteal vein to its point of division into the posterior and anterior tibial veins, which drain the leg.

4.   In your specimen, trace the portal drainage depicted in Figure 25.11. Locate the hepatic portal vein by removing the peritoneum between the first portion of the small intestine and the liver. It appears brown due to coagulated blood, and it is unlikely that it or any of the vessels of this circulation contain latex. In the cat, the portal vein is formed by the union of the **gastrosplenic** and **superior mesenteric veins.** (In the human, the hepatic portal vein is formed by the union of the splenic and superior mesenteric veins.) If possible, locate the following vessels, which empty into the hepatic portal vein:

Gastrosplenic vein: carries blood from the spleen and stomach; located dorsal to the stomach
Superior mesenteric vein: a large vein draining the small and large intestines and the pancreas
Inferior mesenteric vein: parallels the course of the inferior mesenteric artery
Coronary vein: drains the lesser curvature of the stomach
Pancreaticoduodenal veins (anterior and posterior): the anterior branch empties into hepatic portal vein; the posterior branch empties into the superior mesenteric vein (in the human, both of these are branches of the superior mesenteric vein)

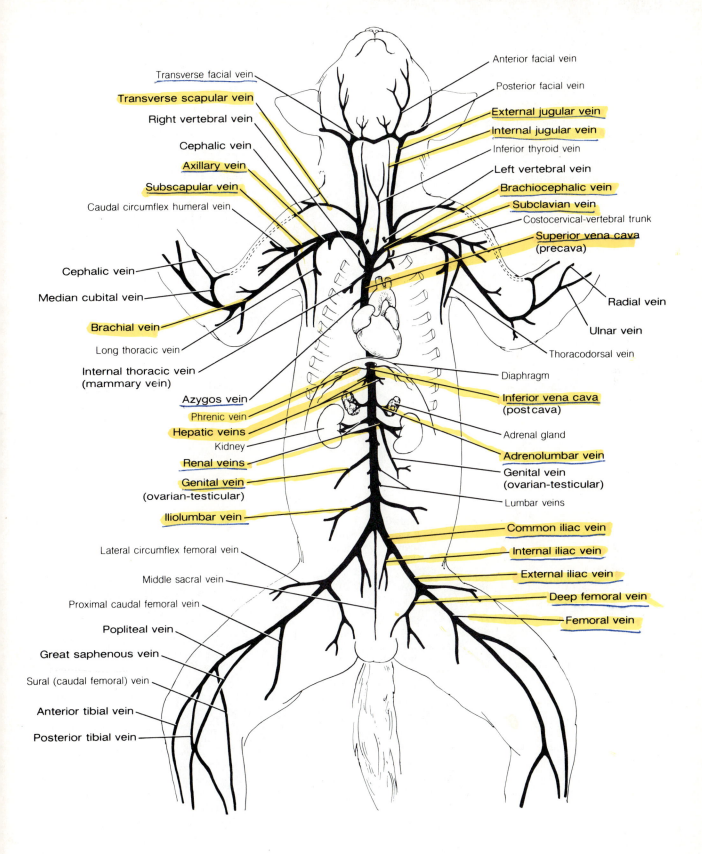

Figure 25.10

Venous system of the cat. See blood vessel dissection, Color Plate E.

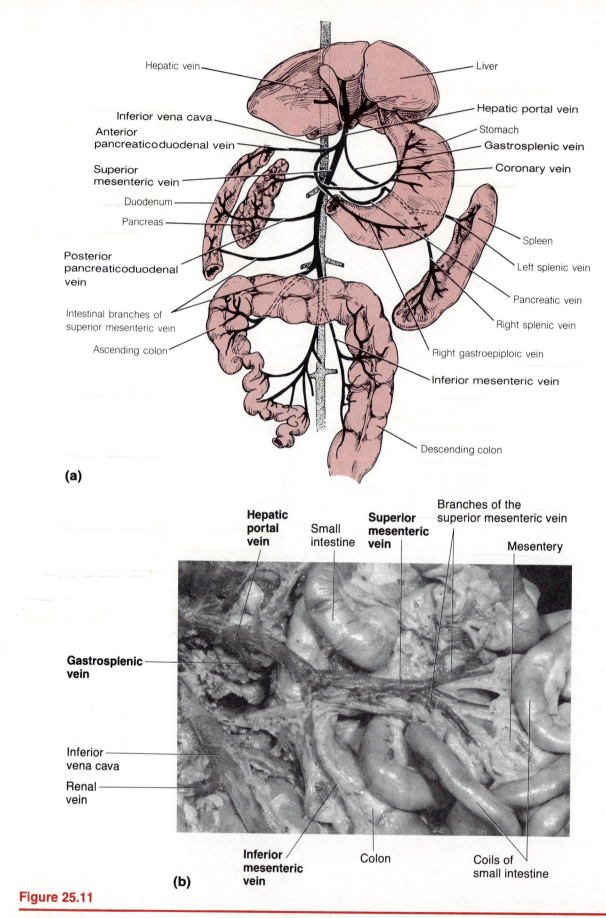

(a)

(b)

Figure 25.11

Hepatic portal circulation of the cat: (a) diagrammatic view; (b) photograph of hepatic portal system of the cat, midline to left lateral view, just below the liver and pancreas. Intestines have been pulled to the left side of the cat. The mesentery of the small intestine has been partially dissected to show the veins of the portal system.

ARTERIAL SUPPLY   Refer to Figure 25.12 and Color Plate E as you continue the dissection.

1.   Return to the thorax, lift the left lung, and follow the course of the descending aorta through the thoracic cavity. The esophagus overlies it along its course. Note the paired intercostal arteries that branch laterally from the aorta in the thoracic region.

2.   Follow the aorta through the diaphragm into the abdominal cavity. Carefully pull the peritoneum away from its ventral surface to make the identification of the following vessels possible.

Celiac trunk: the first abdominal branch of the aorta diverging from the aorta immediately as it enters the abdominal cavity; supplies the stomach, liver, gall bladder, pancreas, and spleen (trace as many of its branches to these organs as possible)

Superior mesenteric artery: immediately posterior to the celiac trunk; supplies the small intestine and most of the large intestine. (Spread the mesenteric attachment of the small intestine to observe the branches of this artery as they run to supply the small intestine.)

Adrenolumbar arteries: paired arteries diverging from the aorta slightly posterior to the superior mesenteric artery and running to supply the muscles of the body wall and adrenal glands

Renal arteries: paired arteries running to the kidneys, which they supply

Genital arteries (testicular or ovarian): paired arteries supplying the gonads

Inferior mesenteric artery: an unpaired thin vessel arising from the ventral surface of the aorta posterior to the genital arteries; supplies the second half of the large intestine

Iliolumbar arteries: paired, rather large arteries that supply the body musculature in the iliolumbar region

3.   The descending aorta ends posteriorly by dividing into three arteries: the two lateral external iliac arteries, which continue through the body wall and pass under the inguinal ligament to the leg, and the single median internal iliac artery. (There is no common iliac artery in the cat.) The internal iliac persists only briefly before dividing into the right and left internal iliac arteries and the caudal artery, which supply the pelvic viscera.

4.   Trace the external iliac artery into the thigh, where it becomes the femoral artery. The femoral artery is most easily identified in the **femoral triangle** at the medial surface of the upper thigh. Follow the femoral artery as it courses through the thigh, along with the femoral vein and nerve, which give off branches to the thigh muscles. (These various branches are indicated on Figure 25.12.) As you approach the knee region, the saphenous artery branches off the fem-

oral artery to supply the medial portion of the leg. The femoral artery then descends deep to the knee to become the popliteal artery in the popliteal region. The popliteal artery in turn gives off two main branches, the **sural artery** and the **posterior tibial artery,** and continues as the **anterior tibial artery.** These branches supply the leg and foot.

Blood Vessels of the Head, Neck, and Upper Extremities

VENOUS SUPPLY   Refer to Figure 25.10 and Color Plate E as you continue the dissection.

1.   Return to the thoracic region, and reidentify the precava as it enters the right atrium. Trace it anteriorly to identify its branches:

Azygos vein: passing directly into its dorsal surface; drains the thoracic intercostal muscles

Internal mammary (thoracic) veins: drain the chest and abdominal walls

Right vertebral vein: drains the spinal cord and brain; usually enters right side of precava approximately at the level of the internal mammary veins but may enter the brachiocephalic vein in your specimen

Right and left brachiocephalic veins: form the precava by their union

2.   Reflect the pectoral muscles, and trace the brachiocephalic vein laterally. Identify the two large veins that unite to form it—the external jugular vein and the subclavian vein.

3.   Follow the **external jugular vein** as it courses anteriorly along the side of the neck to the point where it is joined on its medial surface by the **internal jugular vein.** The internal jugular veins are quite small and often difficult to identify in the cat. Note the difference between cat and human jugular veins. (The internal jugular is considerably larger in humans and drains into the subclavian vein; in the cat, the external jugular is larger, and the interior jugular vein drains into it.) Identify the common carotid artery, since it accompanies the internal jugular vein in this region. Also attempt to find the **sympathetic nerve trunk,** which is located in the same area running lateral to the trachea. Several other vessels drain into the external jugular vein (transverse scapular vein, facial veins, and others). These are not discussed here but are shown on the figure and may be traced if time allows.

4.   Return to the shoulder region and follow the course of the **subclavian vein** as it moves laterally toward the arm. It becomes the **axillary vein** as it passes in front of the first rib and runs through the

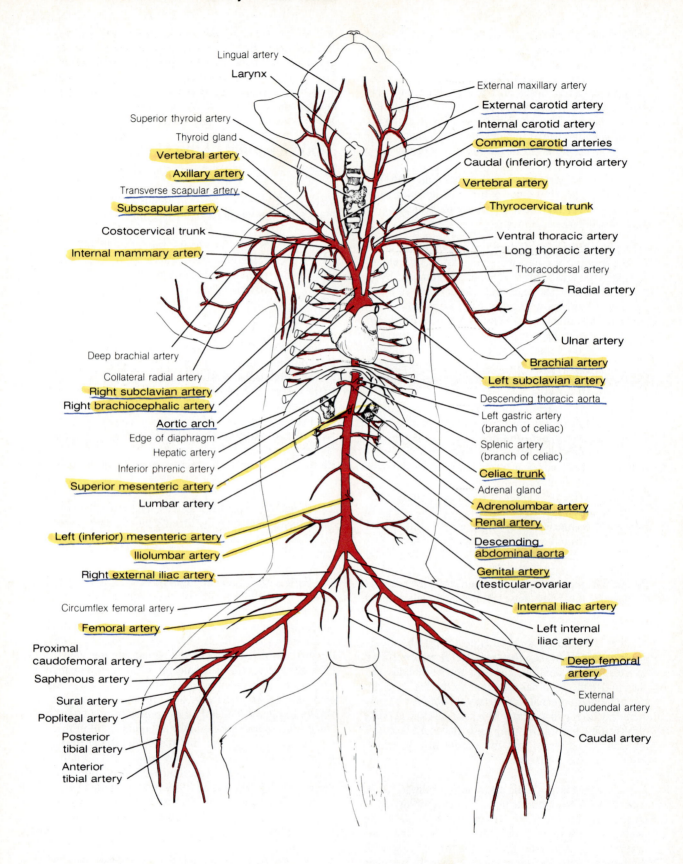

Lingual artery
Larynx
Superior thyroid artery
Thyroid gland
Vertebral artery
Axillary artery
Transverse scapular artery
Subscapular artery
Costocervical trunk
Internal mammary artery

Deep brachial artery
Collateral radial artery
Right subclavian artery
Right brachiocephalic artery
Aortic arch
Edge of diaphragm
Hepatic artery
Inferior phrenic artery
Superior mesenteric artery
Lumbar artery

Left (inferior) mesenteric artery
Iliolumbar artery
Right external iliac artery

Circumflex femoral artery
Femoral artery
Proximal caudofemoral artery
Saphenous artery
Sural artery
Popliteal artery
Posterior tibial artery
Anterior tibial artery

External maxillary artery
External carotid artery
Internal carotid artery
Common carotid arteries
Caudal (inferior) thyroid artery
Vertebral artery
Thyrocervical trunk
Ventral thoracic artery
Long thoracic artery
Thoracodorsal artery
Radial artery

Ulnar artery
Brachial artery
Left subclavian artery
Descending thoracic aorta
Left gastric artery (branch of celiac)
Splenic artery (branch of celiac)
Celiac trunk
Adrenal gland
Adrenolumbar artery
Renal artery
Descending abdominal aorta
Genital artery (testicular-ovariar
Internal iliac artery
Left internal iliac artery
Deep femoral artery
External pudendal artery
Caudal artery

Figure 25.12

Arterial system of the cat. See blood vessel dissection, Color Plate E.

brachial plexus, giving off several branches, the first of which is the **subscapular vein.** The subscapular vein drains the proximal part of the upper arm and shoulder. The four other branches that receive drainage from the shoulder, pectoral, and latissimus dorsi muscles are shown in the figure but need not be identified in this dissection.

5. Follow the axillary vein into the arm, where it becomes the **brachial vein.** You can locate this vein on the medial side of the arm accompanying the brachial artery and nerve. Trace it to the point where it receives the **radial** and **ulnar veins** (which drain the forelimb) at the inner bend of the elbow. Also locate the superficial **cephalic vein** on the dorsal side of the arm; it communicates with the brachial vein via the median cubital vein in the elbow region and then enters the transverse scapular vein in the shoulder.

ARTERIAL SUPPLY   Refer to Figure 25.12 and Color Plate E as you continue the dissection.

1. Reidentify the aorta as it emerges from the left ventricle. As you noted in the dissection of the sheep heart, the first branches of the aorta are the coronary arteries, which supply the myocardium. The coronary arteries emerge from the base of the aorta and can be seen on the surface of the heart. Follow the aorta as it arches (aortic arch), and identify its major branches. In the cat, the aortic arch gives off two large vessels, the **right brachiocephalic artery** and the **left subclavian artery.** The right brachiocephalic artery has three major branches, the right subclavian artery and the right and left common carotid arteries. (Note that in humans, the left common carotid artery is a direct branch off the aortic arch.)

2. Follow the **right common carotid artery** along the right side of the trachea as it moves anteriorly, giving off branches to the neck muscles, thyroid gland, and trachea. At the level of the larynx, it branches to form the **external** and **internal carotid arteries.** The size of the internal carotid is much reduced in the cat and may be difficult to locate. It may even be absent. The distribution of the carotid arteries parallels that in humans.

3. Follow the right **subclavian artery** laterally. It gives off four branches, the first being the tiny vertebral artery, which along with the internal carotid artery provides the arterial circulation of the brain. Other branches of the subclavian artery include the **costocervical trunk** (to the costal and cervical regions), the **thyrocervical trunk** (to the shoulder), and the **internal mammary artery** (serving the ventral thoracic wall). As the subclavian passes in front of the first rib it becomes the **axillary artery.** Its branches, which supply the trunk and shoulder muscles, are the **ventral thoracic artery** (the pectoral muscles), the **long thoracic artery** (pectoral muscles and latissimus dorsi), and the **subscapular artery** (the trunk muscles). As the axillary artery enters the arm, it is called the **brachial artery,** and it travels with the median nerve down the length of the humerus. At the elbow, the brachial artery branches to produce the two major arteries serving the forearm and hand, the **radial** and **ulnar arteries.**

(If the structures of the lymphatic system of the cat are to be studied during this laboratory session, turn to Exercise 26 for instructions to conduct the study. Otherwise, properly clean your dissecting instruments and dissecting pan, and wrap and tag your cat for storage.)

The Lymphatic System

OBJECTIVES

1. To name the components of the lymphatic system.
2. To relate the function of the lymphatic system to that of the blood vascular system.
3. To describe the formation and composition of lymph, and to describe how it is transported through the lymphatic vessels.
4. To describe the structure and function of lymph nodes; and to indicate the localization of T cells (T-lymphocytes), B cells (B-lymphocytes), and macrophages in a typical lymph node.

MATERIALS

Large anatomic chart of the human lymphatic system
Dissection animal, tray and dissection instruments
Prepared microscope slides of lymph nodes
Compound microscope

THE LYMPHATIC SYSTEM

General Description

The lymphatic system consists of a network of successively larger lymph vessels, the lymph nodes, and a number of other lymphoid organs, such as the tonsils, thymus, and spleen. We will focus on the lymphatic vessels and lymph nodes in this section. The overall function of the lymphatic system is twofold. It returns tissue fluid (lymph) to the blood vessels. Because lymph flows only toward the heart, it is a one-way system. In addition, it protects the body by removing foreign material such as bacteria from the lymphatic stream and by serving as a site for lymphocyte multiplication.

Distribution and Function of Lymphatic Vessels and Lymph Nodes

As blood circulates through the body, the hydrostatic and osmotic pressures operating at the capillary beds result in an outward flow of fluid at the arterial end of the bed and in its return at the venous end. However, not all of the lost fluid is returned to the bloodstream by this mechanism; and the fluid that lags behind in the tissue spaces must eventually return to the blood if the vascular system is to operate properly. (If it does not, fluid accumulates in the tissues, producing a condition called edema.) It is the microscopic, blind-ended **lymphatic capillaries,** which ramify through all the tissues of the body, that pick up this leaked fluid (primarily water and a small amount of dissolved proteins) and carry it through successively larger vessels (lymphatic collecting vessels to lymphatic trunks) until the lymph finally returns to the venous system through one of the two large ducts in the thoracic region. The **right lymphatic duct** drains lymph from the right upper extremity, head, and thorax; the large **thoracic duct** receives lymph from the rest of the body. In humans, both ducts empty the lymph into the venous circulation at the junction of the internal jugular vein and the subclavian vein, on their respective sides of the body (Figure 26.1).

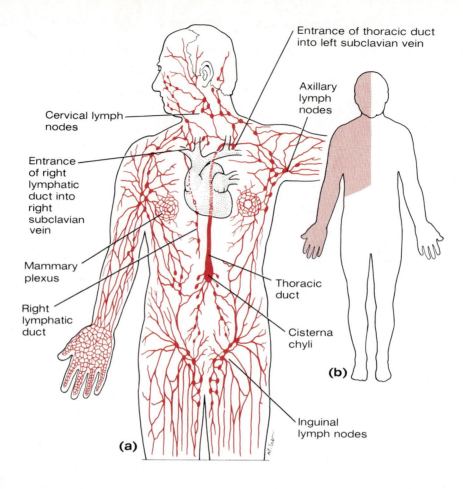

Cervical lymph nodes

Entrance of right lymphatic duct into right subclavian vein

Mammary plexus

Right lymphatic duct

Entrance of thoracic duct into left subclavian vein

Axillary lymph nodes

Thoracic duct

Cisterna chyli

(b)

Inguinal lymph nodes

(a)

Figure 26.1

Lymphatic system: (a) distribution of lymphatic vessels and lymph nodes; (b) dark area represents body area drained by the right lymphatic duct.

Like the veins of the blood vascular system, the lymphatic collecting vessels are thin-walled; and the larger vessels are equipped with valves. Since the lymphatic system is a pumpless system, lymph transport depends largely on the milking action of the skeletal muscles and on pressure changes within the thorax during breathing.

As lymph is transported, it filters through oval or bean-shaped **lymph nodes,** which cluster along the lymphatic vessels of the body. There are hundreds of lymph nodes; but because they are usually embedded in connective tissue, they are not ordinarily seen. Within the lymph nodes are **phagocytic cells (macrophages),** which destroy bacteria, cancer cells, and other foreign matter in the lymphatic stream, thus rendering many harmful substances or cells harmless before the lymph enters the bloodstream. Particularly large collections of lymph nodes are found in the inguinal, axillary, and cervical regions of the body. Although we are not usually aware of the filtering and protective nature of the lymph nodes, most of us have experienced "swollen glands" during an

active infection. This swelling is a manifestation of the trapping function of the nodes.

 Study the large anatomic chart to observe the general plan of the lymphatic system. Note the distribution of the lymphatic vessels and lymph nodes, and the location of the right lymphatic duct and the thoracic duct. Also identify the **cisterna chyli,** the enlarged terminus of the thoracic duct that receives lymph from the digestive viscera.

MAIN LYMPHATIC DUCTS OF THE CAT

 1.   Obtain your cat and a dissecting tray and instruments. Because the lymphatic vessels are extremely thin-walled, it is difficult to locate them in a dissection; however, the large thoracic duct can be localized and identified.

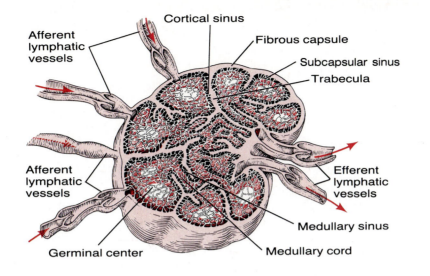

Afferent lymphatic vessels

Cortical sinus

Fibrous capsule

Subcapsular sinus

Trabecula

Efferent lymphatic vessels

Afferent lymphatic vessels

Germinal center

Medullary sinus

Medullary cord

Figure 26.2

Cross section of a lymph node. Note that the afferent vessels outnumber the efferent vessels, which slows the rate of lymph flow. The arrows indicate the direction of the lymph flow.

2. Move the thoracic organs to the side to find the **thoracic duct** traveling through the thorax just to the left of the middorsal line and abutting the dorsal aspect of the descending aorta. It is usually about the size of pencil lead and commonly red-brown with a segmented or beaded appearance caused by the valves within it. Trace it anteriorly to the site where it passes behind the left brachiocephalic vein and then bends and enters the venous system at the junction of the left subclavian and external jugular veins. If the venous system is well injected, some of the blue latex may have slipped past the valves and entered the first portion of the thoracic duct.

3. While in this general region, also attempt to identify the short **right lymphatic duct** draining into the right subclavian vein, and note the collection of lymph nodes in the axillary region.

4. If the cat is triply injected and the lymphatic vessels have been injected with yellow or green latex, trace the thoracic duct posteriorly to identify the cisterna chyli, a saclike enlargement of its distal end, which receives fat-rich lymph from the intestine. This structure begins at the level of the diaphragm and can be localized posterior to the left kidney.

5. Clean the dissecting instruments and tray, and properly wrap and return the cat to storage before continuing with the laboratory exercise.

MICROSCOPIC ANATOMY OF A LYMPH NODE

Obtain a prepared slide of a lymph node and a compound microscope. As you examine the slide, notice the following anatomical features, depicted in Figure 26.2. The node is enclosed within a fibrous **capsule,** from which connective tissue septa (**trabeculae**) extend inward to divide the node into several compartments. Very fine strands of reticular connective tissue issue from the trabeculae, forming the substrate of the gland within which cells are found.

In the outer region of the node, the **cortex,** some of the cells are arranged in globular masses, referred to as **germinal centers.** The germinal centers contain rapidly dividing B cells; the cortex also contains large numbers of circulating T cells and some macrophages. In the internal portion of the gland, the **medulla,** the cells are arranged in cordlike fashion. Most of the medullary cells are macrophages.

Lymph enters the node through a number of afferent vessels, circulates through sinuses within the node, and leaves the node through efferent vessels at the **hilus.** Since each node has fewer efferent than afferent vessels, the lymph flow stagnates somewhat within the node. This allows time for the generation of an immune response and for the phagocytic cells to remove debris from the lymph before it reenters the blood vascular system.

On a separate sheet, draw a pie-shaped section of a lymph node, showing the detail of cells in a germinal center, sinusoids, and afferent and efferent vessels. Label all elements.

Anatomy of the Respiratory System

OBJECTIVES

1. To define the following terms: *respiratory system*, *cellular respiration*, *internal respiration*, *external respiration*, *pulmonary ventilation* (breathing).

2. To label the major respiratory system structures on a diagram (or identify on a model or dissection) and to describe the function of each.

3. To recognize pseudostratified columnar ciliated epithelium and lung tissue on prepared histologic slides, and describe the functions the observed structural modifications serve.

MATERIALS

Human torso model
Respiratory organ model and/or chart of the respiratory system
Larynx model (if available)
Sheep pluck (preserved or, preferably, fresh from the slaughterhouse)
Dissection animals, trays, and instruments
Source of compressed air*
2-foot length of laboratory rubber tubing
Histologic slides of the following (if available): trachea (cross section), lung tissue, both normal and pathologic specimens (e.g., sections taken from lung tissues exhibiting bronchitis, pneumonia, or emphysema)
Compound and dissecting microscopes
Bronchogram (if available)

*If a compressed air source is not available, cardboard mouthpieces that fit the cut end of the rubber tubing should be available for student use. Disposable autoclave bags should also be provided for discarding the mouthpiece.

To carry out their vital processes, body cells require an abundant and continuous supply of oxygen. As the cells use oxygen, they release carbon dioxide, a waste product that the body must eliminate. These oxygen-using cellular processes are collectively referred to as *cellular respiration* and the gas exchanges that occur between the tissue cells and the blood are known as *internal respiration*.

The circulatory and respiratory systems are intimately involved in the acquisition and delivery of oxygen and the removal of carbon dioxide. The structures concerned with pulmonary ventilation, or *breathing* (moving air into and out of the lungs), and with gas exchange between the blood and the air sacs of the lungs (*external respiration*) are collectively referred to as the *respiratory system*. The transport of respiratory gases between the lungs and the cells is accomplished by the structures of the circulatory system, with blood as the transport medium. Should either system fail, the cells begin to die from oxygen starvation and the accumulation of carbon dioxide. If uncorrected, this situation soon causes death of the entire organism.

UPPER RESPIRATORY SYSTEM STRUCTURES

The upper respiratory system structures—nose, pharynx, and larynx—are shown in Figure 27.1 and described below. As you read through the descriptions, identify each structure by referring to this figure.

Air generally passes into the internal nose through the **external nares** (nostrils), into the paired internal **nasal cavity** (divided by the **nasal septum**), and then flows posteriorly over the three pairs of lobelike structures, the **inferior, superior,** and **middle nasal conchae,** which increase the air turbulence. As the air passes through the nasal cavity, it is also warmed, moistened, and filtered by the nasal mucosa. The air that flows directly beneath the upper part of the nasal cavity may chemically stimulate the olfactory receptors located in the mucosa of that region. The nasal cavity is surrounded by the **paranasal sinuses** in the frontal, sphenoid, ethmoid, and maxillary bones. These sinuses act as resonance chambers in speech.

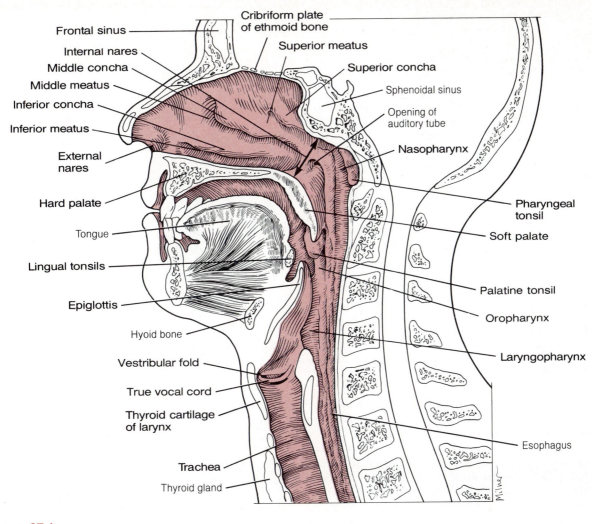

Figure 27.1

Structures of the upper respiratory tract (sagittal section).

▲ Because of the close communication of the sinuses with the nasal passages and the continuity of the mucosa through both areas, nasal infections often invade the sinus areas, causing sinusitis that is difficult to treat. ■

The nasal passages are separated from the oral cavity below by a partition composed anteriorly of the **hard palate** and posteriorly by the **soft palate.**

▲ The genetic defect called cleft palate (failure of the palatine bones and/or the palatine processes of the maxillary bones to fuse medially) causes difficulty in breathing and oral cavity functions such as mastication and speech. ■

Needless to say, in addition to entering the **nasopharynx** through the nasal cavities, air may also enter the body via the mouth and pass through the oral cavity into the **oropharynx** (throat) posteriorly, where the oral and nasal cavities are joined temporarily. The pharynx thus accommodates both ingested food and air. From the oropharynx, air enters the lower respiratory passageways by passing through the

laryngopharynx and **larynx** (voice box) into the **trachea** below.

The larynx (Figure 27.2) consists of nine cartilages, the three most prominent being the large shield-shaped **thyroid cartilage,** whose anterior medial prominence is commonly referred to as the Adam's apple, the inferiorly located, ring-shaped **cricoid cartilage,** whose widest dimension faces posteriorly, and the flaplike **epiglottis,** located superior to the opening of the larynx. The epiglottis, sometimes referred to as the guardian of the airways, forms a lid over the larynx when we swallow. This closes off the respiratory passageways to incoming food or drink, which is routed into the posterior esophagus, or food chute.

● Palpate your larynx by placing your hand on the anterior neck surface approximately halfway down its length. Swallow. Can you feel the cartilaginous larynx rising?

If anything other than air enters the larynx, a cough reflex attempts to expel the substance. Note that this

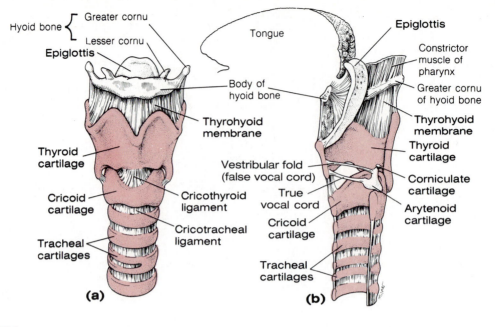

Figure 27.2

Structure of the larynx: (a) anterior view; (b) sagittal section.

reflex is operative only when a person is conscious; thus you should never try to feed or pour liquids down the throat of an unconscious person.

The mucous membrane of the larynx is thrown into two pairs of folds—the upper **vestibular folds,** also called the **false vocal cords,** and the lower **vocal folds,** or **true vocal cords,** which vibrate with expelled air for speech. The vocal cords are attached posterolaterally to the small triangular **arytenoid cartilages.** The slitlike passageway between the folds is called the **glottis.**

LOWER RESPIRATORY SYSTEM STRUCTURES

Air entering the trachea, or windpipe, from the larynx travels down its length (about 11.5 cm) to the level of the sternal angle, where the passageway divides into the right and left **primary bronchi,** which plunge into their respective lungs at an indented area called **hilus** (Figure 27.3a). The right primary bronchus is more vertical, larger in diameter, and shorter than the left primary bronchus; as a result, foreign objects that enter the respiratory passageways are more likely to become lodged in it.

The trachea is lined with ciliated mucus-secreting epithelium as are many of the other respiratory system passageways. The cilia beat in unison and propel mucus (produced by goblet cells) laden with dust particles, bacteria, and other debris away from the lungs and toward the throat, where it can be expectorated or swallowed. The walls of the trachea are reinforced with C-shaped cartilage rings, the incomplete portion being located posteriorly. These

C-shaped cartilages serve a double function: the incomplete parts allow the esophagus to expand anteriorly when a large food bolus is swallowed; the solid portions reinforce the trachea walls to maintain its passageway regardless of the pressure changes that occur during breathing.

The primary bronchi further divide into smaller and smaller branches (the secondary, tertiary, on down), finally becoming the **bronchioles,** which have terminal branches called **respiratory bronchioles** (Figure 27.3b). All but the most minute branches contain cartilaginous reinforcements in their walls, usually in the form of small plates of hyaline cartilage rather than cartilaginous rings. The respiratory bronchioles in turn subdivide into several **alveolar ducts,** which terminate in alveolar sacs that rather resemble clusters of grapes. The walls of the **alveolar sacs,** or individual **alveoli,** are composed of a single very thin layer of squamous epithelium overlying a wispy basal lamina. The external surfaces of the alveoli are literally spider-webbed with a network of pulmonary capillaries (Figure 27.4). It is through the **respiratory membrane,** the fused alveolar and capillary walls, that the gas exchanges occur by simple diffusion—the oxygen passing from the alveolar air into the capillary blood and the carbon dioxide leaving the capillary blood to enter the alveolar air. Respiratory passageways larger than respiratory brachioles simply serve as access or exit routes to and from these gas exchange chambers. Since the larger passageways (trachea and bronchi) have no exchange function, they are termed *anatomical dead space.*

The continuous branching of the respiratory passageways in the lungs is often referred to as the respiratory tree. The comparison becomes much more

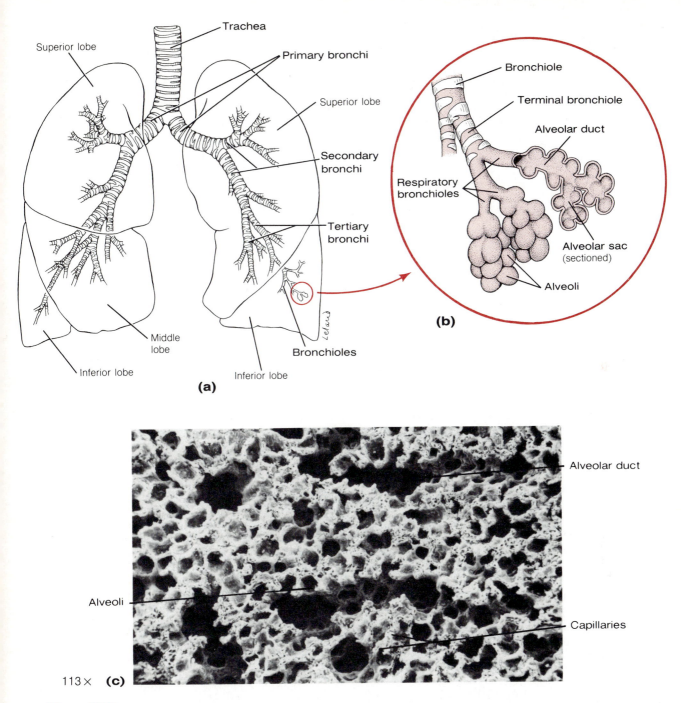

Figure 27.3

Structures of the lower respiratory tract (a). Inset (b) shows enlarged view of alveoli. Photomicrograph (c) shows the spongy nature of lung tissue.

meaningful if you observe a bronchogram. (Do so, if one is available for observation in the laboratory, or refer to Figure 27.5.)

The paired lungs are separated from one another by the structures of the mediastinum (heart, trachea, bronchi, major blood vessels, esophagus). The substance of the lungs, other than the respiratory passageways that make up the bulk of their volume, is primarily elastic connective tissue that allows the lungs to recoil passively during expiration. Each lung is

enclosed in a double-layered sac of serous membrane called the **pleura.** The outer layer, the **parietal pleura,** is attached to the thoracic walls and the **diaphragm;** the inner layer, covering the lung tissue, is the **visceral pleura.** The two pleural layers are separated by a space, which is more of a potential space than an actual one. The production of a lubricating serous fluid by the pleural layers causes the parietal and visceral layers to adhere closely to one another, holding the lungs to the thoracic wall and allowing them

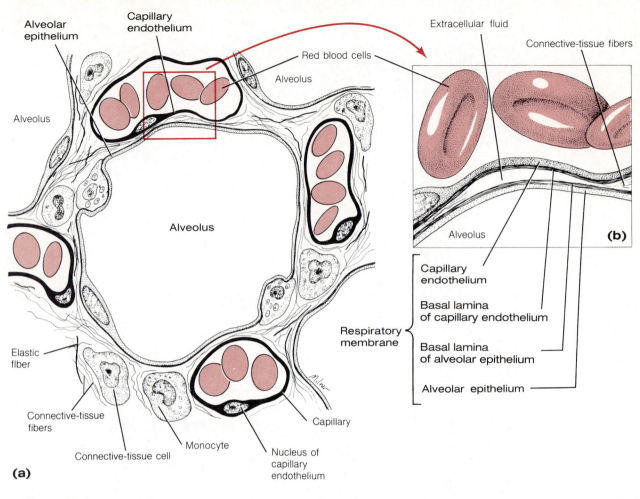

Figure 27.4

Diagrammatic view of the relationship between the alveoli and pulmonary capillaries involved in gas exchange: (a) one alveolus surrounded by capillaries; (b) enlargement of the respiratory membrane.

to move easily against one another in a frictionless environment during the movements of breathing.

Before proceeding, be sure to locate on the torso model, thoracic cavity structures model, or an anatomic chart, all the respiratory structures described.

SHEEP PLUCK DEMONSTRATION

A *sheep pluck* includes the larynx, trachea with attached lungs, the heart and pericardium, and portions of the major blood vessels found in the mediastinum (aorta, pulmonary artery and vein, vena cava). Obtain a sheep pluck and identify the lower respiratory system organs. Once you have completed your observations, insert a hose from an air compressor (vacuum pump) into the trachea and alternately allow air to flow in and out of the lungs. Notice how the lungs inflate. This observation is

educational in a preserved pluck but is a spectacular sight in a fresh one. (If air compressors are not available, the same effect may be obtained by using a length of laboratory rubber tubing to blow into the trachea. Obtain a cardboard mouthpiece and fit it into the cut end of the laboratory tubing before attempting to inflate the lungs. Dispose of the mouthpiece in the autoclave bag immediately after use.)

EXAMINATION OF PREPARED SLIDES OF LUNG AND TRACHEA TISSUE

1. Obtain and examine a cross section of trachea tissue. Identify the smooth muscle wall, the hyaline cartilage supporting rings, and the pseudostratified ciliated epithelium. Also try to identify a few goblet cells in the epithelium. Draw

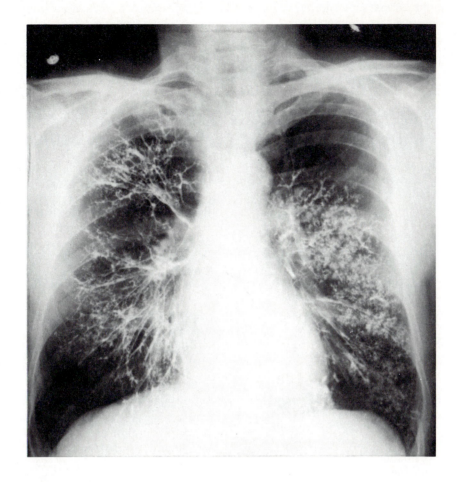

Figure 27.5

Bronchogram showing the extensive branching of the respiratory tree.

a section of the trachea wall and label all tissue layers in the space provided here.

2.   Obtain a slide of lung tissue for examination. The alveolus is the main structural and functional unit of the lung and is the actual site of gas exchange. Identify the thin squamous epithelium of the alveolar walls, a bronchiole, and one of the smaller bronchi. Draw your observations of a small section of the alveolar tissue in the space below. Label the alveoli.

3. Examine slides of pathologic lung tissues, and compare them to the normal lung specimens.

DISSECTION OF THE RESPIRATORY SYSTEM OF THE CAT

1. Obtain your dissection animal. Before beginning this dissection exercise, examine the external nares, oral cavity, and oral pharynx. Use a probe to demonstrate the continuity between the oral pharynx and the nasal pharynx above.

2. After securing the animal to the dissecting tray, expose the respiratory structures by retracting the cut muscle and rib cage. (Do not sever nerves and blood vessels located on either side of the trachea if these have not previously been studied. If you have not previously opened the thoracic cavity, make a medial longitudinal incision through the neck muscles and thoracic musculature to expose and view the thoracic organs.)

3. Using Figure 27.6 and Color Plate D as guides, identify the structures named in items 3 through 6. Examine the trachea, and determine by finger examination whether the cartilage rings are complete or incomplete posteriorly. Locate the thyroid gland inferior to the larynx on the trachea. Free the larynx from the attached muscle tissue, and pull the larynx anteriorly for ease of examination. Identify the thyroid and cricoid cartilages and the flaplike epiglottis. Find the hyoid bone located anterior to the larynx.

Make a longitudinal incision through the ventral wall of the larynx and locate the true and false vocal folds on the inner wall.

4. Locate the large right and left common carotid arteries and the internal jugular veins on either side of the trachea. Also locate a conspicuous white band, the vagus nerve, which lies alongside the trachea, adjacent to the common carotid artery.

5. Examine the contents of the thoracic cavity. Follow the trachea as it bifurcates into two primary bronchi, which plunge into the lungs. Note that there are two pleural cavities containing the lungs and that each lung is composed of many lobes. (In humans there are three lobes in the right lung and two in the left. How does this compare to what is seen in the cat?) Note the pericardial sac containing the heart located in the mediastinum (if it is still present). Examine the pleura, and note its exceptionally smooth texture. Locate the **diaphragm** and the **phrenic nerve** (a conspicuous white thread running along the pericardium to the diaphragm). The phrenic nerve controls the activity of the diaphragm in breathing. Lift one lung and find the esophagus beneath the parietal pleura. Follow it through the diaphragm to the stomach.

6. Make a longitudinal incision in the outer tissue of one lung lobe beginning at a primary bronchus. Attempt to follow part of the respiratory tree from this point down into the smaller subdivisions. Carefully observe the cut lung tissue (under a dissection scope called a stereomicroscope if one is available), noting the richness of the vascular supply and the irregular or spongy texture of the lung.

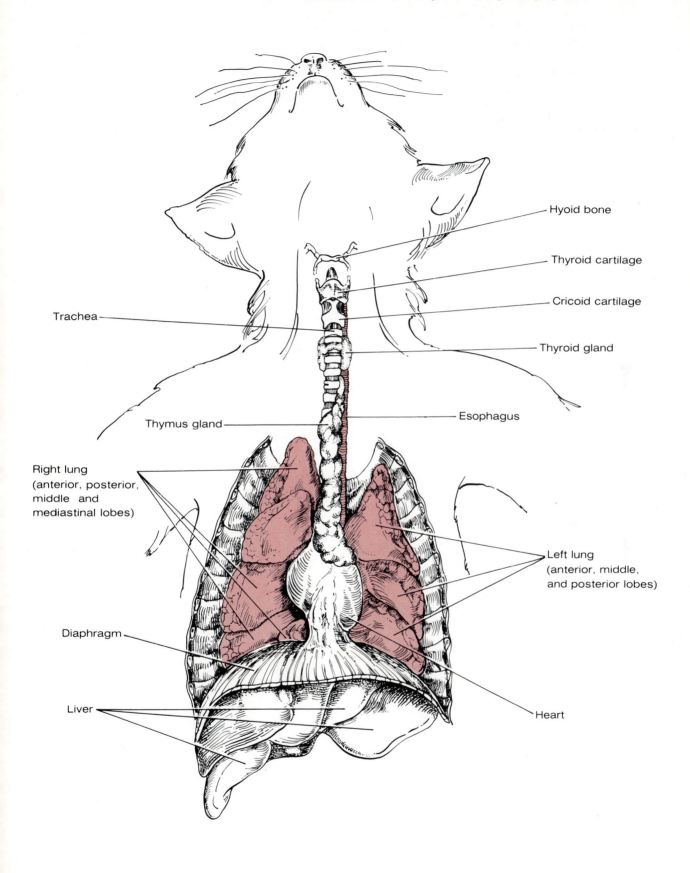

Hyoid bone

Thyroid cartilage

Cricoid cartilage

Trachea

Thyroid gland

Esophagus

Thymus gland

Right lung
(anterior, posterior,
middle and
mediastinal lobes)

Left lung
(anterior, middle,
and posterior lobes)

Diaphragm

Liver

Heart

Figure 27.6

Respiratory system of the cat. See respiratory dissection, Color Plate D.

Anatomy of the Digestive System

OBJECTIVES

1. To state the overall function of the digestive system.

2. To identify on an appropriate diagram, torso model, or dissected cat the organs comprising the alimentary tract; and to name their subdivisions if any.

3. To name and/or identify the accessory digestive organs.

4. To describe the general function of all the alimentary tract organs and accessory organs relative to digestive system activity.

5. To describe the histologic structure of the wall of the alimentary canal and/or label a cross-sectional diagram of the wall with the following terms: mucosa, submucosa, muscularis externa, and serosa, or adventitia.

6. To describe the composition of saliva.

7. To list the major enzymes or enzyme groups produced by each of the following organs: salivary glands, stomach, small intestine, pancreas.

8. To name human deciduous and permanent teeth and describe the anatomy of the generalized tooth.

9. To list and explain the specializations of structure of the stomach and small intestine that contribute to their functional roles.

10. To recognize by microscopic inspection or by viewing an appropriate diagram or photomicrograph the histologic structure of the following organs:

small intestine	pancreas	stomach
salivary glands	tooth	liver

MATERIALS

Dissectible torso model
Anatomical chart of the human digestive system
Dissection animal, trays, instruments, and bone cutters
Model of a villus and liver (if available)
Jaw model or human skull
Prepared microscope slides of the liver, pancreas, mixed salivary glands, longitudinal sections of the gastroesophageal junction and tooth, and cross sections of the stomach, duodenum, and ileum
Hand lens
Compound microscope

The digestive system provides the body with the nutrients, water, and electrolytes essential for metabolic processes and health. The organs of this system are responsible for food ingestion, digestion, absorption, and the elimination of the undigested remains as feces.

The digestive system consists of a hollow tube extending from the mouth to the anus, into which various accessory organs or glands empty their secretions. Food material within this tube, the alimentary canal, is technically outside the body, since it has contact only with the cells lining the tract. For the ingested food to become available to the body cells, it must first be broken down physically (chewing, churning) and chemically (enzymatic hydrolysis) into its smaller diffusible molecules—a process called **digestion.** The digested end products can then pass through the epithelial cells lining the tract into the blood for distribution to the body cells—a process termed **absorption.** In one sense, the digestive tract can be viewed as a disassembly line, in which food is carried from one stage of its digestive processing to the next by muscular activity, and its nutrients are made available to the cells of the body en route.

GROSS ANATOMY OF THE HUMAN DIGESTIVE SYSTEM

The organs of the digestive system are traditionally separated into two major groups: the **alimentary canal,** or **gastrointestinal tract,** and the **accessory digestive organs.** The alimentary canal is approximately 30 feet long in the cadaver but considerably less in a living person and consists of the mouth, pharynx, esophagus, stomach, small and large intestines, and anus. The accessory structures consist of the salivary glands, gallbladder, liver, and pancreas, which secrete their products into the alimentary canal.

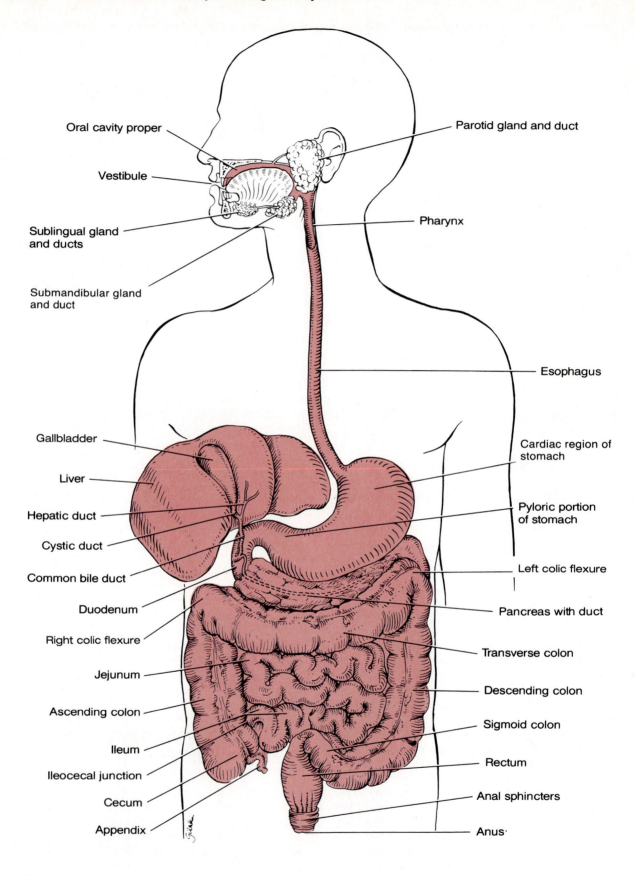

Oral cavity proper

Vestibule

Sublingual gland
and ducts

Submandibular gland
and duct

Parotid gland and duct

Pharynx

Esophagus

Gallbladder

Liver

Hepatic duct

Cystic duct

Common bile duct

Duodenum

Right colic flexure

Jejunum

Ascending colon

Ileum

Ileocecal junction

Cecum

Appendix

Cardiac region of
stomach

Pyloric portion
of stomach

Left colic flexure

Pancreas with duct

Transverse colon

Descending colon

Sigmoid colon

Rectum

Anal sphincters

Anus

Figure 28.1

The human digestive system: alimentary tube and accessory organs. (Liver and gall bladder are reflected superiorly and to the right.)

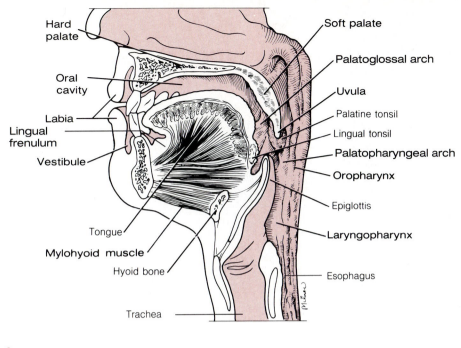

Figure 28.2

Sagittal view of the head showing oral, nasal, and pharyngeal cavities.

The sequential pathway and fate of food as it passes through the digestive tract is described here. Identify each structure on Figure 28.1 and on the torso model.

Alimentary Canal

MOUTH OR ORAL CAVITY   Food enters the digestive tract through the mouth, or oral cavity. Within this mucous membrane-lined cavity are the gums, teeth, tongue, and openings of the ducts of the salivary glands. The **lips (labia)** protect the opening of the chamber anteriorly, the cheeks form its lateral walls, and the palate, its roof. The anterior portion of the palate is referred to as the **hard palate,** since bone (the palatine processes of the maxillae and the palatine bones) underlies it. The posterior **soft palate** is a fibromuscular structure that is unsupported by bone. The **uvula,** a fingerlike projection of the soft palate, extends downward from its posterior margin. The soft palate closes off the oral cavity from the nasal and pharyngeal passages during swallowing. The floor of the oral cavity is occupied primarily by the muscular **tongue,** which is largely supported by the **mylohyoid muscle** and is attached to the hyoid bone, mandible, styloid processes, and pharynx. A membrane called the **lingual frenulum** secures the inferior midline of the tongue to the floor of the mouth. The space between the lips and cheeks and the teeth is the **vestibule;** the area containing the teeth, which is posterior to the alveolar arches, is the **oral cavity** proper. (Figures 28.1 to 28.3 depict the structures of the oral cavity.)

On each side of the mouth at its posterior end are masses of lymphatic tissue, the **palatine tonsils.**

Each lies in a concave area bounded anteriorly and posteriorly by membranes, the **palatoglossal arch** (anterior membrane) and the **palatopharyngeal arch** (posterior membrane). The tonsils, in common with other lymphoid tissues, are part of the body's defense system. Another mass of lymphatic tissue, the **lingual tonsil,** covers the base of the tongue, posterior to the oral cavity proper.

Very often in young children, the palatine tonsils become inflamed and enlarge, partially blocking the entrance to the pharynx posteriorly and making swallowing difficult and painful. This condition is called tonsilitis. ■

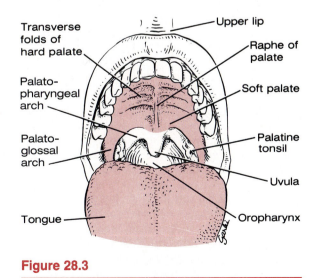

Figure 28.3

Anterior view of the oral cavity.

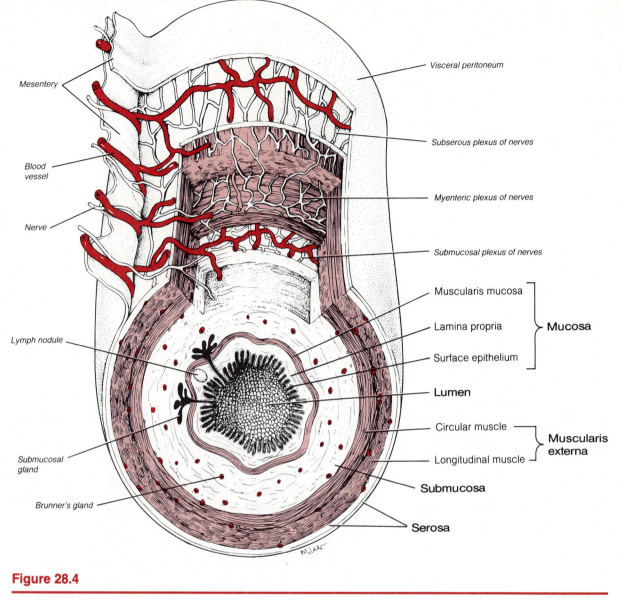

Mesentery

Blood vessel

Nerve

Lymph nodule

Submucosal gland

Brunner's gland

Visceral peritoneum

Subserous plexus of nerves

Myenteric plexus of nerves

Submucosal plexus of nerves

Muscularis mucosa ⎫
Lamina propria ⎬ **Mucosa**
Surface epithelium ⎭

Lumen

Circular muscle ⎫ **Muscularis**
Longitudinal muscle ⎬ **externa**

Submucosa

Serosa

Figure 28.4

Structural pattern of the alimentary canal wall.

Three pairs of salivary glands duct their secretion, saliva, into the oral cavity. One component of saliva, salivary amylase, begins the digestion of starchy foods within the oral cavity. (The salivary glands are discussed in more detail on p. 276.)

As food enters the mouth, it is mixed with saliva and masticated (chewed). The cheeks and lips help hold the food between the teeth during mastication, and the highly mobile tongue manipulates the food for chewing and initiates swallowing. Thus the mechanical and chemical breakdown of food begins before the food has left the oral cavity. As noted in Exercise 21, the surface of the tongue is covered with papillae, many of which contain taste buds, the receptors for taste sensation. So, in addition to its manipulative function, the tongue provides for the enjoyment and appreciation of the food ingested.

PHARYNX   When the tongue initiates swallowing, the food passes posteriorly into the pharynx, a common passageway for food, fluid, and air. The pharynx is often subdivided anatomically into the **nasopharynx** (behind the nasal cavity), the **oropharynx** (behind the oral cavity extending from the soft palate to the epiglottis overlying the larynx), and the **laryngopharynx** (extending from the epiglottis to the base of the larynx), which is continuous with the esophagus.

The walls of the pharynx consist largely of two layers of skeletal muscles: an inner layer of longitudinal muscle (the levator muscles) and an outer layer of circular constrictor muscles, which initiate wavelike contractions that enable the pharynx to propel the food into the esophagus inferiorly.

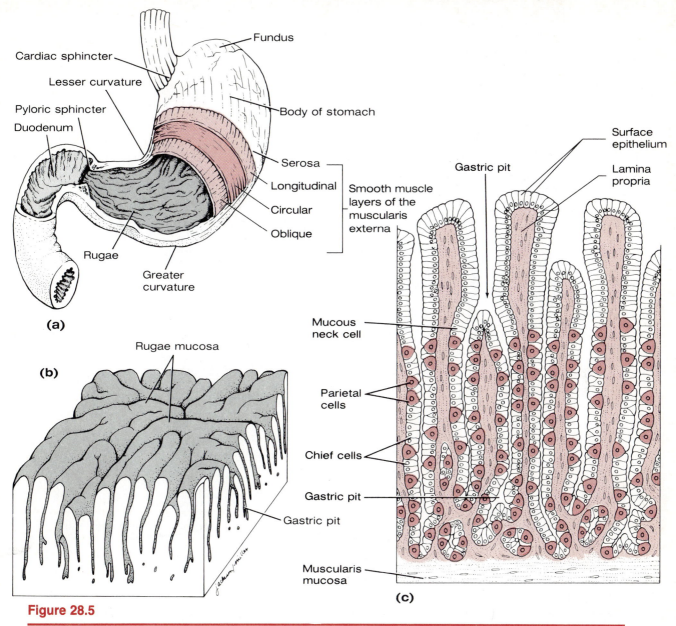

Figure 28.5

Anatomy of the stomach: (a) gross internal and external anatomy; (b) section of the stomach wall showing rugae and gastric pits; (c) enlarged view of gastric pits (longitudinal section).

ESOPHAGUS The **esophagus,** or gullet, extends from the pharynx through the diaphragm to the cardiac sphincter in the superior aspect of the stomach. It is approximately 10 inches long in humans and is essentially a food passageway that conducts food to the stomach in a wavelike peristaltic motion. The esophagus has no digestive or absorptive function. The walls at the superior end of the esophagus contain skeletal muscle, which is replaced by smooth muscle in the area nearing the stomach.

Because the tissue composition of the alimentary tube walls is modified in various areas along its length to serve specific functions, it is worthwhile to briefly describe a cross-sectional view of the general wall structure (Figure 28.4). The walls of the alimentary canal organs from the esophagus to the colon have four characteristic layers:

Mucosa: the innermost layer, which consists of an **epithelium,** a connective tissue basal layer, the **lamina propria,** and a thin smooth muscle layer, the **muscularis mucosa.**

Submucosa: deep to the mucosa and consisting primarily of connective tissue in which blood vessels, nerve endings, and lymphatic tissue are found.

Muscularis externa: commonly composed of two layers of smooth muscle, the inner layer in circular orientation, and the outer layer oriented longitudinally.

Serosa: the outermost layer, consisting of a single layer of serous fluid-producing cells and commonly referred to as the **visceral peritoneum.** (In some areas, this outermost layer may be an adventitia composed of fibrous tissue rather than a serosa.)

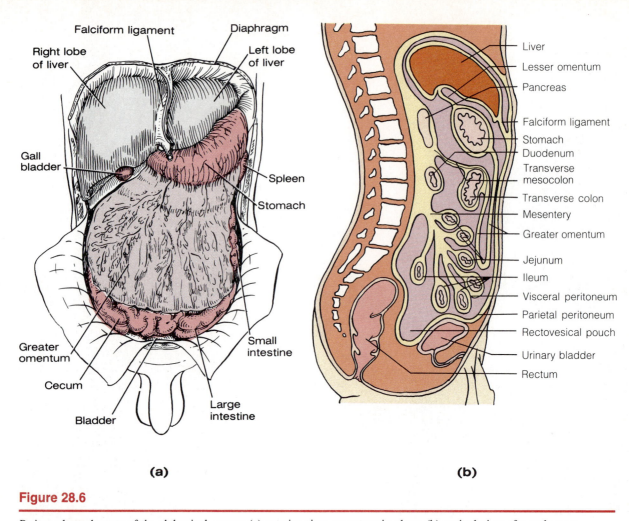

(a)   **(b)**

Figure 28.6

Peritoneal attachments of the abdominal organs: (a) anterior view, omentum in place; (b) sagittal view of a male torso.

STOMACH   The **stomach** (Figures 28.1 and 28.5) is on the left side of the abdominal cavity and is hidden by the liver and diaphragm. Different regions of the saclike stomach are the **cardiac region** (the area surrounding the cardiac sphincter through which food enters the stomach from the esophagus), the **fundus** (the expanded portion of the stomach, superolateral to the cardiac region), the **body** (midportion of the stomach, inferior to the fundus), and the **pylorus** (the terminal part of the stomach, which is continuous with the small intestine through the pyloric sphincter).

The concave medial surface of the stomach is called the **lesser curvature,** and its convex lateral surface is called the **greater curvature.** Extending from these curvatures are two mesenteries called *omenta.* The **lesser omentum** extends from the liver to attach to the lesser curvature of the stomach. The **greater omentum,** a saclike mesentery extends from the greater curvature of the stomach, reflects downward over the abdominal contents to cover them in an apronlike fashion, and then blends with the meso-

colon attaching the transverse colon to the posterior body wall. Figure 28.6 illustrates the omenta as well as the other peritoneal attachments of the abdominal organs.

The stomach is a temporary storage region for food as well as a site for the mechanical and chemical breakdown of food. It contains a third obliquely oriented layer of smooth muscle in its muscularis externa that allows it to churn, mix, and pummel the food, physically reducing it to smaller fragments. Gastric glands of the mucosa secrete hydrochloric acid (HCl) and hydrolytic enzymes (primarily pepsinogen, the inactive form of *pepsin,* a protein-digesting enzyme), which begin the enzymatic, or chemical, breakdown of protein foods. The mucosal glands also secrete a viscous mucus that prevents the stomach itself from being digested by the proteolytic enzymes. Most digestive activity occurs in the pyloric region of the stomach. After the food has been processed in the stomach, it resembles a creamy mass (chyme), which enters the small intestine through the pyloric sphincter.

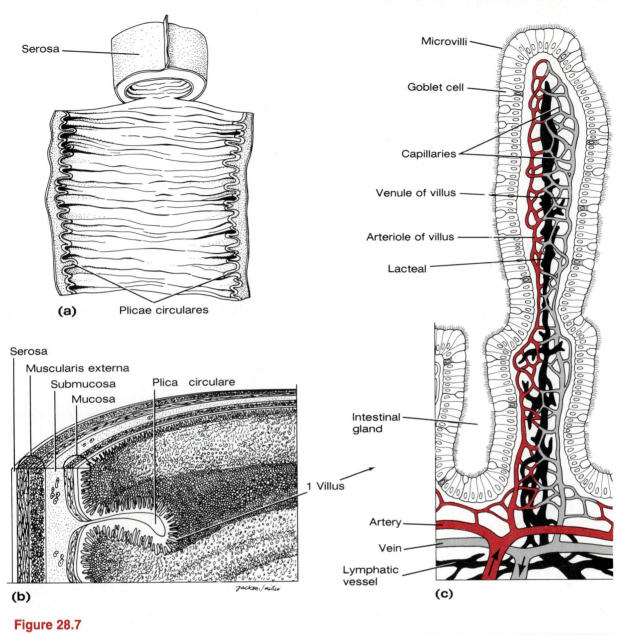

Serosa

(a) Plicae circulares

Serosa
Muscularis externa
Submucosa
Mucosa
Plica circulare

1 Villus

(b)

Microvilli
Goblet cell
Capillaries
Venule of villus
Arteriole of villus
Lacteal

Intestinal gland

Artery
Vein
Lymphatic vessel

(c)

Figure 28.7

Structural modifications of the small intestine: (a) plicae circulares (circular folds) seen on the inner surface of the small intestine; (b) enlargement of one plica circulare to show villi; (c) detailed anatomy of a villus.

SMALL INTESTINE   The **small intestine** is a convoluted tube, 6 to 7 meters (m) long, extending from the pyloric sphincter to the ileocecal valve. The small intestine is suspended by a double layer of peritoneum, the fan-shaped **mesentery,** from the posterior abdominal wall (see Figure 28.6), and it lies, framed laterally and superiorly by the large intestine, in the abdominal cavity. The small intestine has three subdivisions: (1) the **duodenum** extends from the pyloric sphincter for about 25 cm (10 inches) and curves around the head of the pancreas; most of the duodenum lies in a retroperitoneal position. (2) The **jejunum,** continuous with the duodenum, extends for 2.5 to 3 m. Most of the jejunum occupies the umbilical region of the abdominal cavity. (3) The **ileum,**

the terminal portion of the small intestine, is about 4 m long and joins the large intestine at the **ileocecal valve.** It is located inferiorly and somewhat to the right in the abdominal cavity, but its major portion lies in the hypogastric region.

Hydrolytic enzymes ("brush-border" enzymes) bound to the microvilli of the columnar epithelial cells and, more importantly, enzymes produced by the pancreas and ducted into the duodenum via the **pancreatic duct** complete the enzymatic digestion process in the small intestine. Bile (formed in the liver) also enters the duodenum via the **common bile duct** in the same area (see Figure 28.1). At the duodenum, the ducts join to form the bulblike **hepatopancreatic ampulla** and empty their products into the duodenal lumen through the duodenal papilla, an orifice controlled by a muscular valve called the **hepatopancreatic sphincter (of Oddi).**

Nearly all nutrient absorption occurs in the small intestine, where three structural modifications that increase the absorptive area appear—the **microvilli, villi,** and **plicae circulares.** Microvilli are minute projections of the surface plasma membrane of the epithelial lining cells of the mucosa; the villi are the fingerlike projections of the mucosa layer that give it a velvety appearance and texture (Figure 28.7). The plicae circulares are deep folds of the mucosa and submucosa layers that extend partially or totally around the intestine. These structural modifications, which increase the surface area, decrease in frequency and elaboration toward the end of the small intestine. Any residue remaining undigested and unabsorbed at the terminus of the small intestine enters the large intestine through the ileocecal valve. In contrast, the amount of lymphatic tissue in the submucosa of the small intestine (**Peyer's patches**) increases along the length of the small intestine and is very apparent in the ileum. This reflects the fact that the remaining undigested food residue contains large numbers of bacteria that must be prevented from entering the bloodstream.

LARGE INTESTINE   The **large intestine** (see Figure 28.8) is about 5 feet long and extends from the ileocecal valve to the anus. It encircles the small intestine on three sides and consists of the following subdivisions: the **cecum, appendix, colon, rectum,** and **anal canal.**

The blind tubelike appendix (approximately 3 inches long) is a trouble spot in the large intestine. Since it is generally twisted, it provides an ideal location for bacteria to accumulate and multiply. Inflammation of the appendix, or appendicitis, is the result. ■

The colon is divided into several distinct regions. The **ascending colon** travels up the right side of the abdominal cavity and makes a right-angle turn (**right colic flexure**) to cross the abdominal cavity as the **transverse colon.** It then turns (**left colic flexure**) and continues down the left side of the abdominal cavity as the **descending colon,** where it takes an S-shaped course as the **sigmoid colon.** The sigmoid colon, rectum, and the anal canal lie in the pelvis anterior to the sacrum and thus are not considered abdominal cavity structures. The anal canal terminates in the **anus,** the opening to the exterior of the body. The anus, which has an external sphincter of skeletal muscle (the voluntary sphincter) and an internal sphincter of smooth muscle (the involuntary sphincter), is normally closed except during defecation when the undigested remains of the food and bacteria are eliminated from the body as feces.

In the large intestine, the longitudinal muscle layer of the muscularis externa is reduced to three longitudinal muscle bands called the **teniae coli.** Since these bands are shorter than the rest of the wall of the large intestine, they cause the wall to pucker into small pocketlike sacs called **haustra.**

The major functions of the large intestine include the following: it provides a site for the manufacture of some vitamins (B vitamins and vitamin K) by intestinal bacteria; the vitamins are then absorbed into the bloodstream. It also absorbs water (and some electrolytes) from undigested food, thus conserving body water and forming the feces, or stool.

Watery stools, or diarrhea, result from any condition that rushes undigested food residue through the large intestine before it has had sufficient time to absorb the water (as in irritation of the colon by bacteria). Conversely, when food residue remains in the large intestine for extended periods (as with atonic colon or failure of the defecation reflex), excessive water is absorbed and the stool becomes hard and difficult to pass (constipation). ■

Accessory Digestive Organs

SALIVARY GLANDS   The three pairs of major salivary glands (see Figure 28.1) that empty their secretions into the oral cavity are the large **parotid glands,** located anteriorly to the ear and ducting into the mouth over the second upper molar through the parotid duct; the **submandibular glands,** located inside the maxillary arch in the floor of the mouth and ducting under the tongue via the submandibular ducts; and the small **sublingual glands,** located most anteriorly in the floor of the mouth and emptying under the tongue through several small ducts.

Food in the mouth and mechanical pressure (chewing rubber bands or wax) stimulates the salivary glands to secrete saliva. Saliva consists primarily of mucin (a viscous glycoprotein), which moistens the food and helps to bind it together into a mass called a bolus, and a clear serous fluid containing the enzyme, *salivary amylase.* Salivary amylase begins the digestion of starch (a large polysaccharide), breaking it down into disaccharides, or double sugars, and glucose. The secretion of the parotid glands is mainly serous, whereas the submandibular and sublingual glands are mixed glands that produce mucin and serous components.

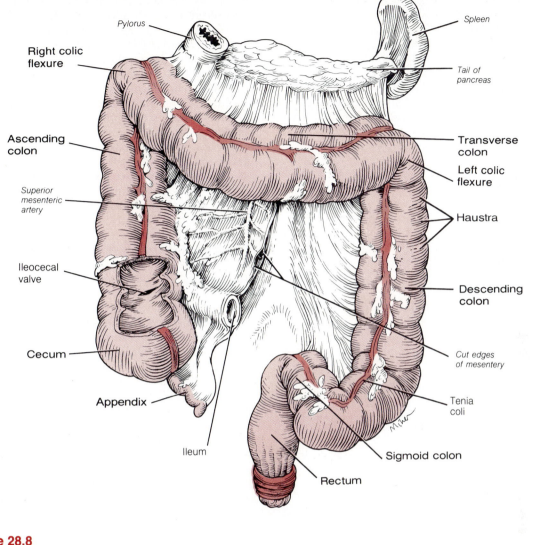

Figure 28.8

The large intestine (section of the cecum removed to show the ileocecal valve).

PANCREAS   The **pancreas** is a soft, triangular gland that extends horizontally across the posterior abdominal wall from the spleen to the duodenum (see Figure 28.1). Like the duodenum, it is a retroperitoneal organ (see Figure 28.6). It produces a whole spectrum of hydrolytic enzymes, which it secretes in an alkaline fluid into the duodenum through the pancreatic duct. Pancreatic juice is very alkaline. Its high concentration of bicarbonate ion (HCO_3^-) neutralizes the acidic chyme entering the duodenum from the stomach, enabling the pancreatic and intestinal enzymes to operate at their optimal pH. (Optimal pH for digestive activity to occur in the stomach is very acidic and results from the presence of HC1; that for the small intestine is slightly alkaline.) The pancreas also has an endocrine function: it produces the hormones, insulin and glucagon (see Exercise 22, Anatomy and Basic Function of the Endocrine Glands).

LIVER AND GALLBLADDER   The **liver** (see Figure 28.1), the largest gland in the body, is located inferior to the diaphragm, more to the right than the left side of the body. As noted, it hides the stomach from view in a superficial observation of abdominal contents. The human liver has four lobes and is suspended from the diaphragm and anterior abdominal wall by the **falciform ligament** (see Figure 28.6a).

The liver is one of the body's most important organs, and it performs many metabolic roles. However, its digestive function is to produce bile, which leaves the liver through the **hepatic duct** and then enters the duodenum through the **common bile duct.** Bile has no enzymatic action but emulsifies fats (spreads thin or breaks up large particles into smaller ones), thus creating a larger surface area for more efficient lipase activity. Without bile, very little fat digestion or absorption occurs.

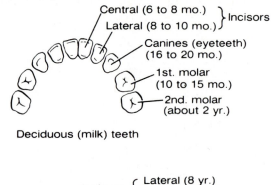

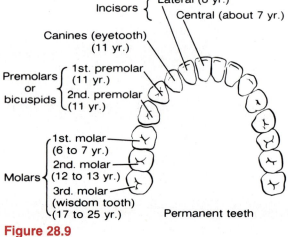

Figure 28.9

Human deciduous and permanent teeth. (Approximate time of teeth eruption shown in parentheses.)

When digestive activity is not occurring in the digestive tract, bile backs up the cystic duct and enters the **gallbladder,** a small, green sac on the inferior surface of the liver. It is stored there until needed for the digestive process. While in the gallbladder, bile is concentrated by the removal of water and some ions. When fat-rich food enters the duodenum, a hormonal stimulus causes the gallbladder to contract, releasing the stored bile and making it available to the duodenum.

If the hepatic or common bile duct is blocked (for example, by wedged gall stones), bile is prevented from entering the small intestine, begins to accumulate, and eventually backs up into the liver. This exerts pressure on the liver cells, and bile begins to enter the bloodstream. As the bile circulates through the body, the tissues become yellow or *jaundiced*.

Blockage of the ducts is just one cause of jaundice; more often it results from actual liver problems such as **hepatitis** (an inflammation of the liver) or **cirrhosis,** a condition in which the liver is severely damaged, becoming hard and fibrous. Cirrhosis is almost guaranteed in those who drink excessive alcohol for many years. ■

TEETH By the age of 21, two sets of teeth have developed (Figure 28.9). The initial set, called the

deciduous, or **milk teeth,** normally appears between the ages of 6 months and 2½ years. The first of these to erupt are the lower central incisors, an event which is usually applauded by the child's parents. The child begins to shed the deciduous teeth around the age of 6. The second set of teeth, the **permanent teeth,** gradually replace them. As the deeper permanent teeth progressively enlarge and develop, the roots of the deciduous teeth are resorbed, leading to their final shedding, and during the sixth to twelfth years, the child has mixed dentition—both permanent and deciduous teeth. Generally, by the age of 12, all of the deciduous teeth have been shed, or exfoliated.

The teeth are classified as **incisors, canines** (eye teeth), **premolars** (bicuspids), and **molars.** Teeth names reflect differences in relative structure and function. The incisors are chisel-shaped and exert a shearing action used in biting. The canines are cone-shaped or fanglike, the latter description being much more applicable to the canines of animals whose teeth are used for the tearing of food. Both the incisors and the canines have single roots. The premolars, or bicuspids, have two cusps (grinding surfaces) and typically two roots. The molars have relatively flat, broad superior surfaces specialized for the fine grinding of food and typically have two or three roots but may have more.

Dentition is described by means of a **dental formula,** which designates the numbers, types, and position of the teeth in one side of the jaw. (Since tooth arrangement is bilaterally symmetrical, it is only necessary to designate one side of the jaw.) The complete dental formula for the deciduous teeth from the medial aspect of each jaw and proceeding posteriorly is as follows:

$$\frac{\text{Upper teeth: 2 incisors, 1 canine, 0 premolars, 2 molars}}{\text{Lower teeth: 2 incisors, 1 canine, 0 premolars, 2 molars}} \times 2$$

This formula is generally abbreviated to read as follows:

$$\frac{2,1,0,2}{2,1,0,2} \times 2 = 20 \text{ (number of deciduous teeth)}$$

The 32 permanent teeth are then described by the following dental formula:

$$\frac{2,1,2,3}{2,1,2,3} \times 2 = 32 \text{ (number of permanent teeth)}$$

Although 32 is designated as the normal number of permanent teeth, not everyone develops a full complement. In many people, the No. 3 molars, commonly called the wisdom teeth, never erupt.

 Identify the four types of teeth (incisors, canines, premolars, and molars) on the jaw model or human skull.

A tooth is commonly considered to consist of two major regions, the **crown** and the **root.** A longitudinal section made through a tooth shows the following basic anatomic plan (Figure 28.10). The crown is the superior portion of the tooth; the portion of the crown visible above the **gum,** or **gingiva,** is referred to as the **clinical crown.** The entire area covered by **enamel** is called the **anatomical crown.** The crevice between the end of the anatomical crown and the upper margin of the gingiva is referred to as the **gingival sulcus** and its apical border is the **gingival margin.** Enamel is the hardest substance in the body and is fairly brittle. It consists of 95% to 97% inorganic calcium salts (chiefly $CaPO_4$) and thus is heavily mineralized. That portion of the tooth embedded in the alveolar portion of the jaw is the root, and the root and crown are connected by a slight constriction, the **neck.** The outermost surface of the root is covered by **cementum,** which is similar to bone in composition and less brittle than enamel. The cementum attaches the tooth to the **periodontal ligament (membrane)** which holds the tooth in the alveolar socket and exerts a cushioning effect. **Dentin,** which comprises the bulk of the tooth, consists of bonelike material and is medial to the enamel and cementum. The **pulp cavity** occupies the central portion of the tooth. **Pulp,** connective tissue liberally supplied with blood vessels, nerves, and lymphatics, occupies this cavity and provides the source of nutrition for the tooth tissues. Specialized cells, **odontoblasts,** reside in the outer margins of the pulp cavity and produce the dentin. Since the pulp contains the nerve supply of the tooth, it also provides tooth sensation. As the pulp cavity extends into the root it becomes the **root canal.** An opening at the root apex, the **apical foramen** provides a route of entry into the tooth for the blood vessels, nerves, and other structures from the tissues beneath.

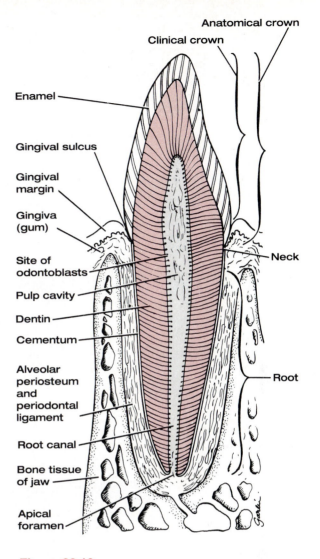

Figure 28.10

Longitudinal section of human canine tooth.

DISSECTION OF THE DIGESTIVE SYSTEM OF THE CAT

 1. Obtain your cat and secure it to the dissecting tray, dorsal surface down. Obtain all necessary dissecting instruments. If you have completed the dissection of the circulatory and respiratory systems, the abdominal cavity is already exposed and many of the digestive system structures have been previously identified. However, duplication of effort generally provides a good learning experience, so all of the digestive system structures will be traced and identified in this exercise.

2. To expose and identify the salivary glands, which secrete saliva into the mouth, remove the skin from one side of the head and clear the connective tissue away from the angle of the jaw, below the ear, and superior to the masseter muscle. Many lymph nodes are in this area and you should remove them if they

obscure the salivary glands. The cat possesses five pairs of salivary glands, but only those glands described in humans are easily localized and identified (Figure 28.11). Locate the parotid gland on the cheek just inferior to the ear. Follow its duct over the surface of the masseter muscle to the angle of the mouth. The submandibular gland is posterior to the parotid, near the angle of the jaw, and the sublingual gland is just anterior to the submandibular gland within the lower jaw. The ducts of the submandibular and sublingual glands run deep and parallel to each other and empty on the side of the frenulum of the tongue. These need not be identified on the cat.

3. To expose and identify the structures of the oral cavity, cut through the mandible with bone cutters just anterior to the angle to free the lower jaw from

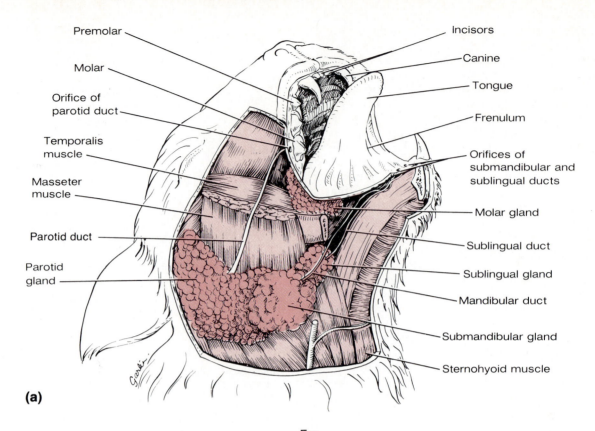

(a)

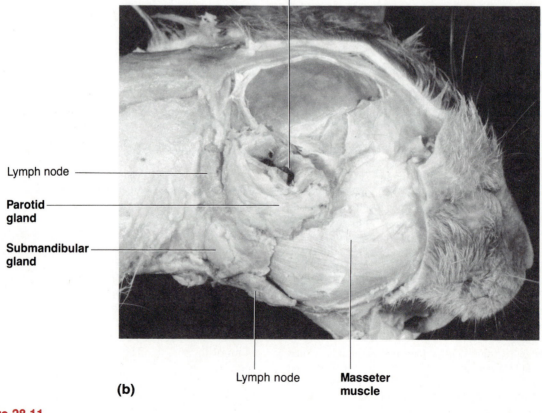

(b)

Figure 28.11

Salivary glands of the cat: (a) diagrammatic view; (b) photograph.

the maxilla. Observe the teeth of the cat. The dental formula for the adult cat is as follows:

$$\frac{3,1,3,1}{3,1,2,1} \times 2 = 30$$

What differences in diet and mode of food-getting explain the difference between cat and human dentition? _____

_____

Identify the hard and soft palates; use a probe to trace the hard palate to its posterior limits. Note the transverse ridges, or *rugae,* on the hard palate, which play a role in holding food in place while chewing.

Do these appear in humans? _____

Does the cat have a uvula? _____

Identify the oropharynx at the rear of the oral cavity and the palatine tonsils on the posterior walls at the junction between the oral cavity and oropharynx. Identify the tongue and rub your finger across its surface to feel the papillae. Some of the papillae, especially at the anterior end of the tongue, should feel sharp and bristly. These are the filiform papillae, which are much more numerous in the cat than in humans. What do you think their function is?

_____

Locate the **frenulum** attaching the tongue to the floor of the mouth. Trace the tongue posteriorly until you locate the **epiglottis,** the flap of tissue that covers the entrance to the respiratory passageway when swallowing occurs. Identify the esophageal opening posterior to the epiglottis.

4. Locate the abdominal alimentary tube structures. (If the abdominal cavity has not been previously opened, make a midline incision from the rib cage to the pubic symphysis and then make four lateral cuts—two parallel to the rib cage and two at the inferior margin of the abdominal cavity so that the abdominal wall can be reflected back while you examine the abdominal contents (Figure 28.12). Observe the shiny membrane lining the inner surface of the abdominal wall, which is the parietal peritoneum. Identify the large reddish brown liver just beneath the diaphragm and the greater omentum covering the abdominal contents. The greater omentum assists in regulating body temperature and its phagocytic cells function in body protection. Note that the greater omentum is riddled with fat deposits. Lift the greater omentum, noting its two-layered structure and attachments, and lay it to the side or remove it to make subsequent organ identifications easier. Does the liver of the cat have the same number of lobes

as the human liver? _____

Lift the liver and examine its inferior surface to locate the gallbladder, a dark, greenish sac embedded in its ventral surface. Identify the falciform ligament, a delicate layer of mesentery separating the main lobes of the liver (right and left median lobes) and attaching the liver superiorly to the abdominal wall. Also identify the thickened area along the posterior edge of the falciform ligament, the **round ligament,** or **ligamentum teres,** a remnant of the umbilical vein of the embryo.

Displace the left lobes of the liver to expose the stomach. Identify the cardiac, fundic, body, and pyloric regions of the stomach. What is the general

shape of the stomach? _____

Locate the lesser omentum, the serous membrane attaching the lesser curvature of the stomach to the liver. Make an incision through the stomach wall to expose its inner surface. Can you see the **rugae?** (When the stomach is empty, its mucosa is thrown into large folds called rugae. As the stomach fills, the rugae gradually disappear and are no longer visible.) Lift the stomach and locate the pancreas, which appears as a greyish or brownish diffuse glandular mass in the mesentery. It extends from the vicinity of the spleen and greater curvature of the stomach and wraps around the duodenum. Attempt to find the pancreatic duct as it empties into the duodenum at a swollen area referred to as the **hepatopancreatic ampulla,** or **ampulla of Vater.** Close to the pancreatic duct, locate the common bile duct and trace its course superiorly to the point where it diverges into the cystic duct (gallbladder duct) and the hepatic duct (duct from the liver). Note that the duodenum assumes a looped position.

Lift the small intestine to note the manner in which it is attached to the posterior body wall by the mesentery. Observe the mesentery closely. What types of structures do you see in this double peritoneal

fold? _____

_____

Other than providing support for the intestine, what

other functions does the mesentery have? _____

_____

Trace the course of the small intestine from its proximal, or duodenal, end to its distal, or ileal, end. Can you see any obvious differences in the external anatomy of the small intestine from one end to the

other? _____

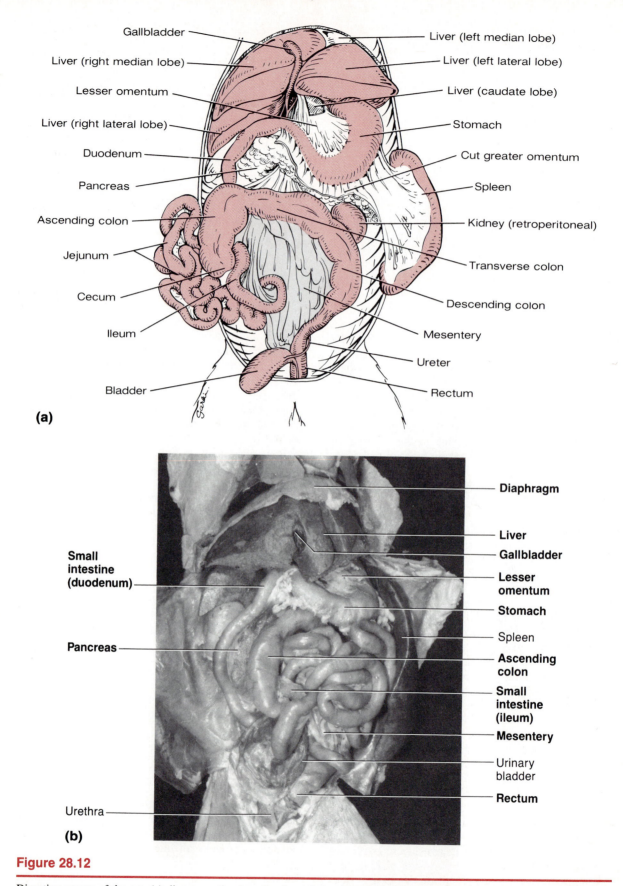

Gallbladder

Liver (right median lobe)

Lesser omentum

Liver (right lateral lobe)

Duodenum

Pancreas

Ascending colon

Jejunum

Cecum

Ileum

Bladder

Liver (left median lobe)

Liver (left lateral lobe)

Liver (caudate lobe)

Stomach

Cut greater omentum

Spleen

Kidney (retroperitoneal)

Transverse colon

Descending colon

Mesentery

Ureter

Rectum

(a)

Small intestine (duodenum)

Pancreas

Urethra

Diaphragm

Liver

Gallbladder

Lesser omentum

Stomach

Spleen

Ascending colon

Small intestine (ileum)

Mesentery

Urinary bladder

Rectum

(b)

Figure 28.12

Digestive organs of the cat: (a) diagrammatic view showing relationship to the left kidney (small intestine, liver, and greater omentum reflected); (b) photograph of digestive organs of cat. The greater omentum has been cut from its attachment to the stomach.

With a scalpel, slice open the distal portion of the ileum and flush out the inner surface with water. Feel the inner surface with your finger tip. How does it

feel? _____

_____

Use a hand lens to see if you can see any villi and to locate the areas of lymphatic tissue called Peyer's patches, which appear as scattered white patches on the inner intestinal surface.

Return to the duodenal end of the small intestine. Make an incision into the duodenum. As before, flush the surface with water, and feel the inner surface. Does it feel any different than the ileal mucosa?

_____

If so, describe the difference. _____

_____

_____

Use the hand lens to observe the villi. What differences do you see between the villi in the two areas

of the small intestine? _____

Make an incision into the junction between the ileum and cecum to locate the ileocecal valve. Observe the cecum, the initial expanded part of the large intestine. (Lymph nodes may have to be removed from this area to observe it clearly). Does the cat

have an appendix? _____

Identify the short ascending, transverse, and descending portions of the colon and the **mesocolon,** a membrane that attaches the colon to the posterior body wall. Trace the descending colon to the rectum, which penetrates the body wall, and identify the anus.

Identify the two portions of the peritoneum, the parietal peritoneum lining the abdominal wall (identified previously) and the visceral peritoneum, which is the outermost layer of the wall of the abdominal organs (serosa).

5. Prepare your cat for storage by wrapping it in paper towels wet with embalming fluid, return it to the plastic bag, and attach your name label. Wash the dissecting tray and instruments before continuing or leaving the laboratory.

MICROSCOPIC ANATOMY OF SELECTED DIGESTIVE SYSTEM AND ACCESSORY ORGANS

 Obtain a microscope and the following slides in preparation for the histologic study: salivary glands (submandibular or sublingual), pancreas, liver, cross sections of the duodenum, ileum, liver, and the stomach, and longitudinal sections of a tooth and the gastroesophageal junction.

Salivary Glands

Examine the glandular tissue under low power and then high power to become familiar with the appearance of a glandular tissue. Note the clustered arrangement of the cells around their ducts. The cells are basically triangular with their pointed ends facing the duct orifice. If possible, differentiate between the serous cells, which produce the clear enzyme-containing fluid and have granules in their cytoplasm, and the mucus-producing cells, which look hollow or have a clear cytoplasm. In many cases the serous crescents are distal to the duct and the mucus-producing cells. (Figure 28.13a may be helpful in this task.) Draw a small portion of the salivary gland tissue and label it appropriately.

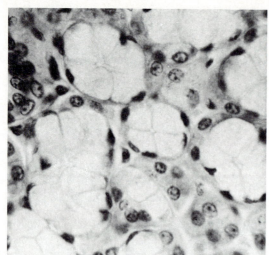

(a)

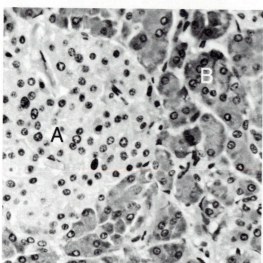

(b)

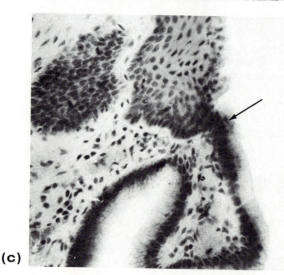

(c)

Pancreas

Observe the pancreas tissue under low power and then high power to distinguish between the lighter-staining, endocrine-producing clusters of cells (islets of Langerhans) and the deeper-staining parenchyma cells (acinar cells), which produce the hydrolytic enzymes and form the major portion of the pancreatic tissue (see Figure 28.13b). Note the arrangement of the exocrine parenchyma cells around their central ducts. Draw a small portion of the exocrine pancreatic tissue in the space provided. Appropriately label your drawing.

Small Intestine

DUODENUM Observe the tissue under low power to identify the four basic layers of the wall of the intestine—mucosa, submucosa, muscularis externa, and serosa. Refer back to Figure 28.4 if you need to refresh your memory on the tissue makeup of these layers. Identify the scattered duodenal (Brunner's) glands (mucus-producing glands) in the submucosa.

What type of epithelium do you see here? _____

_____

Notice the large leaflike villi, which increase the surface area for absorption. Note also the intestinal crypts (of Lieberkühn), invaginated areas of the mucosa between the villi. Sketch and label a small section of the duodenal wall, showing all layers and villi.

Figure 28.13.

Photomicrographs: (a) mixed salivary glands (160X); (b) pancreas, islets of Langerhans = A, acinar cells = B (100X); (c) gastroesophageal junction (100X).

ILEUM   The structure of the ileum is similar to that of the duodenum except that the villi tend to be less elaborate (most of the absorption has occurred by the time the ileum is reached). Observe the villi, and identify the four layers of the wall and the large Peyer's patches.

What tissue comprises Peyer's patches? _____

_____

Stomach

Examine the tissue under low power to locate the muscularis externa; then move to high power to closely examine this layer. How many smooth muscle layers

are visible? _____ How does this correlate with the churning movements performed by the

stomach? _____

_____

_____

_____

Identify the gastric glands and the gastric pits. If the section is taken from the stomach fundus and appropriately stained, you can identify the blue-staining **zymogenic, or chief, cells,** which produce pepsinogen, and the red-staining **parietal cells,** which secrete HCl, in the gastric glands. Draw a small section of the stomach wall and label it appropriately.

Gastroesophageal Junction

Examine the slide under low power, and scan it to localize the junction between the end of the esophagus and the beginning of the stomach, the gastroesophageal junction. Compare your view to that shown in Figure 28.13(c). How does the epithelium of the

esophagus differ from that of the stomach? _____

_____

_____

Why do you suppose this is so? _____

_____

_____

Tooth

Observe a slide of a longitudinal section of a tooth and compare your observations with the structures detailed in Figure 28.10. Identify as many of these structures as possible. Make a rough drawing of the tooth you are observing and label it appropriately.

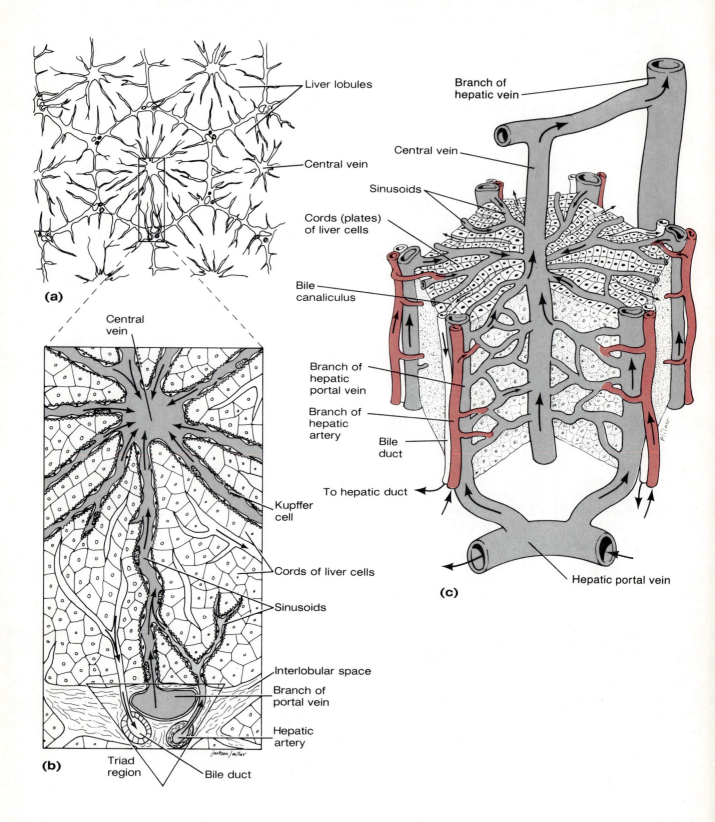

Figure 28.14

Microscopic anatomy of the liver (diagrammatic view); (a) several liver lobules (cross section); (b) enlarged view of a portion of one liver lobule (cross section); (c) portion of one liver lobule (three-dimensional representation). Arrows show direction of bile and blood flow.

The Liver

The liver (Figure 28.14) is very important in the initial processing of the nutrient-rich blood draining the digestive organs. It is composed of structural and functional units called **lobules.** Each lobule is a basically cylindrical structure consisting of cordlike arrangements of parenchyma cells, which radiate outward from a central vein running upward in the longitudinal axis of the lobule. At each of the six corners of the lobule is a **triad** region, so named because three basic structures are always present: a branch of the hepatic artery (the functional blood supply of the liver), a branch of the hepatic portal vein (carrying nutrient-rich blood from the digestive viscera), and a bile duct. Between the liver parenchyma cells are blood-filled spaces, or **sinusoids,** through which blood from the hepatic portal vein and hepatic artery percolates past the parenchyma cells. Special phagocytic cells, **Kupffer cells,** line the sinusoids and remove debris such as bacteria from the blood as it flows past, while the parenchyma cells pick up oxygen and nutrients. Much of the glucose transported to the liver from the digestive system is stored as glycogen in the liver for later use, and amino acids are taken from the blood by the liver cells and utilized to make plasma proteins. The sinusoids empty into the central vein, and the blood ultimately drains from the liver into the hepatic vein. Bile is also continuously being made by the parenchyma cells. It flows through tiny canals, the **bile canaliculi,** which run between adjacent parenchyma cells toward the bile duct branches in the triad regions, where the bile eventually leaves the liver. Note that the direction of blood and bile flow in the liver lobule is exactly opposite.

Examine a slide of liver tissue and identify as many of the structural features illustrated in Figure 28.14 as possible. Also examine a three-dimensional model of the liver if this is available. Draw your observations below.

Anatomy of the Urinary System

<table>
<tr><td>

OBJECTIVES

1. To describe the overall function of the urinary system.

2. To identify on an appropriate diagram, torso model, or dissection specimen, the urinary system organs and to describe the general function of each.

3. To define *micturition,* and to explain the pertinent differences in the control of the two bladder sphincters (internal and external).

4. To compare the course and length of the urethra in males and females.

5. To identify the following regions of the dissected kidney (longitudinal section): hilus, cortex, medulla, medullary pyramids, major and minor calyces, pelvis, renal columns, and capsule layers.

6. To trace the blood supply of the kidney from the renal artery to the renal vein.

7. To define the nephron as the physiological unit of the kidney and to describe its anatomy.

8. To define *glomerular filtration, tubular reabsorption,* and *tubular secretion,* and to note the nephron areas involved in these processes.

9. To recognize microscopic or diagrammatic views of the histologic structure of the kidney and bladder.

</td><td>

MATERIALS

Human dissectible torso model and/or anatomical chart of the human urinary system

3-dimensional model of the cut kidney and of the nephron (if available)

Dissection animal, tray, and instruments

Pig or sheep kidney, doubly or triply injected

Prepared histologic slides of a longitudinal section of kidney and cross sections of the bladder

Compound microscope

</td></tr>
</table>

Metabolism of nutrients by the body produces various wastes (carbon dioxide, nitrogenous wastes, ion excesses, and so on) that must be eliminated from the body if normal function is to continue. Although excretory processes involve several organ systems (the lungs excrete carbon dioxide and the skin glands excrete salts and water), it is the **urinary system** that is primarily concerned with the removal of nitrogenous wastes from the body. In addition to this purely excretory function, the kidney maintains the electrolyte, acid-base, and fluid balance of the blood and is thus a major, if not *the* major, homeostatic organ of the body.

To perform its functions, the kidney acts much like a blood filter. It allows toxins, metabolic wastes, and excess ions to leave the body in the urine, while simultaneously retaining needed substances and

returning them to the blood. Malfunction of the urinary system, particularly of the kidneys, leads to a failure in homeostasis, which, unless corrected, results in death.

GROSS ANATOMY OF THE HUMAN URINARY SYSTEM

The urinary system (Figure 29.1) consists of the paired kidneys and ureters and the single urinary bladder and urethra. The kidneys perform the functions described above and manufacture urine in the process. The remaining organs of the system provide temporary storage reservoirs for urine or transport urine from one body region to another.

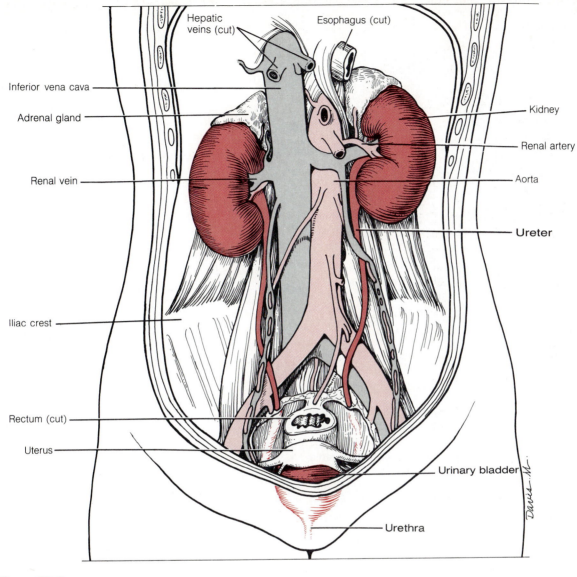

Figure 29.1

Anterior view of the urinary organs of a human female. (Most unrelated abdominal organs have been removed.)

 Examine the human torso model, a large anatomic chart, or a three-dimensional model of the urinary system to locate and study the anatomy and relationships of the urinary organs.

1. Locate the paired **kidneys** on the dorsal body wall in the superior lumbar region. Note that they are not positioned at exactly the same level: the right kidney is slightly lower than the left kidney. Why do you suppose this is so? _____

_____

_____

In the living person, fat deposits hold the kidneys in place in a retroperitoneal position. When the fatty material is reduced or deficient in amount (in cases of rapid weight loss or in very thin individuals), the kidneys are less securely anchored to the body wall and may drop to a lower or more inferior position in the abdominal cavity. This phenomenon is called **ptosis.** ■

2. Observe the **renal arteries** as they diverge from the descending aorta and plunge into the indented medial region (**hilus**) of each kidney. Note also the **renal veins,** which drain the kidneys (circulatory drainage) and the two **ureters,** which drain the urine from the kidneys and conduct it by peristalsis to the bladder for temporary storage.

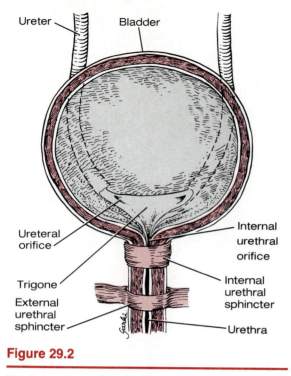

Figure 29.2

Detailed structure of the bladder.

3.   Locate the **urinary bladder,** and observe the point of entry of the two ureters into this organ. Also locate the single **urethra,** which drains the bladder. The triangular region of the bladder, which is delineated by these three openings (two ureteral and one urethral orifice), is referred to as the **trigone** (Figure 29.2). Although the formation of urine by the kidney is a continuous process, urine is usually removed from the body when voiding is convenient. In the meantime the bladder provides temporary storage for urine.

Voiding, or **micturition,** is the process in which urine empties from the bladder. Two sphincter muscles or valves, the **internal urethral sphincter** (more superiorly located) and the **external urethral sphincter** (more inferiorly located) control emptying of the bladder. Ordinarily, the bladder continues to collect urine until about 200 ml have accumulated, at which time the stretching of the bladder wall activates stretch receptors. Impulses transmitted to the central nervous system in turn produce reflex contractions of the bladder wall through parasympathetic nervous system pathways. As the contractions increase in force and frequency, the stored urine is forced past the internal sphincter, which is a smooth muscle involuntary sphincter, into the superior part of the urethra. It is then that a person feels the urge to void. The inferior external sphincter consists of skeletal muscle and is also reinforced by skeletal muscles of the pelvic diaphragm which are voluntarily controlled. If it

is not convenient to void, the opening of this sphincter can be inhibited. Conversely, if the time is convenient, the sphincter may be relaxed and the stored urine flushed from the body. If voiding is inhibited, the reflex contractions of the bladder cease temporarily and urine continues to accumulate in the bladder. After another 200 to 300 ml of urine have been collected, the micturition reflex will again be initiated.

Lack of voluntary control over the external sphincter is referred to as **incontinence.** Incontinence is a normal phenomenon in children 2 years old or younger, as they have not yet gained control over the voluntary sphincter. Past this age, incontinence is generally a result of emotional problems, bladder irritability, or some other pathologic condition of the urinary tract.

4.   Follow the course of the urethra to the body exterior. In the male, it is approximately 20 cm (8 inches) long; it travels the length of the **penis,** and opens at its tip. The urethra of the male has a dual function: it is a urine conduit to the body exterior, and it provides a passageway for the ejaculation of sperm. Thus, in the male, the urethra is part of both the urinary and reproductive systems. In the female, the urethra is very short, approximately 4 cm (1½ inches) long, and it travels downward and slightly forward from the bladder to the external urethral opening or orifice. There are no common urinary–reproductive pathways in the female, and the female urethra serves only to transport urine to the body exterior.

GROSS INTERNAL ANATOMY OF THE PIG OR SHEEP KIDNEY

1.   Obtain a preserved sheep or pig kidney, dissecting pan, and instruments. Observe the kidney to identify the **renal capsule,** a smooth transparent membrane that adheres tightly to the kidney tissue.

2.   Find the ureter, renal vein, and renal artery at the hilus (indented) region. The renal vein has the thinnest wall and will be collapsed. The ureter is the largest of these structures and has the thickest wall.

3.   Make a cut through the longitudinal axis (frontal section) of the kidney and locate the anatomic areas described below and depicted in Figure 29.3.

Kidney cortex: the outer kidney region, which is lighter in color. (If the kidney is double-injected with latex, you will see a predominance of red and blue latex specks in this region indicative of the rich vascular supply.)

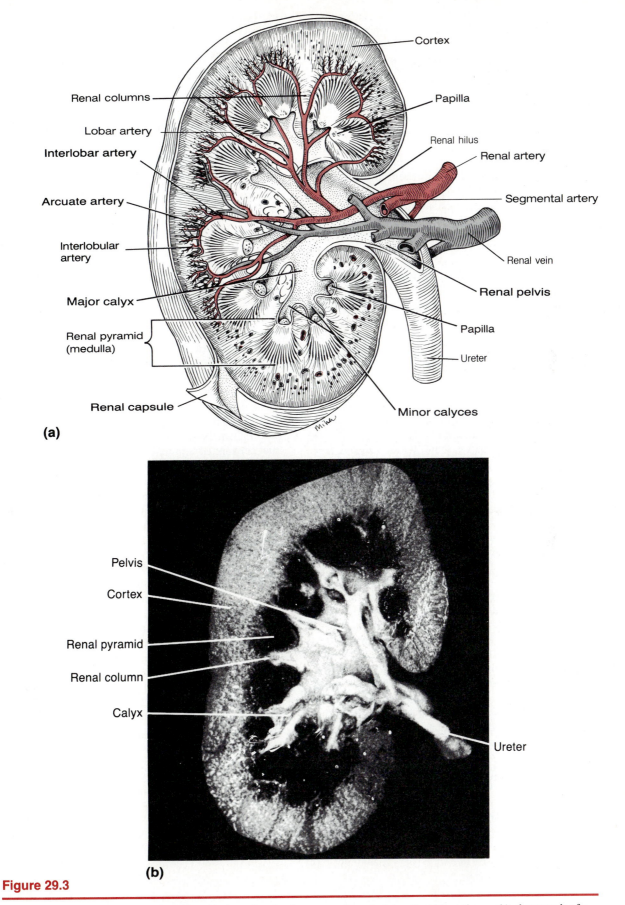

(a)

(b)

Figure 29.3

Frontal section of a kidney: (a) diagrammatic view, showing larger arteries supplying the kidney tissue; (b) photograph of a pig kidney. (Photo courtesy of Ann Allworth)

Medullary region: deep to the cortex; a darker, reddish-brown color. The medulla is segregated into triangular regions that have a striped, or striated, appearance—the **medullary pyramids.** The base of each pyramid faces toward the cortex; its **apex,** or **papilla,** points to the innermost kidney region.

Renal columns: Areas of tissue, more like the cortex in appearance, which segregate and dip downward between the pyramids.

Renal pelvis: Medial to the hilus; a relatively flat, basinlike cavity that is continuous with the **ureter,** which exits from the hilus region. Fingerlike extensions of the pelvis should be visible. The larger, or primary, extensions are called the **major calyces;** subdivisions of the major calyces are the **minor calyces.** Note that the minor calyces terminate in cuplike areas that enclose the apexes of the medullary pyramids and collect urine draining from the pyramidal tips into the pelvis.

4. If the preserved kidney is doubly or triply injected, follow the renal blood supply from the renal artery to the **glomeruli.** The glomeruli appear as little red and blue specks in the cortex region. (See Figures 29.3 and 29.4 and the discussion below.)

Because the kidneys continuously cleanse the blood and adjust its composition, it is not surprising that they have a rich vascular supply. Approximately a fourth of the total blood flow of the body is delivered to the kidneys each minute, by the **renal arteries.** As a renal artery approaches the kidney, it breaks up into several branches called **segmental arteries,** which enter the hilus. Each segmental artery, in turn, divides into several **lobar arteries.** The lobar arteries branch to form **interlobar arteries,** which ascend toward the cortex in the renal column areas. At the top of the medullary region, these arteries give off arching branches, the **arcuate arteries,** which curve over the bases of the medullary pyramids. Small **interlobular arteries** branch off the arcuate arteries and ascend into the cortex, giving off the individual **afferent arterioles,** which provide the capillary networks (**glomeruli** and **peritubular capillary beds**) supplying the nephrons, or functional units, of the kidney. Blood draining from the nephron capillary networks in the cortex enters the **interlobular veins** and then drains through the **arcuate veins** and the **interlobar veins** to finally enter the **renal vein** in the pelvis region. (There are no lobar or segmental veins.)

MICROSCOPIC ANATOMY OF THE KIDNEY AND BLADDER

Obtain prepared slides of kidney and bladder tissue, and a compound microscope.

Kidney

Each kidney contains approximately one million nephrons, which are the anatomical units responsible for the filtration, reabsorption, and secretion activities of the kidney. Figure 29.4 depicts the detailed structure and the relative positioning of the nephrons in the kidney.

Each nephron consists of two major structures: the **glomerulus** (a capillary knot) and the **renal tubule.** During embryologic development, each renal tubule begins as a blind-ended tubule that gradually encloses an adjacent capillary cluster, or glomerulus. The enlarged end of the tubule encasing the glomerulus is the **glomerular (Bowman's) capsule,** and its inner, or visceral, wall consists of highly specialized cells called **podocytes.** Podocytes have long branching processes that interdigitate with those of other podocytes and cling to the endothelial wall of the glomerular capillaries, thus forming a very porous epithelial membrane surrounding the glomerulus. The glomerular-capsule complex is sometimes called the **renal corpuscle.**

The rest of the tubule is approximately 3 cm (1.25 inches) long. As it emerges from the glomerular capsule, it becomes highly coiled and convoluted, drops down into a long hairpin loop, and then again coils and twists before entering a collecting duct, or tubule. In order from the glomerular capsule, the anatomical areas of the tubule are: the **proximal convoluted tubule, loop of Henle** (descending and ascending limbs), and the **distal convoluted tubule.** The wall of the renal tubule is composed almost entirely of cuboidal epithelial cells, with the exception of the descending limb of the loop of Henle, which is simple squamous epithelium. The lumen surfaces of the cuboidal cells in the proximal convoluted tubule have dense microvilli (a cellular modification that greatly increases the surface area exposed to the lumen contents, or filtrate). Microvilli also occur on cells of the distal convoluted tubule but in much reduced numbers.

Most nephrons, called **cortical nephrons,** are located entirely within the cortex. However, parts of the loops of Henle of the **juxtamedullary nephrons** (located close to the cortex—medulla junction), penetrate well into the medulla. The **collecting tubules,** each of which receives urine from many nephrons, run downward through the medullary pyramids, giving them their striped appearance, to empty the final urinary product into the calyces and pelvis of the kidney.

The function of the nephron depends on several unique features of the renal circulation. The capillary vascular supply consists of two distinct capillary

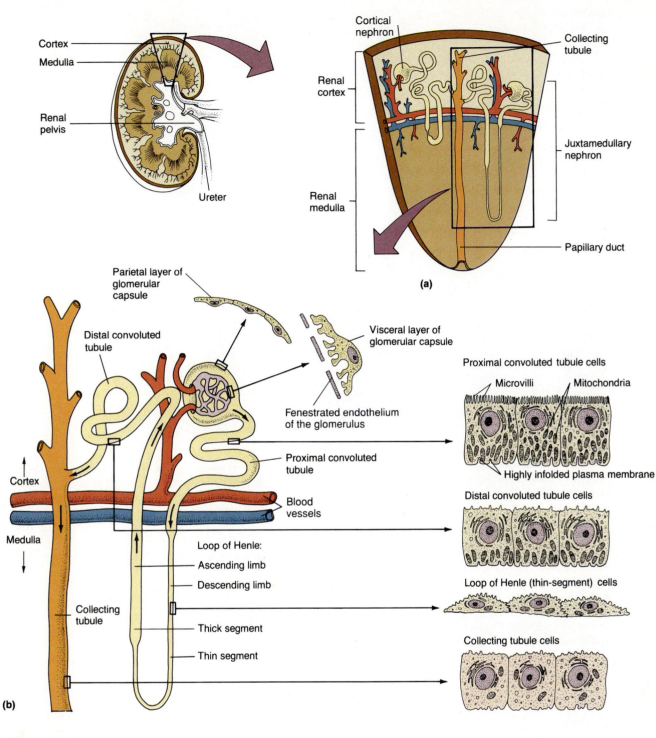

Figure 29.4

Location and structure of nephrons: (a) Wedge-shaped section (lobule) of kidney tissue, indicating the location of nephrons in the kidney; (b) Schematic view of a nephron depicting the structural characteristics of epithelial cells forming its various regions.

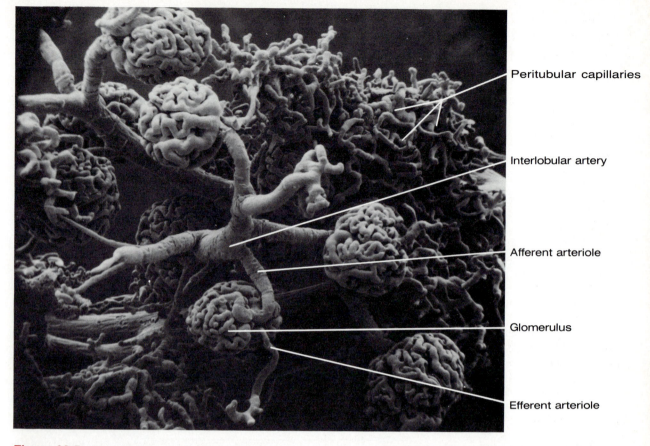

Peritubular capillaries

Interlobular artery

Afferent arteriole

Glomerulus

Efferent arteriole

Figure 29.5

Scanning electron micrograph of the blood vessels associated with glomeruli (206X). (From *Tissues and Organs: A Text-Atlas of Scanning Electron Microscopy* by Richard G. Kessel and Randy H. Kardon. W.H. Freeman and Company. Copyright © 1979.)

beds, the **glomerulus** and the **peritubular capillary bed.** Vessels leading to and from the first capillary bed, the glomerulus, are both arterioles: **the afferent arteriole** feeds the bed while the **efferent arteriole** drains it. The efferent arteriole then breaks up into the second capillary network, the peritubular capillaries (and *vasa recta* in juxtamedullary nephrons), which surrounds the proximal and distal convoluted tubules and the loop of Henle. The peritubular capillaries then drain into an interlobular vein, which leaves the cortex. A scanning electron micrograph showing the relationships of these two capillary beds appears in Figure 29.5.

The glomerular capillary bed has no parallel elsewhere in the body. It is a high-pressure bed along its entire length. This high pressure is a result of two major factors: the bed is fed and drained by arterioles (arterioles are high-resistance vessels as opposed to venules, which are low-resistance vessels), and the afferent feeder arteriole is larger in diameter than the efferent arteriole that drains the bed. The high hydrostatic pressure created by these two anatomic features forces out fluid and blood components smaller than proteins from the glomerulus into the glomerular capsule.

As noted earlier, urine formation is a result of three processes: filtration, reabsorption, and secretion. Filtration is the role of the glomerulus and is largely a passive process in which a portion of the blood passes from the glomerular bed into the glomerular capsule. This filtrate enters the proximal convoluted tubule where tubular reabsorption and secretion begin. During tubular reabsorption, many of the filtrate components move through the tubule cells and return to the blood in the peritubular capillaries. Some of this reabsorption is passive such as that of water, which passes by osmosis, but the reabsorption of most substances depends on active transport processes and is highly selective. Which substances are reabsorbed at any time depends on the composition of the blood and needs of the body at that time. Substances that are almost entirely reabsorbed from the filtrate include water, glucose, and amino acids. Various ions are selectively reabsorbed or allowed to go out in the urine according to what is required to maintain appropriate blood pH and electrolyte composition. Waste products (urea, creatinine, uric acid, and drug metabolites) are reabsorbed to a much lesser degree or not at all. Most (75% to 80%) of the tubular reabsorption occurs in the proximal convoluted tubule: the balance occurs in other tubular areas, primarily the distal parts of the tubule.

Tubular secretion is essentially the reverse process. Substances such as hydrogen and potassium ions and ammonia move either from the blood of the peritubular capillaries through the tubular cells or from the tubular cells into the filtrate to be disposed of in the urine. This process is particularly important for the disposal of selected metabolites not already in the filtrate or as an adjunct method for controlling blood pH.

 Observe a model of the nephron before continuing on with the microscope study of the kidney.

Hold the longitudinal section of the kidney up to the light to identify cortical and medullary areas. Scan the slide under low power. Move the slide so that you can see the cortical area. Identify a glomerulus, which appears as a ball of tightly packed material containing many small nuclei. It is usually delineated by a vacant-appearing region (corresponding to the space between the visceral and parietal layers of the glomerular capsule) that surrounds it. Note that the renal tubules are cut at various angles. Also try to differentiate between the thin-walled loop of Henle portion of the tubules and the cuboidal epithelium of the proximal convoluted tubule, which has microvilli. Draw a glomerulus and a small portion of the surrounding tissue below and label the glomerulus.

Bladder

Scan the bladder tissue. Note the heavy muscular wall (detrusor muscle), which consists of three irregularly arranged muscular layers. The innermost and outermost muscle layers are arranged longitudinally; the middle layer is arranged circularly. Attempt to differentiate the three layers.

Observe the mucosa with its highly specialized transitional epithelium. (Although the organ is the ureter, this type of epithelium is illustrated in Color Plate 10 on the inside of the back cover.) The plump, transitional epithelial cells have the ability to slide over one another, thus decreasing the thickness of the mucosa layer as the bladder fills and stretches to accommodate the increased urine volume. Depending on the degree of stretching of the bladder, the mucosa may be three to eight cell layers thick. In the space at the top of the next column, draw a small section of the bladder wall, and label all regions or tissue areas.

DISSECTION OF THE CAT URINARY SYSTEM

The structures of the reproductive and urinary systems are often considered together as the urogenital system, since they have common embryologic origins. However, the emphasis in this dissection is on identifying the structures of the urinary tract (Figures 29.6 and 29.7) with only a few references to contiguous reproductive structures. The anatomy of the reproductive system is studied in Exercise 30.

 1. Obtain your dissection specimen, and pin or tie its limbs to the dissection tray. Reflect the abdominal viscera (most importantly the small intestine) to locate the kidneys high on the dorsal body wall. Note that the kidneys in the cat, as well as in the human, are retroperitoneal (behind the peritoneum).

2. Carefully remove the peritoneum, and clear away the bed of fat that invests the kidneys. Locate the adrenal (suprarenal) glands lying superiorly and medial to the kidneys.

3. Identify the renal artery (red latex-injected), the renal vein (blue latex-injected), and the ureter at the hilus region of the kidney. (You may find two renal veins leaving one kidney in the cat but not in humans.)

4. Trace the ureters to the urinary bladder, a smooth muscular sac located superiorly to the small intestine. If your cat is a female, be careful not to confuse the ureters with the uterine tubes, which lie superior to the bladder in the same general region. (See Figure 29.7 and Color Plate B.) Observe the sites where the ureters enter the bladder. How would you describe

the entrance point anatomically? _____

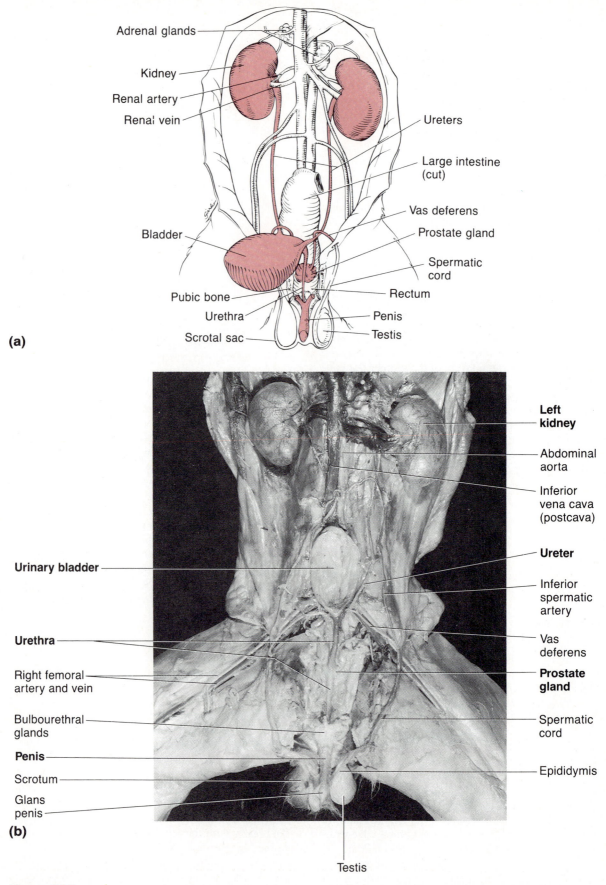

(a)

Adrenal glands
Kidney
Renal artery
Renal vein
Ureters
Large intestine (cut)
Vas deferens
Prostate gland
Spermatic cord
Bladder
Rectum
Pubic bone
Penis
Urethra
Testis
Scrotal sac

(b)

Left kidney
Abdominal aorta
Inferior vena cava (postcava)
Ureter
Inferior spermatic artery
Vas deferens
Prostate gland
Spermatic cord
Epididymis
Testis
Urinary bladder
Urethra
Right femoral artery and vein
Bulbourethral glands
Penis
Scrotum
Glans penis

Figure 29.6

Urinary system of the male cat (reproductive structures also indicated): (a) diagrammatic view; (b) photograph of male urogenital system. See dissection photo, Color Plate A.

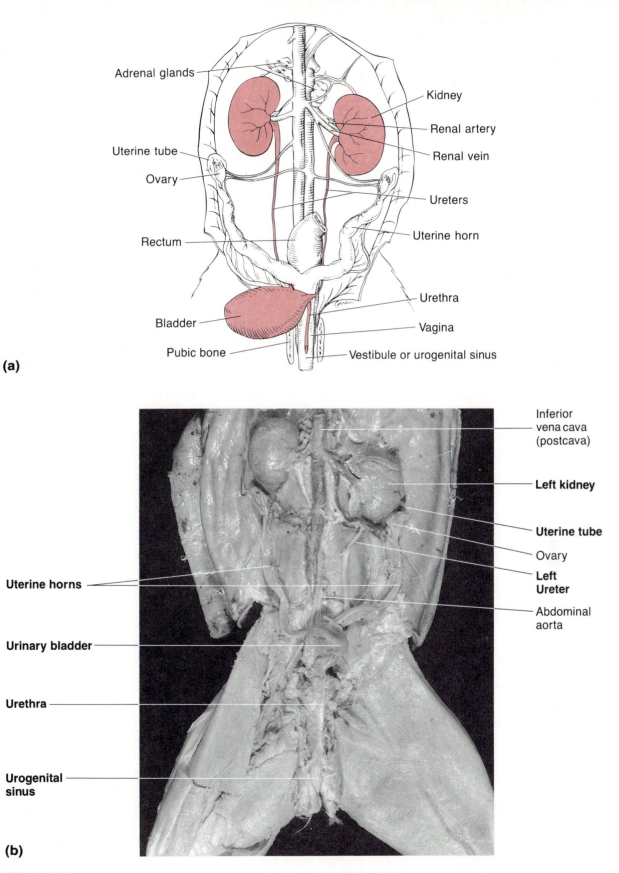

(a)

(b)

Figure 29.7

Urinary system of the female cat (reproductive structures also indicated): (a) diagrammatic view; (b) photograph of female urogenital system. See dissection photo, Color Plate B.

5. Cut through the bladder wall, and examine the region of the urethral exit to see if you can discern any evidence of the internal sphincter.

6. If your cat is a male, identify the prostrate gland (part of the male reproductive system), which encircles the neck of the bladder (see Figure 29.6). Note that the urinary bladder is somewhat fixed in position by ligamentous attachments.

7. Using a probe, trace the urethra as it exits from the bladder to its terminus in the **urogenital sinus*** (opening into the vagina) in the female cat or into the penis of the male. Dissection to expose the urethra along its entire length should not be done at this time because of possible damage to the reproductive structures, which you will study in Exercise 30.

8. Before cleaning up the dissection materials, observe a cat of the opposite sex.

---

*In the human female, the urethra does not empty into the vagina but has a separate external opening located above the vaginal orifice.

Anatomy of the Reproductive System

EXERCISE 30

OBJECTIVES

1. To discuss the general function of the reproductive system.
2. To identify and name the structures of the male and female reproductive systems when provided with an appropriate model or diagram, and to discuss the general function of each.
3. To define *semen,* discuss its composition, and name the organs involved in its production.
4. To trace the pathway followed by a sperm from its site of formation to the external environment.
5. To describe the structure of the penis and relate its structure to its erectile function.
6. To name the exocrine and endocrine products of the testes, indicating the cell types or structures responsible for the production of each.
7. To describe the microscopic structure of the epididymis, and to relate structure to function.
8. To define *ejaculation, erection,* and *gonad.*
9. To discuss the function of the fimbriae and ciliated epithelium of the uterine (fallopian) tubes.
10. To identify homologous structures of the male and female systems.
11. To discuss the microscopic structure of the ovary and be prepared to identify the following ovarian structures: primary follicle, graafian follicle, and corpus luteum, and to state the hormonal products of the last two structures. (At option of instructor.)
12. To define *endometrium, myometrium,* and *ovulation.*
13. To identify the fundus, body, and cervical regions of the uterus.
14. To identify the major reproductive structures of the male and female dissection animal, and to recognize and discuss pertinent differences between the reproductive structures of the human and the dissection animal.

MATERIALS

Models or large laboratory charts of the male and female reproductive tracts
Prepared microscope slides of cross sections of the penis and epididymis.
Compound microscope
Dissection animal, tray, and instruments
Bone clippers

Most simply stated, the biologic function of the reproductive system is to perpetuate the species. Thus the reproductive system is unique, since the other organ systems of the body function primarily to sustain the existing individual.

The essential organs of reproduction—the testes and the ovaries—are those that produce the germ cells. The reproductive role of the male is to manufacture sperm and to deliver them to the female reproductive tract. The female, in turn, produces eggs. If the time is suitable, the combination of sperm and egg produces a fertilized egg, which is the first cell of a new individual. Once fertilization has occurred, the female uterus provides a nurturing, protective environment in which the embryo, later called the fetus, develops until birth.

Although the drive to reproduce is strong in all animals, in humans this drive is also intricately related to nonbiologic factors. Emotions and social considerations often enhance or thwart its expression.

GROSS ANATOMY OF THE HUMAN MALE REPRODUCTIVE SYSTEM

The primary reproductive organs of the male are the **testes,** the male gonads, which have both an exocrine (sperm production) and an endocrine (testosterone production) function. All other reproductive structures are conduits or sources of secretions, which aid

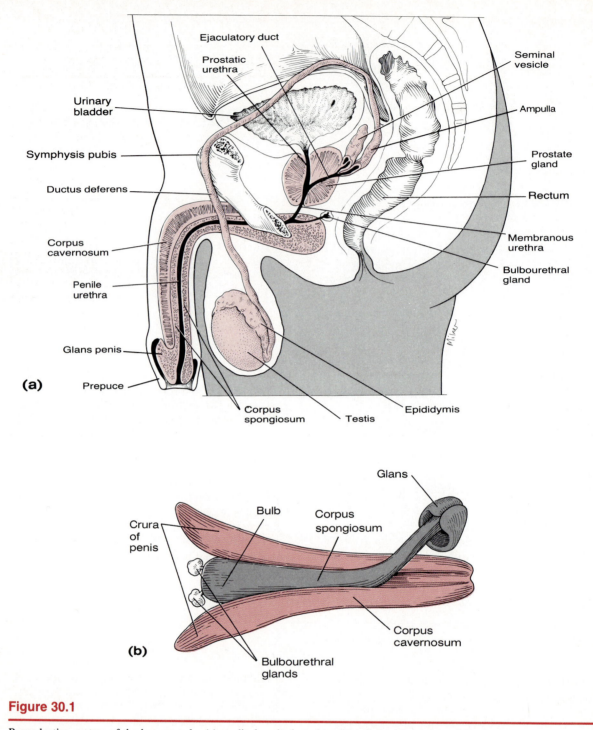

Figure 30.1

Reproductive system of the human male: (a) medical sagittal section; (b) inferior intact view of the penis.

in the safe delivery of the sperm to the body exterior or female reproductive tract.

As the following organs and structures are described, locate them on Figure 30.1, and then identify them on a three-dimensional model of the male reproductive system or on a large laboratory chart.

The paired oval testes lie in the **scrotal sac** outside the abdominopelvic cavity. The temperature there, approximately 34°C (93° F), is slightly lower than body temperature, a requirement for production of viable sperm.

The accessory structures forming the duct system are the epididymis, the ductus deferens, the ejaculatory duct, and the urethra. The **epididymis** is an elongated structure running up the posterolateral aspect of the testis and capping it superiorly. The epididymis forms the first portion of the duct system and provides a site for immature sperm that enter it from the testis to complete their maturation process. The **ductus deferens** (sperm duct) arches upward from the epididymis through the inguinal canal into the pelvic cavity and courses over the superior aspect of the bladder. In life, the ductus deferens (also called the *vas deferens*) is enclosed along with blood vessels and nerves in a connective tissue sheath called the **spermatic cord.** The terminus of the ductus deferens enlarges to form the region called the **ampulla,** which empties into the **ejaculatory duct.** Contraction of the ejaculatory duct propels the sperm through the prostate gland to the **prostatic urethra,** which in turn empties into the **membranous urethra** and then into the **penile urethra,** which runs through the length of the penis to the body exterior.

The accessory glands include the prostate gland, the paired seminal vesicles, and bulbourethral glands. These glands produce **seminal fluid,** the liquid medium in which sperm leave the body. The **seminal vesicles,** which produce about 60% of the seminal fluid, are close to the terminus of the ductus deferens. They produce a viscous secretion containing fructose (a simple sugar) and other substances that nourish the sperm passing through the tract. The duct of each seminal vesicle merges with a ductus deferens to form an ejaculatory duct; thus sperm and seminal fluid enter the urethra together. The **prostate** encircles the urethra just inferior to the bladder. It secretes a milky alkaline fluid into the urethra, which is believed to have a role in activating the sperm.

Hypertrophy of the prostate gland, commonly seen in old age, constricts the urethra, a troublesome condition that makes urination difficult. ■

The **bulbourethral glands (Cowper's glands)** are tiny, pear-shaped glands inferior to the prostate. They produce a clear, mucus-containing alkaline solution that drains into the membranous urethra. This secretion is believed to wash the residual urine out of the urethra when ejaculation of semen (sperm plus seminal fluid) occurs. The relative alkalinity of all of these glandular secretions may also have a role in buffering the sperm against the acidity of the female reproductive tract.

The **penis,** part of the external genitalia of the male along with the scrotal sac, is the copulatory organ of the male and is designed to deliver sperm into the female reproductive tract. It consists of a shaft, which terminates in an enlarged tip, the **glans.** The skin covering the penis is loosely applied, and it reflects downward to form a circular fold of skin, the **prepuce,** or **foreskin,** around the proximal end of the glans. Internally, the penis consists primarily of three elongated cylinders of erectile tissue, which become engorged with blood during sexual excitement. This causes the penis to become rigid and enlarged so that it may more adequately serve as a penetrating device. This event is called **erection.** The paired dorsal cylinders are the **corpora cavernosa.** The single ventral **corpus spongiosum** surrounds the penile urethra.

MICROSCOPIC ANATOMY OF SELECTED REPRODUCTIVE ORGANS

Testis

Each testis is covered by a dense connective tissue capsule called the **tunica albuginea** (literally, white tunic). Extensions of this sheath enter the testis, dividing it into a number of lobes, each of which houses one to four highly coiled **seminiferous tubules,** the sperm-forming factories (Figure 30.2). The seminiferous tubules of each lobe converge to empty the sperm into another tubular region, the **rete testis,** at the mediastinum of the testis. Sperm traveling through the rete testis then enter the epididymis, located on the exterior aspect of the testis, as previously described. Lying between the seminiferous tubules and softly padded with connective tissue are the **interstitial cells,** which produce testosterone, the hormonal product of the testis.

Epididymis

Obtain a cross section of the epididymis. Look for sperm in the lumen. Examine the composition of the tubule wall carefully. Identify the smooth muscle layer and the pseudostratified columnar epithelium bearing stereocilia. What do you think the function of the smooth muscle is?

_____

_____

Draw what you observe below and label appropriately.

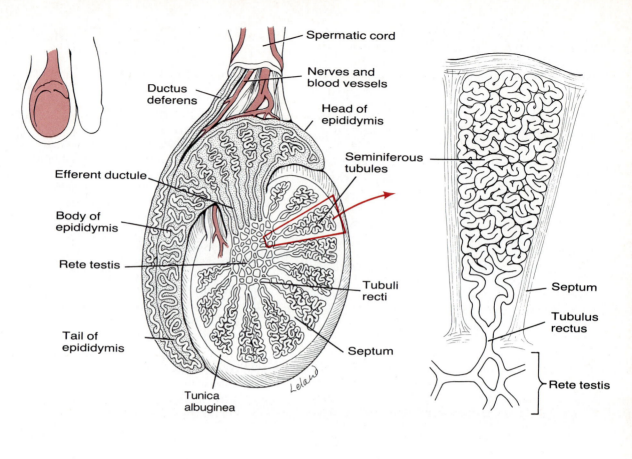

Figure 30.2

Longitudinal-section view of the testis showing seminiferous tubules.

Penis

Obtain a cross section of the penis. Scan the tissue under low power to identify the urethra and the cavernous bodies. Observe the lumen of the urethra carefully. What type of epithelium do you

see? _____

Explain the function of this type of epithelium.

_____

_____

_____

_____

Draw a diagram of the cross section of the penis below. Label the corpus cavernosum, corpus spongiosum, and urethra.

GROSS ANATOMY OF THE HUMAN FEMALE REPRODUCTIVE SYSTEM

The **ovaries** (female gonads) are the primary reproductive organs of the female. Like the testes of the male, the ovaries produce both an exocrine product (the eggs, or ova) and endocrine products (estrogens and progesterone). The other accessory structures of the female reproductive system transport, house, nurture, or otherwise serve the needs of the reproductive cells and/or the developing fetus.

The reproductive structures of the female are generally considered in terms of internal organs and external organs, or external genitalia. As you read the descriptions of these structures, locate them on Figures 30.3 and 30.4 and then on the female reproductive system model or large laboratory chart.

The **external genitalia (vulva)** consist of the mons pubis, labia majora and minora, the clitoris, urethral and vaginal orifices, the hymen, and the greater vestibular glands. The **mons pubis** is a rounded fatty eminence overlying the pubic symphysis. Running inferiorly and posteriorly from the mons pubis are two elongated, pigmented, hair-covered skin folds, the **labia majora,** which enclose two smaller hair-free folds, the **labia minora.** The labia majora are homologous to the scrotum of the male. The labia minora, in turn, enclose a region called the **vestibule,** which contains many structures: the clitoris, most anteriorly, followed by the urethral orifice and the vaginal orifice. The diamond-shaped region between the anterior terminus of the labia minora folds, the ischial tuberosities laterally, and the anus posteriorly is called the **perineum.**

The **clitoris** is a small protruding structure, homologous to the male penis, and like its counterpart is composed of highly sensitive, erectile tissue. It is hooded by skin folds of the anterior labia minora, referred to as the **prepuce of the clitoris.** The urethral orifice, which lies posterior to the clitoris, is the outlet for the urinary system and has no reproductive function in the female. The vaginal opening is partially closed by a thin fold of mucous membrane called the **hymen** and is flanked by the mucus-secreting **greater vestibular (Bartholin's) glands,** which lubricate the distal end of the vagina during coitus. (These glands are not depicted in the illustrations.)

The internal female organs include the vagina, uterus, uterine tubes, ovaries, and the ligaments and supporting structures that suspend these organs in the pelvic cavity. The **vagina** extends for approximately 8–10 cm (3–4 inches) from the vestibule to the uterus superiorly. It serves as the copulatory organ and the birth canal and permits passage of the menstrual flow. The pear-shaped **uterus,** situated between the bladder and the rectum, is a highly muscular organ with its narrow end, the **cervix,** directed inferiorly.

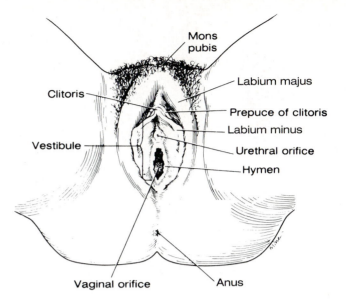

Figure 30.3

External genitalia of the human female.

The major portion of the uterus is referred to as the **body,** and its superior rounded region above the entrance of the uterine tubes is called the **fundus.** The fertilized egg is implanted in the uterus, which houses the embryo or fetus during its development. In some cases, the fertilized egg may be implanted in a uterine tube or even on the abdominal viscera, creating an **ectopic pregnancy.** Such implantations are usually unsuccessful and may even endanger the mother's life, since the uterine tubes cannot accommodate the increasing size of the fetus. ■

The superior part of the **endometrium,** the thick mucosal lining of the uterus, sloughs off periodically (usually about every 28 days) in response to cyclic changes in the levels of ovarian hormones in the woman's blood. This sloughing-off process, which is accompanied by bleeding, is referred to as the **menstrual flow,** or **menses.** The **uterine,** or **fallopian, tubes** enter the superior region of the uterus and extend laterally for about 10 cm (4 inches) toward the **ovaries** in the peritoneal cavity. The distal ends of the tubes are funnel-shaped and have fingerlike projections called **fimbriae.** Unlike the male duct system, there is no actual contact between the female gonad and the initial part of the female duct system—the uterine tube.

Because of this open passageway between the female reproductive organs and the peritoneal cavity, reproductive system infections, such as gonorrhea, can spread to cause widespread inflammations of the pelvic viscera, a condition called *pelvic inflammatory disease* or PID. ■

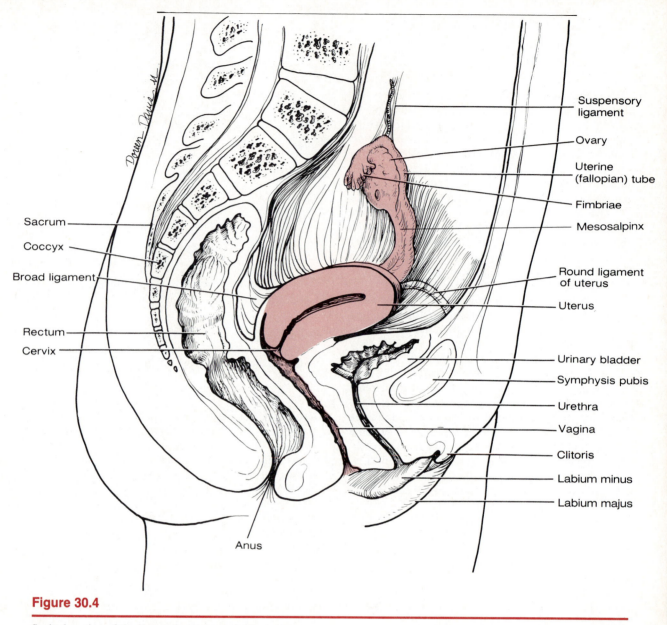

Figure 30.4

Sagittal section of the human female reproductive system.

The flattened almond-shaped ovaries lie adjacent to the uterine tubes but are not connected to them. As an egg is expelled from the ovary, an event called **ovulation,** it enters the pelvic cavity. The waving fimbriae of the uterine tubes create fluid currents that, if successful, draw the egg into the lumen of the uterine tube, where it begins its passage to the uterus, propelled by the cilia of the tubule walls. The usual and most desirable site of fertilization is the uterine tube, since the journey to the uterus takes 3 to 4 days and an egg is usually nonviable after 24 hours. Thus, sperm must swim upward through the vagina and uterus and into the uterine tubes to reach the egg. This must be an arduous journey, since they must swim against the downward current created by ciliary action—rather like swimming against the tide!

The internal female organs are for the most part retroperitoneal, except the ovaries, which are covered by only a thin layer of epithelium. They are supported and suspended somewhat freely by ligamentous folds of peritoneum. The peritoneum takes an undulating course: from the pelvic cavity floor it moves superiorly over the top of the bladder, reflects over the anterior and posterior surfaces of the uterus, and then over the rectum, and up the posterior body wall. The fold that encloses the uterine tubes and uterus and secures them to the lateral body walls is referred to as the **broad ligament.** The portion of the broad ligament specifically anchoring the uterus is called the **mesometrium** and that anchoring the uterine tubes, the **mesosalpinx.** The **round ligaments,** fibrous cords that run from the uterus to the labia

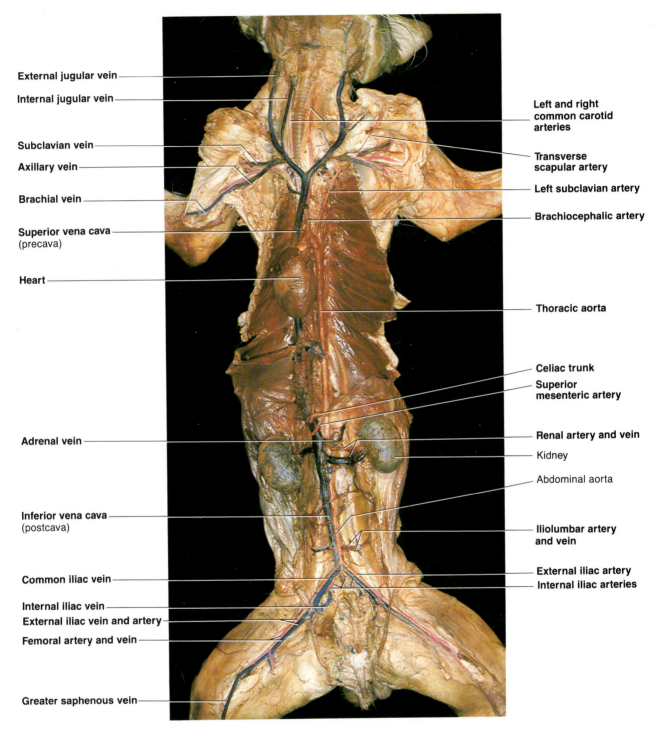

External jugular vein

Internal jugular vein

Subclavian vein

Axillary vein

Brachial vein

Superior vena cava
(precava)

Heart

Adrenal vein

Inferior vena cava
(postcava)

Common iliac vein

Internal iliac vein

External iliac vein and artery

Femoral artery and vein

Greater saphenous vein

Left and right
common carotid
arteries

Transverse
scapular artery

Left subclavian artery

Brachiocephalic artery

Thoracic aorta

Celiac trunk

Superior
mesenteric artery

Renal artery and vein

Kidney

Abdominal aorta

Iliolumbar artery
and vein

External iliac artery

Internal iliac arteries

Color Plate E

Cat dissected to reveal major blood vessels. See pages 248–255.

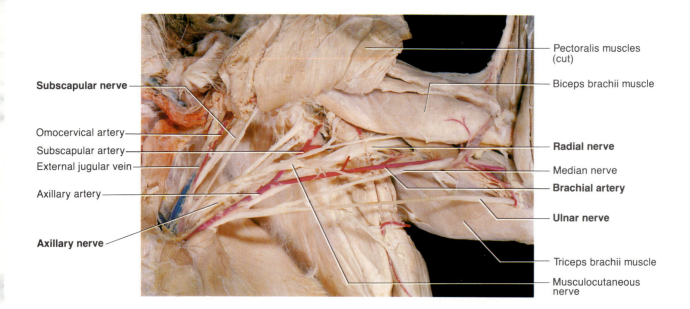

Subscapular nerve

Omocervical artery
Subscapular artery
External jugular vein

Axillary artery

Axillary nerve

Pectoralis muscles
(cut)

Biceps brachii muscle

Radial nerve

Median nerve

Brachial artery

Ulnar nerve

Triceps brachii muscle

Musculocutaneous
nerve

Color Plate C

Brachial plexus and major blood vessels of the left arm of the cat. See pages 175–178.

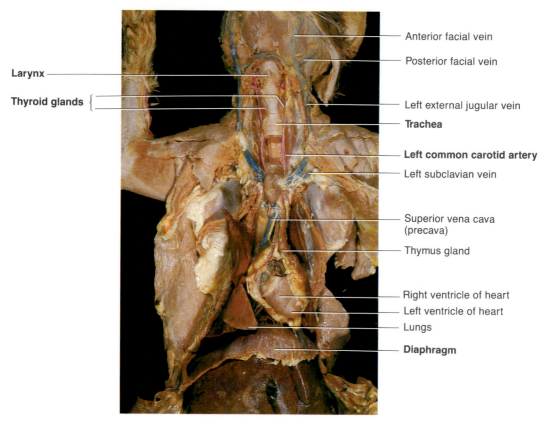

Larynx

Thyroid glands {

Anterior facial vein

Posterior facial vein

Left external jugular vein

Trachea

Left common carotid artery

Left subclavian vein

Superior vena cava
(precava)

Thymus gland

Right ventricle of heart
Left ventricle of heart
Lungs

Diaphragm

Color Plate D

Respiratory system of the cat. See pages 266–267.

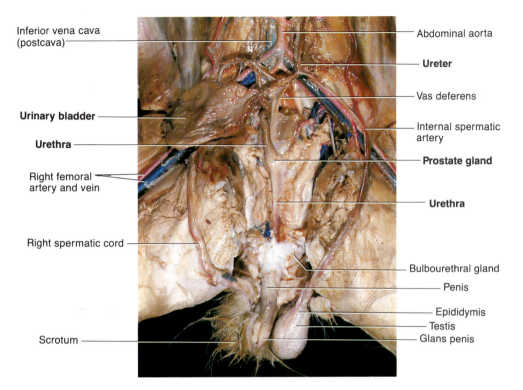

Inferior vena cava (postcava)

Abdominal aorta

Ureter

Vas deferens

Urinary bladder

Internal spermatic artery

Urethra

Prostate gland

Right femoral artery and vein

Urethra

Right spermatic cord

Bulbourethral gland

Penis

Epididymis

Testis

Scrotum

Glans penis

Color Plate A

Reproductive organs of the male cat. See pages 306–310.

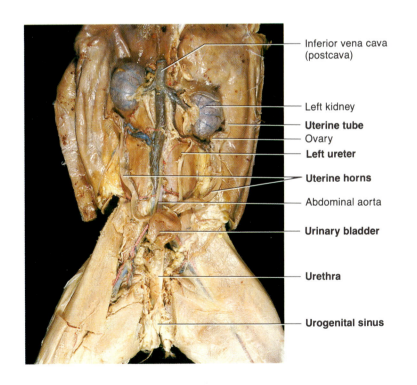

Inferior vena cava (postcava)

Left kidney

Uterine tube

Ovary

Left ureter

Uterine horns

Abdominal aorta

Urinary bladder

Urethra

Urogenital sinus

Color Plate B

Urogenital system of the female cat, with emphasis on the urinary system. See pages 295–298.

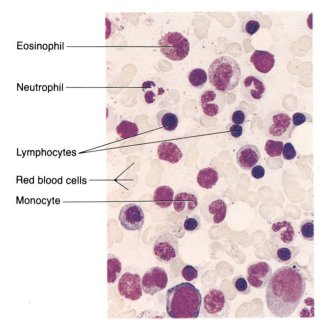

Eosinophil

Neutrophil

Lymphocytes

Red blood cells

Monocyte

Color Plate F

Bone-marrow smear showing different types of human blood cells. See pages 218–221.

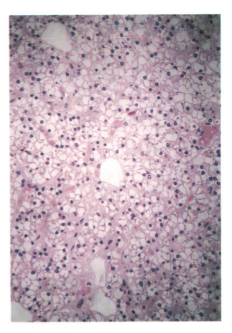

Color Plate G

Photomicrograph of human parathyroid tissue. See pages 210–213.

Neural tissue of posterior pituitary (neurohypophysis)

Glandular tissue of anterior pituitary (adenohypophysis)

Color Plate H

Photomicrograph of human pituitary tissue. See pages 208–214.

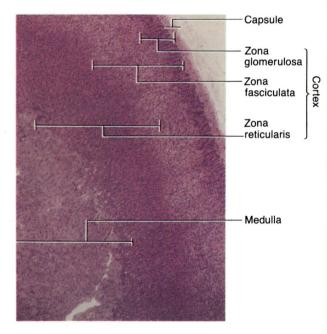

Capsule

Zona glomerulosa

Zona fasciculata

Zona reticularis

Cortex

Medulla

Color Plate I

Photomicrograph of human adrenal tissue. See pages 211–215.

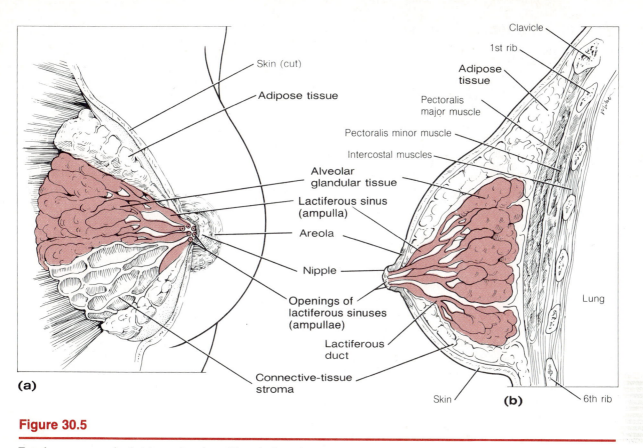

(a)

(b)

Figure 30.5

Female mammary gland: (a) anterior view; (b) sagittal section.

majora, also aid in attachment of the uterus to the body wall. The ovaries are supported posteriorly by a thickened fold of peritoneum, the **mesovarium** portion of the broad ligament, medially by the **ovarian ligament** (extending from the uterus to the ovary), and laterally by the **suspensory ligaments.**

The mammary glands or breasts exist, of course, in both sexes, but they have a reproduction-related function only in females. Since the function of the mammary glands is to produce milk to nourish the newborn infant, their importance is more closely associated with events that occur when reproduction has already been accomplished. Periodic stimulation by the female sex hormones, especially estrogens, increases the size of the female mammary glands at puberty. During this period, the duct system becomes more elaborate, and fat is deposited—fat deposition being the more important contributor to increased breast size.

The rounded, skin-covered mammary glands lie anterior to the pectoral muscles of the thorax, attached to them by connective tissue. Slightly below the center of each breast is a pigmented area, the **areola,** which surrounds a centrally protruding **nipple** (Figure 30.5).

Internally each mammary gland consists of 15 to 20 **lobes** separated by connective tissue and adipose, or fatty, tissue, which radiate around the nipple. Within each lobe are smaller chambers called **lobules,** containing the clusters of **alveolar glands** that produce the milk during lactation. The alveolar glands of each lobule pass the milk into a number of **lactiferous ducts,** which join to form an expanded storage chamber, the **lactiferous sinus,** or **ampulla,** as they approach the nipple. The ampullae open to the outside at the nipple.

DISSECTION OF THE REPRODUCTIVE SYSTEM OF THE CAT

Obtain your cat, a dissection tray, and the necessary dissection instruments. After you have completed the study of the reproductive structures of your specimen, observe a cat of the opposite sex. (The following instructions assume that the abdominal cavity has been opened in previous dissection exercises.)

Male Reproductive System

Refer to Figure 30.6 and Color Plate A as you identify the male structures.

1. Identify the penis and note the prepuce covering the glans. Carefully cut through the skin overlying the penis to expose the cavernous tissue beneath, then cross section the penis to observe the relative positioning of the three cavernous bodies.

2. Identify the scrotal sac, and then carefully make a shallow incision through the scrotum to expose the testes. Note that the scrotum is divided internally.

3. Lateral to the medial aspect of the scrotal sac, locate the spermatic cord, which contains the spermatic artery, vein, and nerve, as well as the ductus deferens, and follow it up through the inguinal canal into the abdominal cavity. (It is not necessary to cut through the pelvic bone; a slight tug on the spermatic cord in the scrotal sac region will reveal its position in the abdominal cavity.) Carefully loosen the spermatic cord from the connective tissue investing it, and follow its course as it travels superiorly in the pelvic cavity, loops over the ureter,* and then courses downward to pass posteriorly to the bladder and enters the prostate gland.

4. Note that the prostate gland is comparatively smaller in the cat than in the human and it is more distal to the bladder. (In the human, the prostate is immediately adjacent to the base of the bladder.) Carefully slit open the prostate gland to follow the entrance of the ductus deferens into the urethra, which exits from the bladder midline. The male cat urethra, like the human, serves as both a urinary and sperm duct. In the human, the ductus deferens is joined by the duct of the seminal vesicle to form the ejaculatory duct, which enters the prostate. Seminal vesicles are not present in the cat.

5. Trace the urethra to the proximal end of the cavernous tissue of the penis. Carefully split the proximal portion of the penis along a sagittal plane to reveal Cowper's, or the bulbourethral, glands lying beneath it.

6. Once again, turn your attention to the testis. Cut it from its attachment to the spermatic cord and carefully slit open the **tunica vaginalis** capsule enclosing it. Identify the epididymis running along one side of

*This position of the spermatic cord and ductus deferens is due to the fact that during fetal development, the testis was in the same relative position as the ovary is in the female. In its descent, it passes laterally and ventrally to the ureter.

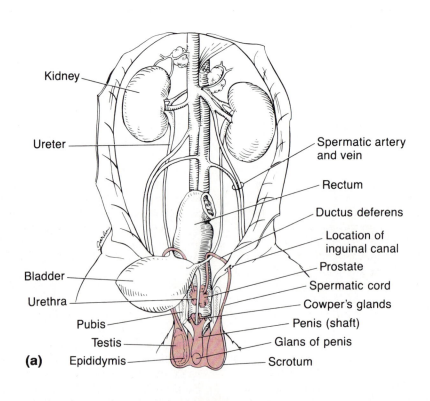

(a)

Kidney

Ureter

Spermatic artery and vein

Rectum

Ductus deferens

Location of inguinal canal

Prostate

Spermatic cord

Cowper's glands

Penis (shaft)

Glans of penis

Scrotum

Bladder

Urethra

Pubis

Testis

Epididymis

Figure 30.6

Reproductive system of the male cat: (a) diagrammatic view.

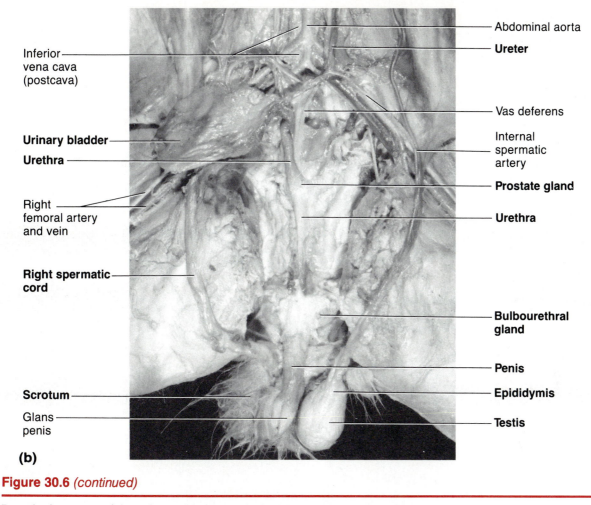

Inferior vena cava (postcava)

Abdominal aorta

Ureter

Urinary bladder

Urethra

Vas deferens

Internal spermatic artery

Prostate gland

Right femoral artery and vein

Urethra

Right spermatic cord

Bulbourethral gland

Penis

Scrotum

Epididymis

Glans penis

Testis

(b)

Figure 30.6 *(continued)*

Reproductive system of the male cat: (b) photograph. See dissection photo, Color Plate A.

the testis. Make a longitudinal cut through the testis and epididymis. Can you see the tubular nature of the epididymis and the rete testis portion of the testis

with the naked eye? _____

Female Reproductive System

Refer to Figure 30.7 and Color Plate B showing a dissection of the urogenital system of the female cat as you identify the structures described below.

1. Unlike the pear-shaped simplex, or one-part, uterus of the human, the uterus of the cat is Y-shaped (bipartite or bicornuate) and consists of a **uterine body** from which two **uterine horns** (cornua) diverge. Such an enlarged uterus enables the animal to produce litters. Examine the abdominal cavity and identify the bladder and the body of the uterus lying just dorsal to it.

2. Follow one of the uterine horns as it travels superiorly in the body cavity, noting the thin mesentery (the broad ligament), which helps anchor it and the other reproductive structures to the body wall. Approximately halfway up the length of the uterine horn, it should be possible to identify the more important round ligament, a cord of connective tissue extending laterally and posteriorly from the uterine horn to the region of the body wall that would correspond to the inguinal region of the male.

3. Identify the uterine tube and ovary at the distal end of the uterine horn just caudal to the kidney. Observe how the funnel-shaped end of the uterine tube curves around the ovary. As in the human, the distal end of the tube is fimbriated, or fringed, and the tube is lined with ciliated epithelium. The uterine tubes of the cat are tiny and much shorter than in the human. Identify the ovarian ligament, a short thick cord that extends from the uterus to the ovary and anchors the ovary to the body wall. Also observe

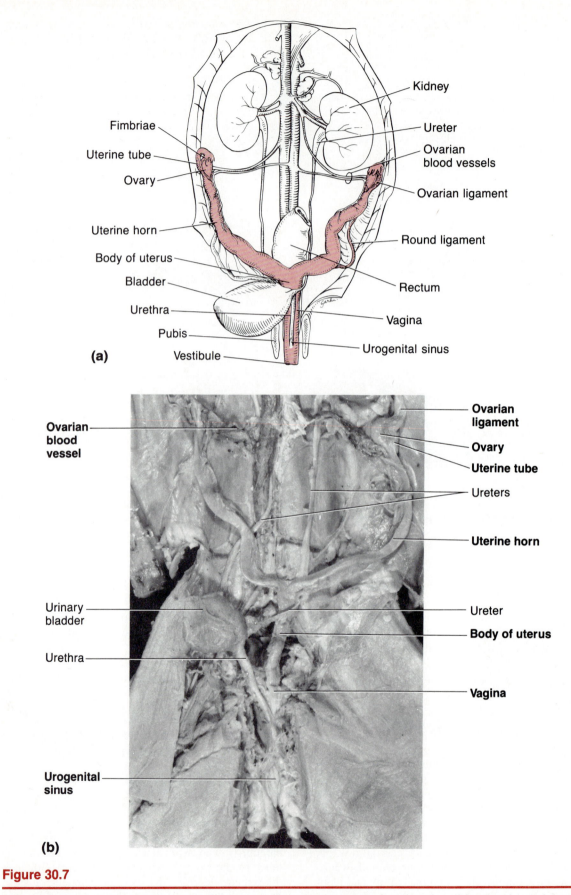

(a)

(b)

Figure 30.7

Reproductive system of the female cat: (a) diagrammatic view; (b) photograph. See dissection of female urogenital system, Color Plate B.

the ovarian artery and vein passing through the mesentery to the ovary and uterine structures.

4. Return to the body of the uterus and follow it caudad to the bony pelvis. Use bone clippers to cut through the median line of the pelvis (the pubic symphysis), cutting carefully so you do not damage the underlying urethra. Expose the pelvic region by pressing the thighs dorsally. Follow the uterine body caudally to the vagina, and note the point where the urethra draining the bladder and the vagina enter a common chamber, the **urogenital sinus.** How does this anatomic arrangement compare to that seen in

the human female? _____

_____

5. Observe the vulva of the cat, which is similar to the human vulva. Identify the raised labia majora surrounding the urogenital opening.

6. To reveal the length of the vagina, which is difficult to ascertain by external inspection, slit through the vaginal wall just superior to the urogenital sinus and cut toward the body of the uterus with scissors. Reflect the cut edges, and identify the muscular cervix of the uterus. Approximately how long is the vagina of the cat? (Measure the distance between the

urogenital sinus and the cervix.) _____

7. When you have completed your observations of both the male and female cats, clean your dissecting instruments and tray and properly wrap the cat for storage.

EXERCISE 31

Survey of Embryonic Development

OBJECTIVES

1. To define *fertilization* and *zygote*.
2. To define *cleavage* and discuss its function.
3. To define and discuss the *gastrulation* process.
4. To name the three primary germ layers and discuss the importance of each.
5. To differentiate between the blastula and gastrula forms of the sea urchin and human when provided with appropriate models or diagrams.
6. To identify a blastocyst and/or chorionic vesicle.
7. To identify the following structures of a human chorionic vesicle when provided with an appropriate diagram, and to state the function of each.

   inner cell mass   trophoblast
   amnion   allantois
   yolk sac   chorionic villi
8. To describe the process and timing of implantation in the human.
9. To define *decidua basalis* and *decidua capsularis*.
10. To state the germ-layer origin of several body organs and organ systems of the human: circulatory system, nervous system, skeleton, skeletal muscles, and lining of the digestive and respiratory tracts.
11. To describe developmental direction.
12. To describe the gross anatomy and general function of the human placenta.

MATERIALS

Prepared slides of sea urchin development (zygote through larval stages)
Compound microscope
Human development models or plaques (if available)
Life Before Birth, Educational Reprint #27, available from Time-Life Educational Materials, Box 834, Radio City P.O., New York, N.Y. 10019
Pregnant cat, rat, or pig uterus (one per laboratory session) with uterine wall dissected to allow student examination
Dissecting instruments
Model of pregnant human torso
Preserved human embryos (if available)
Fresh or formalin-preserved placenta (obtained from a clinical agency
Microscope slide of placenta tissue

EARLY EMBRYOLOGY OF THE SEA URCHIN AND THE HUMAN

Because reproduction is such a familiar event, we tend to lose sight of the wonder of the process. One part of that process, the development of the embryo, is the concern of embryologists who study the changes in structure that occur from the time of fertilization until the time of birth.

Early development in all animals involves three basic types of activities, which are integrated to ensure the formation of a viable offspring: (1) an increase in cell number and subsequent cell growth; (2) cellular specialization; and (3) morphogenesis, the formation of functioning organ systems.

This exercise provides a rather broad overview of the changes in structure that take place during embryonic development in the sea urchin. The pattern of changes in this marine animal provides a basis for comparison with developmental events in the human.

Microscopic Study of
Sea Urchin Development

1.   Obtain a compound microscope and a set of slides depicting embryonic development of the sea urchin. Draw simple diagrams of your observations.

2.   Observe the **zygote,** or fertilized egg, which appears as a single cell immediately surrounded by a fertilization membrane and a jellylike membrane. After an egg is penetrated by a sperm, the egg and the sperm nuclei fuse to form a single nucleus. This process is called **fertilization.** Within 2 to 5 minutes after sperm penetration, a fertilization membrane forms beneath the jelly coat to prevent the entry of additional sperm. Label the fertilization and jelly membranes and the zygote.

2-cell stage   4-cell stage

Fertilized egg

8-cell stage   16-cell stage

3.   Observe the cleavage stages. Once fertilization has occurred, the zygote begins to divide, forming a mass of successively smaller and smaller cells, called **blastomeres.** This series of mitotic divisions without intervening growth periods is referred to as **cleavage,** and it results in a multicellular embryonic body. As the division process continues, a solid ball of cells forms. (At the 16-cell stage, it is called **morula,** and the embryo resembles a raspberry in form.) Then the cell mass hollows out to become the embryonic form called the **blastula,** which is a ball of cells surrounding a central cavity. The blastula is the final product of cleavage.

The cleavage stage of embryonic development provides a large number of building blocks (cells) with which to fashion the forming body. (If this is a little difficult to understand, consider trying to build a structure with a huge block of granite rather than with small bricks.)

Diagram the 2-, 4-, 8-, and 16-cell stages as you observe them.

Identify and sketch the blastula stage of cleavage—a ball of cells with an apparently lighter center, owing to the presence of the central cavity.

Blastula

4. Identify the early **gastrula** form (which follows the blastula in the developmental sequence). The gastrula looks as if one end of the blastula has been indented or pushed into the central cavity, forming a two-layered embryo. In time, a third layer of cells appears between the initial two cell layers. Thus, as a result of gastrulation, a three-layered embryo forms, each layer corresponding to a primary germ layer from which certain body tissues develop. The innermost layer, the **endoderm,** and the middle layer, the **mesoderm,** form the internal organs; the outermost layer, the **ectoderm,** forms the surface tissues of the body.

Draw a gastrula below. Label the ectoderm and endoderm. If you can see the third layer of cells, the mesoderm, budding off between the other two layers, label that also.

Gastrula

5. Gastrulation in the sea urchin is followed by the appearance of the free-swimming larval form, in which the three germ layers have differentiated into the various tissues and organs of the animal's body. The larvae exist for a few days in the unattached form and then settle to the ocean bottom to attach and develop into the sessile adult form. If time allows, observe the larval form on the prepared slides. Observations need not be recorded.

Developmental Stages of the Human

1. Go to the demonstration area where the models of human development are on display. (If these are not available, use Figure 31.1 for this study.) Observe the human development models and respond to the questions posed below.

Is the observed human cleavage process similar to that in the sea urchin? _____

Why do you suppose this is so? _____

_____

_____

_____

_____

2. Observe the blastula, which, in the human, is called the **blastocyst** or **chorionic vesicle.** Unlike what occurs in the sea urchin, only a portion of the blastula cells in the human contribute to the formation of the embryonic body—those seen on the top of the blastocyst forming the so-called **inner cell mass (ICM).** The rest of the blastocyst enclosing the central cavity and overriding the ICM is referred to as the **trophoblast.** The trophoblast becomes an extraembryonic membrane called the **chorion,** which forms the fetal portion of the *placenta.*

3. Observe the implanting blastocyst shown on the model or in the figure. By approximately the seventh day after ovulation, the developing embryo is in the blastocyst stage and is floating free in the uterine cavity. About that time, the blastocyst becomes attached to the uterine wall over the ICM area, and implantation begins. The trophoblast cells secrete enzymes that erode the uterine mucosa at the point of attachment to reach the vascular supply in the submucosa. By the fourteenth day after ovulation, implantation is completed and the uterine mucosa has grown over the burrowed-in embryo. The portion of the endometrium deep to the ICM, which is destined to take part in placenta formation, is called the **decidua basalis;** that surrounding the rest of the blastocyst is called the **decidua capsularis.** Identify these regions.

By the time implantation has been completed, embryonic development has progressed to the **gastrula** stage, and the three primary germ layers are present and are beginning to differentiate. Within the next 6 weeks, virtually all of the body organ systems will have been laid down at least in rudimentary form by the germ layers. The ectoderm gives rise to the epidermis of the skin and the nervous system; the endoderm forms the mucosa of the digestive and respiratory tracts and associated structures; and the mesoderm forms virtually everything lying between the two (skeleton, walls of the digestive organs, urinary system, muscular and circulatory systems, and others). By the ninth week of development, the embryo is referred to as a **fetus.** From this point on, the major activities are growth and tissue and organ specialization. All the groundwork has been completed by the eighth week.

4. Again observe the blastocyst or chorionic vesicle to follow the formation of the extraembryonic membranes and the placenta (see Figure 31.1). Note the villus extensions of the trophoblast. By the time

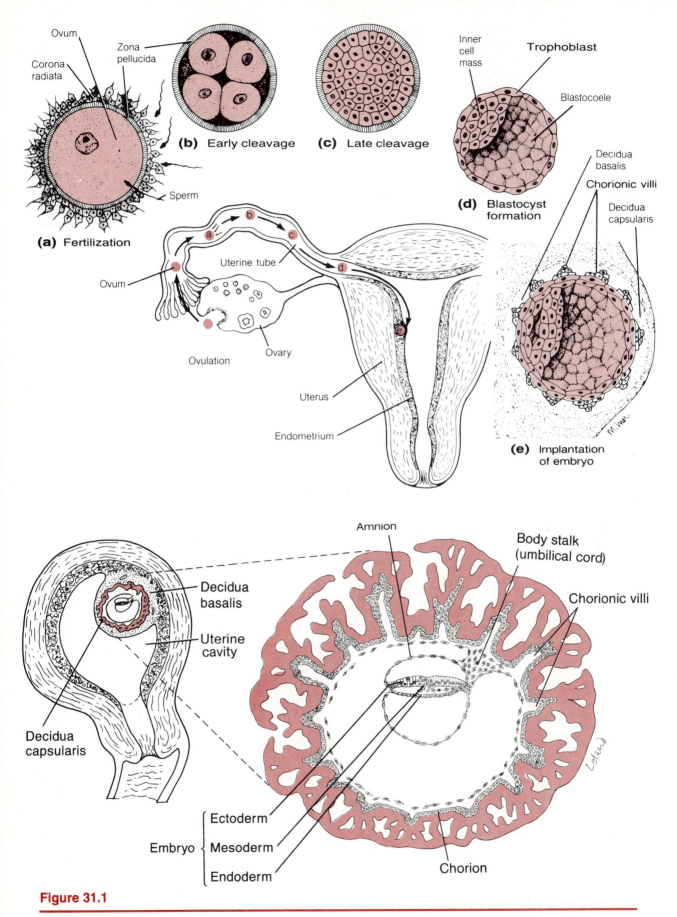

(a) Fertilization

Ovum

Corona radiata

Zona pellucida

Sperm

(b) Early cleavage

(c) Late cleavage

Inner cell mass

Trophoblast

Blastocoele

(d) Blastocyst formation

Ovum

Uterine tube

Ovulation

Ovary

Uterus

Endometrium

Decidua basalis

Chorionic villi

Decidua capsularis

(e) Implantation of embryo

Amnion

Body stalk (umbilical cord)

Chorionic villi

Decidua basalis

Uterine cavity

Decidua capsularis

Ectoderm

Embryo Mesoderm

Endoderm

Chorion

Figure 31.1

Early embryonic development of the human: top (a–e), from fertilization to blastocyst implantation in the uterus; below, embryo of approximately 22 days. Embryonic membranes and germ layers present.

implantation has been completed, the trophoblast has differentiated into the **chorion,** and the large elaborate villi extending from the chorion are lying in the blood-filled sinusoids in the uterine tissue. This composite of uterine tissue and **chorionic villi** is called the **placenta,** and all exchanges to and from the embryo occur through the chorionic membranes.

Three embryonic membranes (originating in the ICM) have also formed by this time: the amnion, the allantois, and the yolk sac. Identify each. The **amnion** encases the young embryonic body in a fluid-filled chamber that protects the embryo from mechanical trauma and prevents adhesions during rapid embryonic growth. The **yolk sac** in humans has lost its original function, which was to pass nutrients to the embryo after digesting the yolk mass, primarily because the placenta has taken over that task. (Also, the human egg has very little yolk.) However, the yolk sac is not totally useless since the embryo's first blood cells originate here, and the primordial germ cells migrate from it into the embryo's body to seed the gonadal tissue. The **allantois,** which protrudes from the posterior end of the yolk sac, is also a largely redundant structure in humans because of the placenta. In birds and reptiles, it is a repository for embryonic wastes; in the human, it is the structural basis on which the mesoderm migrates to form the body stalk, or **umbilical cord,** which attaches the embryo to the placenta.

5. Go to the demonstration area to view the photographic series, *Life Before Birth*. This series, by Lennart Nilsson, illustrates human development in a way you will long remember. After viewing it, respond to the following questions.

In your own words, what do the chorionic villi look

like? _____

_____

_____

What organs or organ systems appear *very* early in

embryonic development? _____

_____

_____

_____

Does development occur in a rostral to caudal (head

to toe) direction, or vice versa? _____

_____

Does development occur in a distal–proximal direc-

tion, or vice versa? _____

Does spontaneous movement occur *in utero?* _____

_____

How does the mother recognize this? _____

_____

_____

The very young embryo has been described as resembling "an astronaut suspended and floating in space." Do you think this definition is appropriate?

Why or why not? _____

_____

_____

_____

What is vernix caseosa? _____

_____

_____

What is lanugo? _____

_____

_____

IN UTERO DEVELOPMENT

1. Go to the appropriate demonstration area and observe the fetuses in the Y-shaped animal uterus. Identify the following fetal or fetal-related structures:

Placenta. (a composite structure formed from the uterine mucosa and the fetal chorion).

Describe its appearance. _____

_____

_____

Umbilical cord. Note its relationship to the pla-

centa and fetus. _____

_____

_____

Amniotic sac. Identify the transparent amnion surrounding a fetus. Open one amniotic sac and note the amount, color, and consistency of the fluid.

_____

_____

Remove a fetus and observe the degree of development of the head, body, and extremities. Is

the skin thick or thin? _____

What is the basis for your response? _____

_____

_____

_____

2. Observe the model of a pregnant human torso. Identify the placenta. How does it differ in shape

from the animal placenta observed? _____

_____

_____

Identify the umbilical cord. In what region of the uterus does implantation usually occur, as indicated by the position of the placenta?

_____

_____

_____

What might be the consequence if it occurred

lower? _____

_____

_____

Why would a feet-first position (breech presentation) be less desirable than the positioning in

the model? _____

_____

_____

3. If human fetuses are available for demonstration, examine them, making mental comparisons with the appearance of the dissected animal fetuses and the illustrations in the *Life Before Birth* series. *Do not pick up the demonstration cases or jiggle the embryos around.* Embryonic tissues are exceedingly soft and tend to disintegrate easily, even when properly preserved.

GROSS AND MICROSCOPIC ANATOMY OF THE PLACENTA

The placenta is a remarkable temporary organ. It is composed of maternal and fetal tissues and is responsible for providing nutrients and oxygen to the embryo and fetus while removing carbon dioxide and metabolic wastes.

1. Note that the placenta on display has two very different-appearing surfaces—one smooth and the other spongy, roughened, and torn-looking.

Which is the fetal side? _____

Basis of your conclusion? _____

_____

Identify the umbilical cord. Within the cord, identify the umbilical vein and two umbilical arteries. What is the function of the umbilical

vein? _____

_____

The umbilical arteries? _____

_____

Are any of the fetal membranes still attached?

_____

If so, which? _____

_____

2. Obtain a microscope slide of placental tissue. Observe the tissue carefully, comparing it to Figure 31.2. Identify the intervillus spaces (maternal sinusoids), which are blood-filled in life. Identify the villi, and notice their rich vascular supply. Draw a small representative diagram of your observations below the figure and label it appropriately.

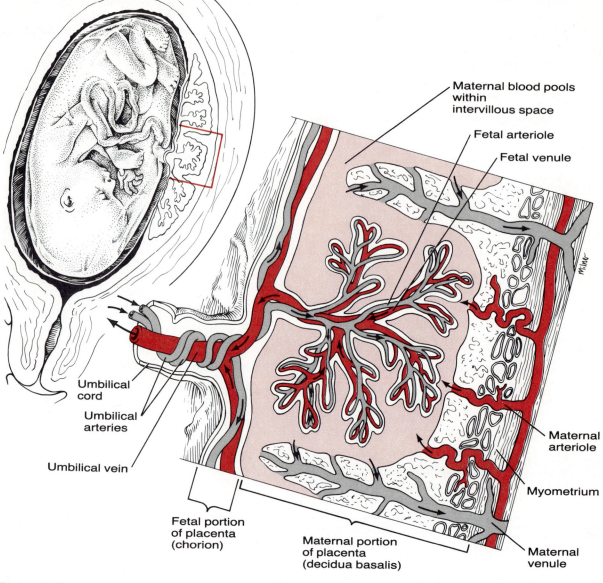

Figure 31.2

Diagrammatic representation of the structure of the placenta.

Review Sheets

Review Sheet

EXERCISE 2

Organ Systems Overview

1. Use the key below to indicate the body system(s) that perform the following functions for the body.

   Key: a. cardiovascular   e. integumentary   i. reproductive
          b. digestive   f. lymphatic   j. respiratory
          c. endocrine   g. muscular   k. skeletal
          d. immune   h. nervous   l. urinary

   _____ rids the body of nitrogen-containing wastes

   _____ is affected by removal of the thyroid gland

   _____ provides support and levers on which the muscular system acts

   _____ includes the heart

   _____ causes the onset of the menstrual cycle

   _____ protects underlying organs from drying out and from mechanical damage

   _____ protects the body; destroys bacteria and tumor cells

   _____ breaks down ingested food into its building blocks

   _____ removes carbon dioxide from the blood

   _____ delivers oxygen and nutrients to the tissues

   _____ moves the limbs; facilitates facial expression

   _____ conserves body water or eliminates excesses

   _____ facilitates conception and childbearing

   _____ controls the body with chemical molecules called hormones

   _____ is damaged when you cut your finger or get a severe sunburn

2. Name the organ system to which each of the following sets of organs (or body structures) belongs:

   blood vessels, heart _____

   trachea, bronchi, alveoli _____

   testis, vas deferens, urethra _____

   adrenal glands, pancreas, pituitary _____

   esophagus, large intestine, rectum _____

kidneys, bladder, ureters _____

bone, cartilages, tendons _____

3. Using the key below, place the following organs in their proper body cavity.

   Key: a. abdominopelvic   b. cranial   c. spinal   d. thoracic

   _____ 1. stomach   _____ 5. liver   _____ 9. lungs

   _____ 2. small intestine _____ 6. spinal cord   _____ 10. brain

   _____ 3. large intestine _____ 7. bladder   _____ 11. rectum

   _____ 4. spleen   _____ 8. heart

4. Using the organs listed in item 3 above, record by number which would be found in the abdominal regions listed below.

   _____ hypogastric region   _____ epigastric region

   _____ right lumbar region   _____ left iliac region

   _____ umbilical region   _____ left hypochondriac region

5. The five levels of organization of a living body are cell, _____ , _____ ,

   _____ , and organism.

6. Define *organ:* _____

   _____

   _____

7. During the course of this laboratory exercise, a rat was dissected. What is the value of observing the anatomy of a rat (or any other small mammal) when human anatomy is the actual topic of study?

   _____

   _____

STUDENT NAME _____

LAB TIME/DATE _____

Review Sheet

EXERCISE 3   # The Microscope

1. Label all indicated parts of the microscope.

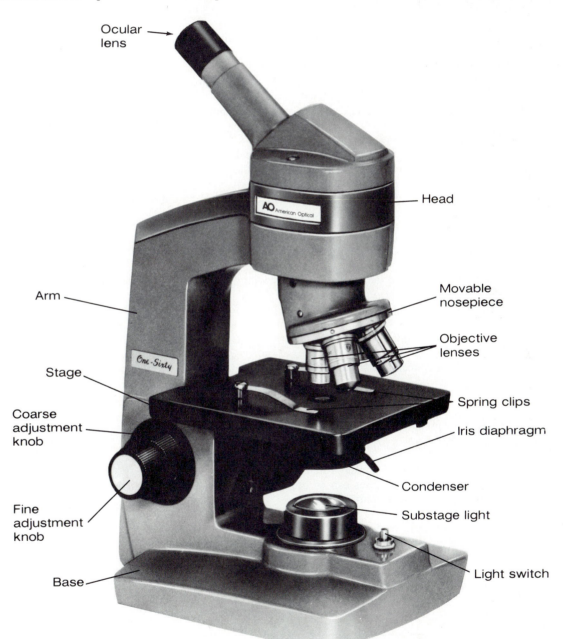

Ocular lens

Head

Arm

Movable nosepiece

Objective lenses

Stage

Spring clips

Coarse adjustment knob

Iris diaphragm

Condenser

Fine adjustment knob

Substage light

Base

Light switch

Care and Structure of the Compound Microscope

2. The following statements are true or false. If true, insert *T* on the answer blank. If false, correct the underlined word or phrase by inserting the correct answer.

_____   The microscope lens may be cleaned <u>with any soft tissue</u>.

_____   The coarse adjustment knob may be used in focusing <u>with all three objectives</u>.

_____   The microscope should be stored with the <u>oil immersion</u> lens in position over the stage.

_____   When beginning to focus, the <u>l.p.</u> lens should be used.

_____   Always focus <u>toward</u> the specimen in low power.

_____   A cover slip should always be used with <u>the high-dry and oil lenses</u>.

_____   The greater the amount of light delivered to the objective lens, the <u>less</u> the resolution.

3. Match the microscope structures given in column B with the statements that identify or describe them (column A).

Column A	Column B
_____ platform on which the slide rests for viewing	a. coarse adjustment knob
_____ lens located at the superior end of the body tube	b. condenser
_____ secure(s) the slide to the stage	c. fine adjustment knob
_____ delivers a concentrated beam of light to the specimen	d. iris diaphragm
	e. mechanical stage or spring clips
_____ used for precise focusing once initial focusing has been done	f. movable nosepiece
_____ carries the objective lenses; rotates so that the different objective lenses can be brought into position over the specimen	g. objective lenses
	h. ocular
_____ used to increase or decrease the amount of light passing through the specimen	i. stage

4. Explain the proper technique for transporting the microscope. _____

_____

_____

5. Define the following terms:

   real image: _____

_____

virtual image: _____

6. Define *total magnification:* _____

_____

_____

7. Define *resolution:* _____

Viewing Objects Through the Microscope

1. Complete or respond to the following statements:

_____ The distance from the bottom of the objective lens in use to the specimen is called the _____ .

_____ The resolution of the human eye is _____ μm.

_____ The resolution of the optical microscope is _____ μm.

_____ The area of the specimen seen when looking through the microscope is the _____ .

_____ If a microscope has a 10× ocular and the total magnification at a particular time is 950×, the objective lens in use at that time is _____ × .

_____ If after focusing in low power only the fine adjustment need be used to focus the specimen at the higher powers, the microscope is said to be _____ .

_____ If the field size using a 10× ocular and a 15× objective is 1.5 mm, the approximate field size with a 30× objective is _____ mm or _____ μm.

_____ If the size of the high-power field is 1.2 mm, an object that occupies approximately one-third of that field has an estimated size of _____ mm.

_____ Assume there is an object on the left side that you want to bring into the center of the field (that is, toward the apparent right). In what direction would you move your side?

_____ If the object is in the top of the field and you want to move it downward into the center, you would move the slide _____ .

2. You have been asked to prepare a slide with the letter *k* on it. Draw its appearance in the l.p. field in the adjacent circle.

k

3. Say you are observing an object in the l.p. field. When you switch to h.p., it is no longer in your field of view. Why might this occur? _____

_____

_____

What should be done initially to prevent this from happening? _____

_____

4. Do the following factors increase or decrease as one moves to higher magnifications with the microscope?

   resolution _____   amount of light needed _____

   working distance _____   depth of field _____

5. A student has the high-dry lens in position and appears to be intently observing the specimen. The instructor, noting a working distance of about 1 cm, knows the student isn't actually seeing the specimen.

   How so? _____

6. Why is it important to be able to use your microscope to perceive depth when studying slides?

   _____

   _____

7. If you are observing a slide of tissue two cell-layers thick, how can you determine which layer is superior?

   _____

   _____

   _____

8. Describe the proper procedure for preparing a wet mount.

   _____

   _____

   _____

   _____

   _____

9. Give two reasons why the light should be dimmed when viewing living or unstained material. _____

   _____

   _____

   _____

Review Sheet

EXERCISE 4

The Cell—
Anatomy and Division

Anatomy of the Composite Cell

1. Define the following:

   organelle: _____

   _____

   cell: _____

2. Although cells have differences that reflect their specific functions in the body, what functional capabilities

   do all cells exhibit? _____

   _____

3. Identify the following cell parts:

   _____   external boundary of cell; confines cell contents, regulates entry
   and exit of materials

   _____   contains digestive enzymes of many varieties; "suicide sacs" of the
   cell

   _____   scattered throughout the cell; controls release of energy from
   foodstuffs

   _____   slender projections of the plasma membrane that increase its
   surface area

   _____   stored glycogen granules, crystals, pigments, and so on

   _____   membranous system, consisting of flattened sacs and vesicles;
   packages proteins for export

   _____   control center of the cell; necessary for cell division and cell life

   _____   two rod-shaped bodies near the nucleus; "spin" the mitotic spindle

   _____   dense, darkly-staining nuclear body; possible packaging site for
   ribosomes

   _____   membranous system involved with synthesis of lipid-based
   hormones

   _____   membranous system; involved in intracellular transport of
   proteins and synthesis of membrane lipids

_____ attached to membrane systems or scattered in the cytoplasm; synthesize proteins

_____ threadlike structures in the nucleus; contain genetic material (DNA)

4. Label all indicated parts of the following diagram.

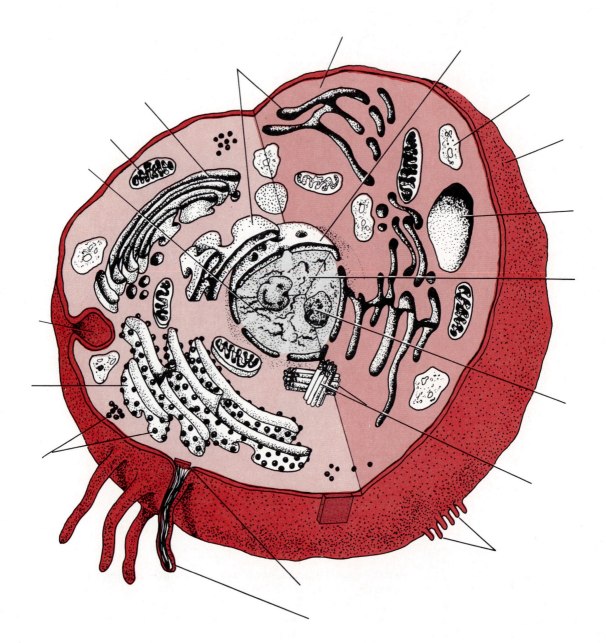

Observing Differences and Similarities in Cell Structure

1. List one important structural characteristic (a) of each of the following cell types observed in the laboratory, then give the function (b) that structure complements or ensures.

    squamous epithelium 1a. _____

                                                    b. _____

    sperm 2a. _____

                                                    b. _____

    smooth muscle 3a. _____

                                                    b. _____

    red blood cell 4a. _____

                                                    b. _____

2. What is the significance of the red blood cell being anucleate (without a nucleus)? _____

    _____

    Did it ever have a nucleus? _____ When? _____

    _____

3. What are selective stains, and what is the basis of their action? _____

    _____

    _____

4. What does Janus green stain for? _____

    Why can't you see these structures without Janus green? _____

Cell Division: Mitosis and Cytokinesis

1. What is the importance of mitotic cell division? _____

    _____

    _____

2. Complete or respond to the following statements:

    Division of the _a_ is referred to as mitosis. Cytokinesis is division of the _b_. The major structural difference between chromatin and chromosomes is that the latter is _c_. Chromosomes attach to the spindle fibers by undivided structures called _d_. If a cell undergoes mitosis but not cytokinesis, the product is _e_. The structure that acts as a scaffolding for chromosomal attachment and movement is called the _f_. _g_ is the period of cell life when the cell is not involved in division. Two cell populations in the body that do not undergo cell division are _h_ and _i_. The implication of an inability of a cell population to divide is that when some of its members die, they are replaced by _j_.

    a. _____

    b. _____

    c. _____

    d. _____

    e. _____

    f. _____

    g. _____

    h. _____

    i. _____

    j. _____

3. Using the key, categorize each of the events described below according to the phase in which it occurs.

Key: a. prophase   b. anaphase   c. telophase   d. metaphase   e. none of these

_____ Chromatin coils and condenses to form deeply staining bodies.

_____ Centromeres break and chromosomes begin migration toward opposite poles of the cell.

_____ The nuclear membrane and nucleoli become reestablished.

_____ Chromosomes cease their poleward movement.

_____ Chromosomes align on the equator of the spindle.

_____ Nucleoli and nuclear membrane disappear.

_____ The spindle forms through the migration of the centrioles.

_____ Chromosomal material replicates.

_____ Centrioles replicate.

_____ Chromosomes first appear to be duplex structures.

_____ Chromosomes attach to the spindle fibers.

_____ Cleavage furrow forms.

_____ The nuclear membrane(s) is absent.

4. Identify the phases of mitosis depicted in the following diagrams:

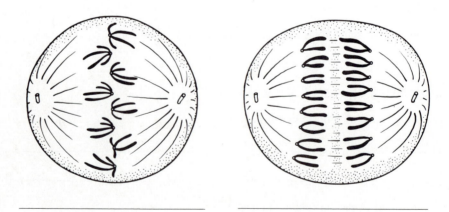

_____   _____

5. What is the physical advantage of the chromatin coiling and condensing to form short chromosomes at the onset of mitosis? _____

_____

_____

Review Sheet

EXERCISE 5

The Cell—Transport Mechanisms and Cell Permeability

Choose all answers that apply to items 1 and 2, and place their letters in the appropriately numbered response blanks to the right.

1. Brownian motion
   a. reflects the kinetic energy of the smaller solvent molecules
   b. reflects the kinetic energy of the larger solute molecules
   c. is ordered and predictable
   d. is random and erratic

1. _____

2. Kinetic energy
   a. is higher in larger molecules
   b. is lower in larger molecules
   c. increases with increasing temperature
   d. decreases with increasing temperature
   e. is reflected in the speed of molecular movement

2. _____

3. Referring to the laboratory experiment using dialysis sacs 1 through 4 to study diffusion through nonliving membranes:

   Sac 1: 40% glucose suspended in distilled water

   Did glucose pass out of the sac? _____

   Test used to determine presence of glucose was _____

   Did the sac weight change? _____

   If so, explain the reason for its weight change: _____

   _____

   _____

   Sac 2: 40% glucose suspended in 40% glucose

   Was there net movement of glucose in either direction? _____

   Explanation: _____

   Sac weight change? _____ Explanation: _____

   _____

   Sac 3: 10% NaCl in distilled water

   Net movement of NaCl out of the sac? _____

Test used to determine the presence of NaCl: _____

Direction of net osmosis? _____

Sac 4: Boiled starch in distilled water

Net movement of starch out of the sac? _____

Test used to determine the presence of starch? _____

Direction of net osmosis? _____

4. What single characteristic of the semipermeable membranes used in the laboratory determines the substances that can pass through them? _____

In addition to this characteristic, what other factors influence the passage of substances through living membranes? _____

_____

5. A semipermeable sac containing 4% NaCl, 9% glucose, and 10% albumin is suspended in a solution with the following composition: 10% NaCl, 10% glucose, and 40% albumin. Assume that the sac is permeable to all substances except albumin. State whether each of the following will (a) move into the sac (b) move out of the sac, or (c) not move.

glucose _____ water _____ albumin _____ NaCl _____

6. The diagrams below represent three microscope fields containing red blood cells. Arrows show the direction of net osmosis. Which field contains a hypertonic solution? _____ The cells in this field are said to be _____ . Which field contains an isotonic bathing solution? _____

Which field contains a hypotonic solution? _____ What is happening to the cells in this field? _____

_____

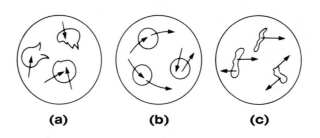

(a)   (b)   (c)

7. What determines whether a transport process is active or passive? _____

_____

8. Use the key terms to characterize each of the statements below. (More than one key choice may apply.)

Key: a. diffusion, dialysis c. permease system e. pinocytosis
   b. diffusion, osmosis d. phagocytosis f. filtration

_____ require ATP (cellular energy)

_____ driven by kinetic energy of the molecules

_____ driven by hydrostatic (fluid) pressure

_____ follow a concentration gradient

_____ proceeds against a concentration gradient; requires a carrier

_____ engulf foreign substances

_____ moves water through a semipermeable membrane

_____ transports amino acids, some sugars, and Na^+ through the plasma membrane

_____ provides for cellular uptake of solid or large particles from the cell exterior

_____ moves small or lipid-soluble solutes through the plasma membrane

9. Define the following terms:

Brownian motion: _____

_____

_____

diffusion: _____

_____

_____

osmosis: _____

_____

_____

dialysis: _____

_____

_____

filtration: _____

_____

_____

active transport: _____

_____

_____

phagocytosis: _____

_____

_____

pinocytosis: _____

_____

_____

Review Sheet

EXERCISE 6

Classification of Tissues

Tissue Structure and Function—General Review

1. Define *tissue*: ___a group of cells with a specialized function___
   _____

2. Use the key choices to identify the major tissue types described below:

   Key: a. connective tissue   b. epithelium   c. muscle   d. nervous tissue

   ___B___ forms membranes—mucous, serous, and cutaneous

   ___A___ allows for movement of limbs, and for organ movements within the body

   ___D___ transmits electrochemical impulses

   ___A___ supports body organs

   ___B___ cells may absorb and/or secrete substances

   ___D___ basis of the major controlling system of the body

   ___C___ major function of the cells of this tissue type is to shorten

   ___B___ forms hormones

   ___A___ packages and protects body organs

   ___A___ characterized by having large amounts of nonliving matrix

   ___A___ most widely distributed in the body

   ___D___ forms brain and spinal cord

Epithelial Tissue

1. Describe the general characteristics of epithelial tissue. ___① tightly packed cells___
   ___w/ little extracellular matrix ② cells always (membranes)___
   ___have one free surface ③ basement membrane___
   ___④ avascular ⑤ can regenerate___
   _____

2. On what bases are epithelial tissues classified? ___shape + # of cells___
   _____

3. What are the major functions of epithelium in the body? (Give examples.) _____

protection - skin

absorption - digestive tract

secretion - secretion of glands

filtration - kidney tubules

4. How is the function of epithelium reflected in its arrangement? _____

_____

_____

5. Where is ciliated epithelium found? _upper respiratory tract_

fallopian tubes

What role does it play? _sweep away/along materials_

_____

_____

6. Transitional epithelium is actually stratified squamous epithelium, but there is something special about it.

How does it differ structurally from other stratified squamous epithelia? _the cells_

at basement are cuboidal

How does this reflect its function in the body? _allows stretching_

of tissue

_____

7. How do the endocrine and exocrine glands differ in structure and function? _____

_____

_____

_____

8. Respond to the following with the key choices:

Key: a. pseudostratified ciliated columnar c. simple cuboidal e. stratified squamous
      b. simple columnar   d. simple squamous f. transitional

_____A_____ lining of the esophagus

_____B_____ lining of the stomach and small intestine

_____D_____ lung tissue, alveolar sacs

_____C_____ collecting tubules of the kidney

_____E_____ epidermis of the skin

_____F_____ lining of bladder; peculiar cells that have the ability to slide over each other

_____D_____ forms the thin serous membranes; a single layer of flattened cells

RS20

Connective Tissue

1. What are the general characteristics of connective tissues? ① _vascular except for cartilage_ ② _have many cell types_ ③ _non-living matrix between cells_

2. What functions are performed by connective tissue? _protect, support, bind other tissues_

3. How are the functions of connective tissue reflected in its structure? _____

4. Using the key, choose the best response to identify the connective tissues described below:

   Key: a. adipose connective tissue   e. fibrocartilage
                 b. areolar connective tissue   f. hemopoietic tissue
                 c. dense connective tissue   g. hyaline cartilage
                 d. elastic cartilage   h. osseous tissue

   _C_   provides great strength through parallel bundles of collagenic fibers; found in tendons

   _A_   acts as a storage depot for fat

   _C_   composes the dermis of the skin _Dense irregular_

   _E_   makes up the intervertebral disks

   _H_   forms the bony skeleton

   _B?_   composes the basement membrane and packages organs; includes a gel-like matrix with all categories of fibers and diverse cell types

   _G_   forms the embryonic skeleton and the surfaces of bones at the joints; reinforces the trachea

   _D_   provides an elastic framework for the external ear

   _G_   structurally amorphous matrix heavily invaded with fibers; appears glassy and smooth

   _H_   contains cells arranged concentrically around a nutrient canal; matrix hard owing to calcium salts

   _A_   provides insulation for the body

5. Why are adipose cells called "signet ring" cells? _the cell nucleus pushed over to one side_

Muscle Tissue

1. The three types of muscle tissue exhibit similarities as well as differences. Check the appropriate space in the chart below to indicate which muscle types exhibit each characteristic.

Characteristic	Skeletal	Cardiac	Smooth
Voluntarily controlled	✓		
Involuntarily controlled		✓	✓
Has a banded appearance	✓	✓	
Has a single nucleus in each cell		✓	✓
Multinucleate	✓		
Found attached to bones	✓		
Allows you to direct your eyeballs	✓		
Found in the walls of the stomach, uterus, and arteries			✓
Contains spindle-shaped cells			✓
Contains cylindrical cells with branching ends		✓	
Contains long, nonbranching cylindrical cells	✓		
Displays intercalated disks		✓	
Concerned with locomotion of the body as a whole	✓		
Changes the internal volume of an organ as it contracts			✓
Tissue of the circulatory pump		✓	

Nervous Tissue

1. What two physiologic characteristics are highly developed in nervous tissue? _____

    _____

2. In what ways are nerve cells similar to other cells? _____

    _____

    How are they different? _____

3. Sketch a neuron, recalling in your diagram the most important aspects of its structure. Below the diagram, describe how its particular structure relates to its function in the body.

Review Sheet

EXERCISE 7

The Integumentary System

Basic Structure of the Skin

1. Complete the following statements by writing the appropriate word or phrase on the correspondingly numbered blanks:

   The two basic tissues of which the skin is composed are dense irregular connective tissue, which makes up the dermis, and __a__, which forms the epidermis. The water-proofing protein found in the epidermal cells is called __b__. Melanin and __c__ contribute to skin color. A localized concentration of melanin is referred to as a __d__.

   a. _____

   b. _____

   c. _____

   d. _____

2. Four protective functions of the skin are _____,

   _____, _____,

   and _____ .

3. Label the skin structures and areas indicated in the accompanying diagram.

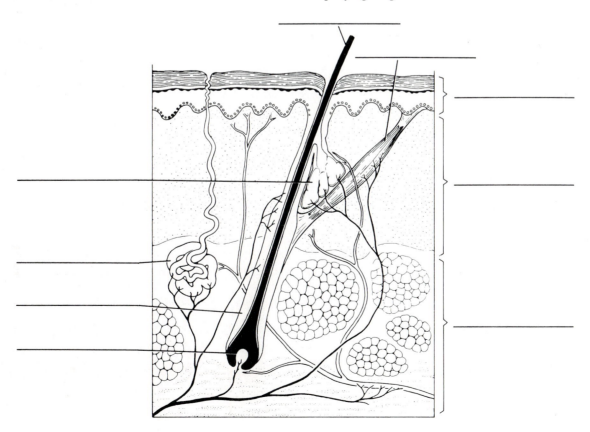

4. Using the key choices, choose all responses that apply to the following descriptions.

Key: a. stratum corneum   d. stratum lucidum   g. epidermis as a whole
    b. stratum germinativum e. papillary layer   h. dermis as a whole
    c. stratum granulosum   f. reticular layer

_____ translucent cells containing keratin

_____ dead cells

_____ dermis layer responsible for fingerprints

_____ vascular region

_____ major skin area where derivatives (nails, hair) arise

_____ epidermal region exhibiting rapid cell division

_____ scalelike dead cells full of keratin that constantly slough off

_____ site of elastic and collagenic fibers

_____ site of melanin formation

_____ actually two strata—the basale and spinosum

5. What substance is manufactured in the skin (but is not a secretion) to play a role elsewhere in the body?

_____

6. What sensory receptors are found in the skin? _____

_____

_____

7. A nurse tells the doctor that a patient is cyanotic. What is cyanosis? _____

What does its presence imply? _____

8. What is the mechanism of a suntan? _____

_____

9. What is a decubitus ulcer? _____

Why does it occur? _____

_____

10. Some injections hurt more than others. On the basis of what you have learned about skin structure, can you determine why this is so? _____

_____

Appendages of the Skin

1. Using key choices, respond to the following descriptions.

    Key: a. arrector pili   d. hair follicles   f. sweat gland—apocrine
            b. cutaneous receptors   e. sebaceous glands   g. sweat gland—eccrine
            c. hair

    _____ A blackhead is an accumulation of oily material that is produced by _____

    _____ Tiny muscles attached to hair follicles, which pull the hair upright during fright or cold.

    _____ Most numerous variety of perspiration gland.

    _____ Sheath formed of both epithelial and connective tissues.

    _____ Less numerous variety of perspiration gland; secretion (often milky in appearance) contains proteins and other substances that promote bacterial growth.

    _____ Found everywhere on body except palms of hands, soles of feet, and lips.

    _____ Primarily dead/keratinized cells.

    _____ Specialized nerve endings that respond to temperature, touch, and so on.

    _____ Become more active at puberty.

2. How does the skin help in regulating body temperature? (Name two different mechanisms.) _____

    _____

    _____

    _____

    _____

Plotting the Distribution of Sweat Glands

1. With what substance in the bond paper does the iodine painted on the skin react? _____

    _____

2. Which skin area—the forearm or palm of hand—has more sweat glands? _____

    Which other body areas would, if tested, prove to have a high density of sweat glands? _____

    _____

3. What organ system controls the activity of the eccrine sweat glands? _____

    _____

STUDENT NAME ________________________

LAB TIME/DATE ________________________

Review Sheet

EXERCISE 8

Classification of Membranes

1. Complete the following chart:

Membrane	Tissue type (epithelial/connective)	Common locations	Functions
mucous			
serous			
synovial			
epidermis			

2. Respond to the following statements by choosing an answer from the key.

Key: a. epidermis b. mucous c. serous d. synovial

_______________ membrane type associated with skeletal system structures

_______________ always formed of simple squamous epithelium

_______________, _______________ membrane types *not* found in the ventral body cavity

_______________ the only membrane type in which goblet cells are found

_______________ the only *external* membrane

Review Sheet

EXERCISE 9

Bone Classification, Structure, and Relationships: An Overview

Bone Markings

1. Match the terms in column B with the appropriate description in column A:

Column A	Column B
_____ sharp, slender process*	a. condyle
_____ small rounded projection*	b. crest
_____ narrow ridge of bone*	c. epicondyle
_____ large rounded projection*	d. fissure
_____ structure supported on neck†	e. foramen
_____ armlike projection†	f. fossa
_____ rounded, convex projection†	g. head
_____ narrow depression or opening‡	h. meatus
_____ canallike structure‡	i. ramus
_____ opening through a bone‡	j. sinus
_____ shallow depression†	k. spine
_____ air-filled cavity	l. trochanter
_____ large, irregularly shaped projection*	m. tubercle
_____ raised area of a condyle*	n. tuberosity

Classification of Bones

1. The four major anatomic classifications of bones are long, short, flat, and irregular. Which category has

   the least amount of spongy bone relative to its total volume? _____

2. Classify each of the bones in the next chart into one of the four major categories by checking the appropriate column. Use appropriate references as necessary.

   *A site of muscle attachment.

   †Takes part in joint formation.

   ‡A passageway for nerves or blood vessels.

RS31

	Long	Short	Flat	Irregular
humerus				
metacarpal				
frontal				
calcaneus				
rib				
vertebra				
radius				

Gross Anatomy of the Typical Long Bone

1. Using the terms to the right, characterize the following statements:

    _____ site of spongy bone in the adult   a. diaphysis

    _____ site of compact bone in the adult   b. endosteum

    _____ site of hemopoiesis in the adult   c. epiphyseal disk

    _____ major submembranous site of osteoclasts   d. epiphysis

    _____ scientific name for bone shaft   e. periosteum

    _____ site of fat storage in the adult   f. red marrow cavity

    _____ site of longitudinal growth in the child   g. yellow marrow cavity

2. What differences between compact and spongy bone can be seen with the naked eye?_____
    _____

3. What is the function of the periosteum? _____
    _____

Microscopic Structure of Compact Bone

1. Trace the route taken by blood through a bone, starting with the periosteum and ending with an osteocyte

    in a lacuna. Periosteum $\Longrightarrow$ _____ $\longrightarrow$ _____

    _____ $\longrightarrow$ _____ $\longrightarrow$ ___osteocyte.

2. Several descriptions of bone structure are given in column B. Identify the structure involved by choosing
    the appropriate term from column A and placing the corresponding letter in the correct blank.

    Column A   **Column B**

    a. Central canal   _____ concentric layers of calcified matrix

    b. concentric lamellae   _____ site of osteocytes

    c. lacunae   _____ longitudinal canal carrying blood vessels, lymphatics, and nerves

d. canaliculi   _____   nonliving, structural part of bone

e. matrix   _____   minute canals connecting lacunae

3. On the photomicrograph of bone below, identify all structures named in column A in question 2 above.

Chemical Composition of Bone

1. What is the function of the organic matrix in bone? _____

2. Name the important organic bone components. _____

3. Calcium salts form the bulk of the inorganic material in bone. What is the function of the calcium salts?

_____

4. Which is responsible for bone structure? (circle the appropriate response)

     inorganic portion   organic portion   both contribute

Review Sheet

EXERCISE 12

The Fetal Skeleton

1. Are the same skull bones seen in the adult found in the fetal skull? _____

2. How does the relative size of the fetal face compare to its cranium? _____

   _____

   How does this compare to the adult skull? _____

   _____

   _____

3. Why are there outward conical projections in some of the fetal cranial bones? _____

   _____

4. What is a fontanel? _____

   What is its fate? _____

   What is the function of the fontanels in the fetal skull? _____

   _____

5. Describe how the fetal skeleton compares with the adult skeleton in the following areas:

   vertebrae _____

   _____

   coxal bone _____

   _____

   carpals and tarsals _____

   sternum _____

   frontal bone _____

   patella _____

   rib cage _____

   _____

6. How does the size of the fetus's head compare to the size of its body? _____

   _____

Review Sheet

EXERCISE 13

Articulations and Body Movements

Types of Joints

1. Use key responses to identify the joint types described below.

   Key: a. cartilaginous   b. fibrous   c. synovial

   _____ typically allows a slight degree of movement

   _____ includes the pubic symphysis and joints between the vertebral bodies

   _____ essentially immovable joints

   _____ sutures are the most remembered example

   _____ characterized by cartilage connecting the bony portions

   _____ all have a fibrous capsule lined with a synovial membrane surrounding a joint cavity

   _____ all are freely movable or diarthrotic

   _____ bone regions are united by fibrous connective tissue

   _____ include the hip, knee, and elbow joints

2. Match the joint subcategories in column B with their descriptions in column A, and place an asterisk (*) beside all choices that are examples of synovial joints.

Column A	**Column B**
_____ joint between skull bones	a. ball and socket
_____ joint between the axis and atlas	b. condyloid
_____ hip joint	c. gliding
_____ intervertebral joints (between articular processes)	d. hinge
_____ joint between forearm bones and wrist	e. pivot
_____ elbow	f. saddle
_____ interphalangeal joints	g. suture
_____ intercarpal joints	h. symphysis
_____ joint between tarsus and tibia/fibula	i. synchrondrosis
_____ joint between skull and vertebral column	j. syndesmosis
_____ joint between jaw and skull	k. symphysis
_____ joints between proximal phalanges and metacarpal bones	

_____ epiphyseal plate of a child's long bone

_____ a multiaxial joint

_____ , _____ biaxial joints

_____ , _____ uniaxial joints

3. What characteristics do all joints have in common? _____

_____

4. Describe the structure and function of the following structures or tissues in relation to a synovial joint
   and label the structures indicated by lines in the diagram to the right.

   ligament _____

   _____

   tendon _____

   _____

   hyaline cartilage _____

   _____

   synovial membrane _____

   _____

   bursa _____

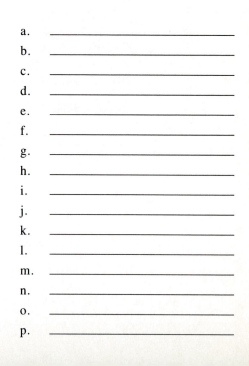

5. What structural joint changes are common to the elderly? _____

_____

_____

Body Movements

1. Complete the following statements:

   The movable attachment of a muscle is called its __a__,
   and its stationary attachment is called the __b__.
   Winding up for a pitch (as in baseball) can properly be
   called __c__. To keep your seat when riding a horse, the
   tendency is to __d__ your thighs. In running, the action
   at the hip joint is __e__ in reference to the leg moving
   forward and __f__ in reference to the leg in the
   posterior position. In kicking a football, the action at
   the knee is __g__. In climbing stairs, the hip and knee
   of the forward leg are both __h__. You have just touched
   your chin to your chest. This is __i__ of the neck.
   Using a screwdriver with a straight arm requires __j__
   of the arm. Consider all the movements of which the
   arm is capable. One often used for strengthening all
   the upper arm and shoulder muscles is __k__. Movement
   of the head that signifies "no" is __l__. Standing on
   your toes as in ballet requires __m__ of the foot. Action
   that moves the distal end of the radius across the ulna
   is __n__. Raising the arms laterally away from the body
   is called __o__ of the arms. Walking on one's heels is
   __p__.

   a. _____

   b. _____

   c. _____

   d. _____

   e. _____

   f. _____

   g. _____

   h. _____

   i. _____

   j. _____

   k. _____

   l. _____

   m. _____

   n. _____

   o. _____

   p. _____

Review Sheet

EXERCISE 14

Microscopic Anatomy, Organization, and Classification of Skeletal Muscle

Skeletal Muscle Cells and Their Packaging into Muscles

1. What capability is most highly expressed in muscle tissue? _____

_____

2. The diagram illustrates a small portion of a muscle myofibril. Using letters from the key, correctly identify each structure indicated by a line. Also add a bracket to delineate the extent of one sarcomere.

Key: a. actin filament   d. myosin filament
       b. A band   e. Z line
       c. I band

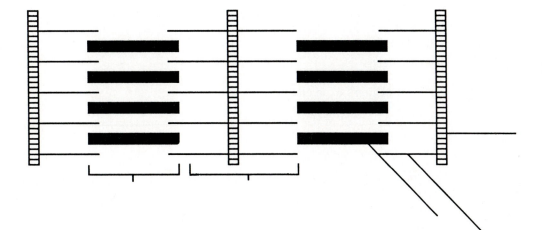

3. Use the letters of the items on the right to correctly identify the structures described on the left.

_____ connective tissue ensheathing a bundle of muscle cells

_____ another term for deep fascia

_____ contractile unit of muscle

_____ a muscle cell

_____ thin reticular connective tissue investing each muscle cell

_____ cell membrane of the muscle cell

_____ a long filamentous organelle found within muscle cells with a banded appearance

_____ actin or myosin-containing structure

_____ cordlike extension of connective tissue beyond the muscle, serving to attach it to a bone

a. endomysium

b. epimysium

c. myofiber

d. myofilament

e. myofibril

f. perimysium

g. sarcolemma

h. sarcomere

i. sarcoplasm

j. tendon

4. Why are the connective tissue wrappings of skeletal muscle important? (Give at least three reasons.)

_____

_____

_____

5. Why are indirect—that is, tendinous—muscle attachments to bone seen more often than direct attachments? _____

_____

_____

6. How does an aponeurosis differ from a tendon? _____

_____

The Neuromuscular Junction

1. Complete the following statements:

The junction between a motor neuron's axon and the muscle cell membrane is called a myoneural junction or a __a__ junction. A motor neuron and all of the skeletal muscle cells it stimulates is called a __b__. The end plates of each motor axon have numerous projections called __c__. The actual gap between the axon and the muscle cell is called a __d__. Within the axonal end plates are many small vesicles containing a neurotransmitter substance called __e__. When the __f__ reaches the ends of the axon, the neurotransmitter is released and diffuses to the muscle cell membrane to combine with receptors there. The combination of the neurotransmitter with the muscle membrane receptors causes the membrane to become permeable to sodium, which results in the influx of sodium ions and __g__ of the membrane. Then contraction of the muscle cell occurs. Before the muscle cell can be stimulated to contract again, __h__ must occur.

a. _____

b. _____

c. _____

d. _____

e. _____

f. _____

g. _____

h. _____

Classification of Skeletal Muscles

1. Several criteria were given relative to the naming of muscles. Match the criteria (column B) to the muscle cell names (column A). Note that more than one criterion may apply in some cases.

Column A	**Column B**
_____ gluteus maximus	a. action of the muscle
_____ adductor magnus	b. shape of the muscle
_____ biceps femoris	c. location of the muscle's origin and/or insertion
_____ abdominis transversus	d. number of origins
_____ extensor carpi ulnaris	e. location of muscle relative to a bone or body region
_____ trapezius	f. direction in which the muscle fibers run relative to some imaginary line
_____ rectus femoris	g. relative size of the muscle
_____ external oblique	

2. When muscles are discussed relative to the manner in which they interact with other muscles, the terms shown in the key are often used. Match the key terms with the appropriate definitions.

Key: a. antagonist   b. fixator   c. prime mover   d. synergist

_____ agonist

_____ postural muscles for the most part

_____ reverses and/or opposes the action of a prime mover

_____ stabilizes a joint so that the prime mover may act at more distal joints

_____ performs the same movement as the prime mover

_____ immobilizes the origin of a prime mover

Review Sheet

EXERCISE 15

Gross Anatomy of the Muscular System

Muscles of the Head and Neck

1. Identify the major muscles described next:

_____ used in smiling

_____ used to suck in your cheeks

_____ used in winking

_____ used in pouting (pulls the corners of the mouth downward)

_____ raises your eyebrows

_____ used to form the vertical frown crease on the forehead

_____ "kissing muscle"

_____ prime mover of jaw closure

Muscles of the Trunk

1. Identify the major muscles described next:

_____ a major spine flexor

_____ prime mover for pulling the arm posteriorly

_____ prime mover for shoulder flexion

_____ , _____ , assume major responsibility for forming the abdominal girdle (three pairs of muscles)

_____

_____ pulls the shoulder backward and downward

_____ prime mover of shoulder abduction

_____ , _____ important in shoulder adduction; antagonists of the shoulder abductor (two muscles)

_____ moves the scapula forward and downward

_____ small, inspiratory muscles between the ribs; elevate the ribs

_____ extends the head

_____ pulls scapulae medially

Arm Muscles

1. Identify the muscles described next:

_____ places the palm upward

_____ flexes the forearm and supinates the hand

_____, _____ forearm flexors; no role in supination (two muscles)

_____ elbow extensor

_____ power wrist flexor and hand adductor

_____ flexes wrist and distal phalanges

_____ pronate the hand (two muscles)

_____ flexes the thumb

_____, _____, extend and abduct the wrist (three muscles)

_____

_____ extend the wrist and digits

Muscles of the Lower Extremity

1. Identify the muscles described next:

_____ moves the thigh laterally to take the "at ease" stance

_____ used to extend the hip when climbing stairs

_____ "toe dancer's" muscles—calf pair

_____ inverts the foot

_____ allow you to draw your legs to the midline of your body, as when standing at attention

_____ "tailor's muscle"

_____, _____, extends thigh, and flexes knee (three muscles)

_____

_____ extends knee and flexes thigh

Muscle Descriptions: General Review

1. Completion: identify the muscles described below:

_____, _____, _____ are commonly used for intramuscular injections (three muscles).

Dissection and Identification of Cat Muscles

1. Many human muscles are modified from those of the cat (or any quadruped) as a result of the requirements of an upright posture. The following questions refer to these differences:

How does the human trapezius muscle differ from the cat's? _____

_____

_____

How does the deltoid differ? _____

_____

_____

How does the extent orientation of the human sartorious muscle differ from its relative position in the cat?

_____

_____

_____

Explain these differences in terms of differences in function. _____

_____

_____

The human rectus abdominis is definitely divided by four transverse tendons. These tendons are absent or difficult to identify in the cat's homologue. How do these tendons affect the human upright posture?

_____

_____

_____

Review Sheet

EXERCISE 16

Histology of Nervous Tissue

1. The cellular unit of the nervous system is the neuron. What is the major function of this cell type? ____

   _____

2. Name four types of neuroglia and list at least four functions of these cells. (You will need to consult your textbook for this.)

   Types   **Functions**

   _____   _____

   _____   _____

   _____   _____

   _____   _____

3. Match each statement with a response chosen from the key.

   Key: a. afferent neuron   f. ganglion   k. peripheral nervous
           b. association neuron   g. neuroglia   system
           c. autonomic nervous   h. neurotransmitters   l. Schwann cells
              system   i. nerve   m. somatic nervous system
           d. central nervous system   j. nuclei   n. synapse
           e. efferent neuron   o. tract

   _____ the brain and spinal cord collectively

   _____ specialized cells that myelinate the fibers of neurons found in the peripheral nervous system

   _____ specialized cells that myelinate the fibers of neurons found in the CNS

   _____ junction or point of close contact between neurons

   _____ a bundle of nerve processes inside the central nervous system

   _____ neuron serving as part of the conduction pathway between sensory and motor neurons

   _____ spinal and cranial nerves and ganglia

   _____ collection of nerve cell bodies found outside the CNS

   _____ neuron that conducts impulses away from the CNS to muscles and glands

   _____ neuron that conducts impulses toward the CNS from the body periphery

   _____ system that controls the involuntary activities of the body

_____ chemicals released by neurons that stimulate other neurons, muscles, or glands

_____ system involved in the voluntary activities of the body, such as activation of the skeletal muscles

Neuron Anatomy

1. Match the following anatomic terms (column B) with the appropriate description or function (column A).

Column A	**Column B**
_____ contains enlarged region of the cell body from which the axon originates	a. axon
_____ releases neurotransmitters	b. axonal end bulb
_____ conducts an electrical current toward the cell body	c. axon hillock
_____ increases the speed of impulse transmission and insulates the nerve fibers	d. dendrite
_____ is site of the nucleus	e. myelin sheath
_____ may be involved in the transport of substances within the neuron	f. neuronal cell body
_____ essentially rough endoplasmic reticulum, important metabolically	g. neurofibril
_____ conducts impulses away from the cell body	h. Nissl body

2. Draw a "typical" neuron in the space below. Include and label the following structures on your diagram: cell body, nucleus, Nissl bodies, dendrites, axon, axon collaterals, myelin sheath, and nodes of Ranvier.

3. How is one-way conduction at synapses assured? _____

_____

4. What anatomic characteristic determines whether a particular neuron is classified as unipolar, bipolar, or

multipolar? _____

Make a simple line drawing of each type here.

Unipolar neuron   Bipolar neuron   Multipolar neuron

5. Describe how the Schwann cells form the myelin sheath and the neurilemma encasing the nerve processes.

(You may want to diagram the process.) _____

_____

_____

_____

_____

_____

Structure of a Nerve

1. What is a nerve? _____

_____

_____

2. State the location of each of the following connective tissue coverings:

endoneurium _____

perineurium _____

epineurium _____

3. What is the value of the connective tissue wrappings found in a nerve? _____

_____

4. Define *mixed nerve:* _____

_____

5. Identify all indicated parts of the nerve section.

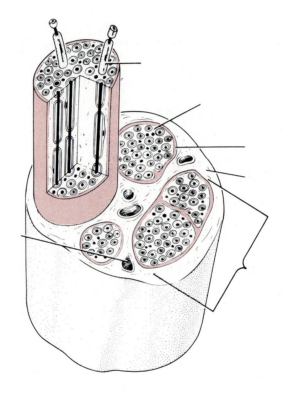

Review Sheet

EXERCISE 17

Gross Anatomy of the Brain and Cranial Nerves

The Human Brain

1. Match the letters on the diagram of the human brain (right lateral view) to the proper terms listed at the left.

_____ frontal lobe

_____ parietal lobe

_____ temporal lobe

_____ precentral gyrus

_____ parieto-occipital sulcus

_____ postcentral gyrus

_____ lateral sulcus

_____ central sulcus

_____ cerebellum

_____ medulla

_____ occipital lobe

_____ pons

2. In which of the cerebral lobes would the following functional areas be found?

auditory area _____   olfactory area _____

primary motor area _____   visual area _____

primary sensory area _____   Broca's area _____

3. Which of the following structures are *not* part of the brain stem? (Circle the appropriate response or responses.)

cerebral hemispheres   pons   midbrain   cerebellum   medulla   diencephalon

4. Completion: insert responses in corresponding blanks on the right.

A __a__ is an elevated ridge of cerebral tissue. The convolutions seen in the cerebrum are important because they increase the __b__. Gray matter is composed of __c__. White matter is composed of __d__. A fiber tract that provides for communication between different parts of the same cerebral hemisphere is called a(n) __e__, while one that carries impulses to and from the cerebrum from and to lower CNS areas is called a(n) __f__ tract. The lentiform nucleus along with the amygdaloid and caudate nuclei are collectively called the __g__.

a. _____
b. _____
c. _____
d. _____
e. _____
f. _____
g. _____

5. Identify the structures on the sagittal view of the human brain by matching the lettered areas to the proper terms at the left.

_____ cerebellum

_____ cerebral aqueduct

_____ cerebral hemisphere

_____ cerebral peduncle

_____ choroid plexus

_____ corpora quadrigemina

_____ corpus callosum

_____ fornix

_____ fourth ventricle

_____ hypothalamus

_____ mammillary bodies

_____ massa intermedia

_____ medulla oblongata

_____ optic chiasma

_____ pineal body

_____ pituitary gland   _____ septum pellucidum

_____ pons   _____ thalamus

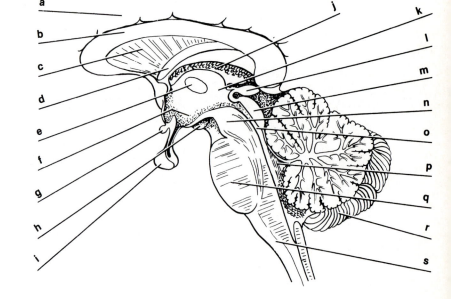

6. Using the letters from the diagram in item 5, match the appropriate structure with the descriptions given below:

_____ site of regulation of body temperature and water balance; most important autonomic center

_____ consciousness depends on the function of this part of the brain

_____ located in the midbrain, contains reflex centers for vision and audition

_____ responsible for the regulation of posture and coordination of complex muscular movements

_____ important synapse site for afferent fibers traveling to the sensory cortex

_____ contains autonomic centers regulating blood pressure, heart rate, and respiratory rhythm, as well as coughing, sneezing, and swallowing centers

_____ large commisure connecting the cerebral hemispheres

_____ fiber tract involved with olfaction

_____ connects the third and fourth ventricles

_____ encloses the third ventricle

7. Embryologically, the brain arises from the rostral end of a tubelike structure that quickly becomes divided into three major regions. Groups of structures that develop from the embryonic brain are listed below. Designate the embryonic origin of each group as the hindbrain, midbrain, or forebrain.

_____ the diencephalon, including the thalamus, optic chiasma, and hypothalamus

_____ the medulla, pons, and cerebellum

_____ the cerebral hemispheres

8. What is the importance of the fact that the human cerebral hemispheres are highly convoluted? _____

_____

9. What is the function of the basal nuclei? _____

_____

_____

10. What is the corpus striatum, and how is it related to the fibers of the internal capsule? _____

_____

_____

11. A brain hemorrhage within the region of the right internal capsule results in paralysis of the left side of the body. Explain why the left side (rather than the right side) is affected. _____

_____

12. Explain why trauma to the base of the brain is often much more dangerous than trauma to the frontal lobes. (_Hint:_ Think about the relative functioning of the cerebral hemispheres and the brain stem structures. Which contain centers more vital to life?) _____

_____

_____

RS63

13. In "split brain" experiments, the main commissure connecting the cerebral hemispheres is cut. First,

name this commissure: _____

Then, describe what results (in terms of behavior) can be anticipated in such experiments. (Use an appropriate reference if you need help with this one!)

_____

_____

_____

_____

Meninges of the Brain

1. Identify the meningeal (or associated) structures described below.

_____ outermost meninx covering the brain, composed of tough fibrous connective tissue

_____ innermost meninx covering the brain, delicate and highly vascular

_____ structure instrumental in returning cerebrospinal fluid to the venous blood in the dural sinuses

_____ structure that forms the cerebrospinal fluid

_____ middle meninx, like a cobweb in structure

_____ its outer layer forms the periosteum of the skull

_____ a dural fold that attaches the cerebrum to the crista galli of the skull

_____ a dural fold separating the cerebrum from the cerebellum

Cerebrospinal Fluid

1. Fill in the following flowsheet, which relates to the circulation of cerebrospinal fluid from its formation site (assume that this is one of the lateral ventricles) to the site of its reabsorption into the venous blood.

Lateral ventricle _ _ _ _ _ _ _ _ _____ _ _ _ _ _ _ _ _ _ _ _ _ _

Third ventricle _ _ _ _ _ _ _ _ _____ _ _ _ _ _ _ _ _ _ _ _ ►

_____ _ _ _ _ ►single _____foramen _ _ _ _ _ _ _ ►

_ ► two _____foramina _ ⁄

_____ surrounding the brain and cord _ _ _ _ _ _ _ _ _ _ _ ►
(and central canal of the cord)

Arachnoid villi _ _ _ _ _ _ _ _ ► _____ containing venous blood

Cranial Nerves

1. Provide the name and number of the cranial nerves involved in each of the following activities, sensations, or disorders.

_____ shrugging the shoulders

_____ smelling a flower

_____ raising the eyelids; focusing the lens of the eye for accommodation; and pupillary constriction

_____ slows the heart; increases the mobility of the digestive tract

_____ involved in Bell's palsy (facial paralysis)

_____ chewing food

_____ listening to music; seasickness

_____ secretion of saliva; tasting well-seasoned food

_____ involved in "rolling" the eyes (three nerves—provide numbers only)

_____ feeling a toothache

_____ reading Playgirl or Playboy magazine

_____ purely sensory in function (three nerves—provide numbers only)

Dissection of the Sheep Brain

1. In your own words, describe the relative hardness of the sheep brain tissue as observed when cutting into it.

_____

_____

Since formalin hardens all tissue, what conclusions might you draw about the relative hardness and texture

of living brain tissue? _____

_____

2. How does the relative size of the cerebral hemispheres compare in sheep and human brains? _____

_____

What is the significance? _____

_____

3. What is the significance of the fact that the olfactory bulbs are much larger in the sheep brain than in the

human brain? _____

_____

Application of Knowledge

You have been given all the information you need to identify the brain region involved in the situations described below. See how well your nervous system has integrated this information, and name the region most likely to be involved in each situation.

1. Following a train wreck, a man with an obvious head injury was observed stumbling about the scene. An inability to walk properly and a loss of balance were quite obvious. What brain region was injured?

   _____

2. A young woman was brought into the emergency room with extremely dilated pupils. Her friends stated

   that she had overdosed on cocaine. What cranial nerve was stimulated by the drug? _____

3. A young man is admitted to the hospital after receiving serious burns resulting from standing with his back too close to a bonfire. He is muttering that he never felt the pain—otherwise, he would have

   smothered the flames by rolling on the ground. What part of his CNS might be malfunctional? _____

   _____

4. An elderly gentleman had just suffered a stroke. He is able to understand verbal and written language, but when he tries to respond, his words come out garbled. What cortical region has been damaged by the

   stroke? _____

Review Sheet
_____

EXERCISE 18

The Spinal Cord, Spinal Nerves, and the Autonomic Nervous System
_____

Anatomy of the Spinal Cord

1. Match the descriptions given below to the proper anatomic term:

   a. cauda equina   b. conus medullaris   c. filium terminale   d. foramen magnum

   _____ most superior boundary of the spinal cord

   _____ meningeal extension beyond the spinal cord terminus

   _____ spinal cord terminus

   _____ collection of spinal nerves traveling in the vertebral canal below the terminus of the spinal cord

2. Choose the proper answer from the following key to respond to the following descriptions relative to spinal cord anatomy.

   Key: a. afferent   b. efferent   c. both afferent and efferent   d. interneuron

   _____ neuron type found in dorsal horn   _____ fiber type in ventral root

   _____ neuron type found in ventral horn   _____ fiber type in dorsal root

   _____ neuron type in dorsal root ganglion   _____ fiber type in spinal nerve

3. Where in the vertebral column is a lumbar puncture generally done? _____

   Why is this the site of choice? _____

   _____

4. The spinal cord is enlarged in two regions, the _____ and the _____

   region. What is the significance of these enlargements? _____

   _____

5. How does the position of the gray and white matter differ in the spinal cord and the cerebral hemispheres?

   _____

   _____

   _____

   _____

6. Choose the name of one tract, from the following key, that might be damaged when the following conditions are observed. (More than one choice may apply.)

_____ uncoordinated movement

_____ lack of voluntary movement

_____ tremors, jerky movements

_____ diminished pain perception

_____ diminished sense of touch

Key:

a. fasciculus gracilis
b. fasciculus cuneatus
c. lateral corticospinal tract
d. ventral corticospinal tract
e. tectospinal tract
f. rubrospinal tract
g. lateral spinothalamic tract
h. ventral spinothalamic tract
i. dorsal spinocerebellar tract
j. vestibulospinal tract
k. olivospinal tract
l. ventral spinocerebellar tract

7. Use an appropriate reference to describe the functional significance of an upper motor neuron and a lower motor neuron:

upper motor neuron _____

_____

lower motor neuron _____

_____

Will contraction of a muscle occur if the lower motor neurons serving it have been destroyed? _____

If the upper motor neurons serving it have been destroyed? _____ Using an appropriate reference,

differentiate between flaccid and spastic paralysis and note the possible causes of each. _____

_____

_____

_____

_____

_____

_____

Spinal Nerves and Nerve Plexuses

1. In the human there are 31 pairs of spinal nerves named according to the region of the vertebral column from which they issue. The spinal nerves are named below; note the vertebral level at which they emerge by number.

cervical nerves _____   sacral nerves _____

lumbar nerves _____   thoracic nerves _____

2. The ventral rami of spinal nerves C_1 through T_1 and T_{12} through S_4 take part in forming _____,

which serve the _____ of the body. The ventral rami of T_1 through T_{12} run between the ribs to serve the _____. The posterior rami of the spinal nerves serve _____

_____.

3. What would happen if the following structures were damaged or transected? (Use key choices for responses.)

Key: a. loss of motor function   b. loss of sensory function   c. loss of both motor and
                                                                            sensory function

_____ dorsal root of a spinal nerve

_____ ventral root of a spinal nerve

_____ anterior ramus of a spinal nerve

4. Define *plexus:* _____

_____

5. Name the major nerves that serve the following body areas:

_____ head, neck, shoulders (name plexus only)

_____ diaphragm

_____ posterior thigh

_____ leg and foot (name two)

_____ most anterior forearm muscles

_____ arm muscles

_____ abdominal wall (name plexus only)

_____ anterior thigh

_____ medial side of the hand

The Autonomic Nervous System

1. For the most part, sympathetic and parasympathetic fibers serve the same organs and structures. How can they exert antagonistic effects? (After all, nerve impulses are nerve impulses—aren't they?)

_____

_____

2. You are alone in your home late in the evening, and you hear an unfamiliar sound in your backyard. List four physiologic events promoted by the sympathetic nervous system that would aid you in coping with this rather frightening situation.

_____

_____

_____

_____

3. The following chart states a number of conditions. Use a check mark to show which division of the autonomic nervous system is involved in each.

Sympathetic division	Condition	Parasympathetic division
	secretes norepinephrine, adrenergic fibers	
	secretes acetylcholine, cholinergic fibers	
	long preganglionic axon, short postganglionic axon	
	short preganglionic axon, long postganglionic axon	
	arises from cranial and sacral nerves	
	arises from spinal nerves T_1 to L_3	
	normally in control	
	fight or flight system	
	has more specific control (look it up!)	

4. Often after surgery, people are temporarily unable to urinate, and bowel sounds are absent. What division of the ANS is affected by the anesthesia? _____

5. Name three structures that receive sympathetic innervation but not parasympathetic innervation. _____

_____

_____

6. The pelvic nerve contains (circle one):

a. preganglionic sympathetic fibers   c. preganglionic parasympathetic fibers

b. postganglionic sympathetic fibers   d. postganglionic parasympathetic fibers

Review Sheet

EXERCISE 19

Special Senses: Vision

Anatomy of the Eye

1. Three accessory eye structures contribute to the formation of tears and/or aid in lubrication of the eyeball. Name each and then name its major secretory product. Indicate which has antibacterial properties by circling the correct secretory product.

Accessory structures	Product
_____	_____
_____	_____
_____	_____

2. The eyeball is wrapped in adipose tissue within the orbit. What is the function of the adipose tissue?

_____

What seven bones form the bony orbit? (Think! If you can't remember, check a skull or your text.)

_____   _____   _____

_____   _____

_____   _____

3. Why does one often have to blow one's nose after having a good cry? _____

_____

4. Identify the extrinsic eye muscle predominantly responsible for the actions described below.

_____ turns the eye laterally

_____ turns the eye medially

_____ turns the eye up and laterally

_____ turns the eye inferiorly

_____ turns the eye superiorly

_____ turns the eye down and laterally

5. What is a sty? _____

   Conjunctivitis? _____

6. Match the key responses with the descriptive statements on the left.

   _____ attaches the lens to the ciliary body

   _____ fluid filling the anterior segment of the eye

   _____ the "white" of the eye

   _____ retinal area devoid of photoreceptors

   _____ modification of the choroid, which controls the shape of
   the crystalline lens

   _____ nutritive (vascular) tunic of the eye

   _____ drains the aqueous humor of the eye

   _____ tunic containing the rods and cones

   _____ substance occupying the posterior segment of the eyeball

   _____ forms most of the heavily pigmented vascular tunic that
   prevents light-scattering within the eye

   _____, _____ smooth muscle structures (intrinsic eye muscles)

   _____ area of acute or discriminatory vision

   _____ form (by filtration) the aqueous humor

   _____, _____, _____, _____ refractory media of the eye

   _____ anteriormost portion of the fibrous tunic—your "window
   on the world"

   _____ composed of tough, white fibrous connective tissue

   Key:

   a. aqueous humor

   b. canal of Schlemm

   c. choroid coat

   d. ciliary body

   e. ciliary processes of
   the ciliary body

   f. cornea

   g. fovea centralis

   h. iris

   i. lens

   j. optic disc

   k. retina

   l. sclera

   m. suspensory ligaments

   n. vitreous humor

7. The iris is composed primarily of two smooth muscle layers, one arranged radially and the other circularly.

   Which of these dilates the pupil? _____

8. You would expect the pupil to be dilated in which of the following circumstances? (Circle the correct response(s).)

   a. in brightly lighted surroundings   c. during focusing for near vision

   b. in dimly lighted surroundings   d. in observing distant objects

9. The intrinsic eye muscles are under the control of which of the following? (Circle the correct response.)

   autonomic nervous system   somatic nervous system

Dissection of the Cow (Sheep) Eye

1. What modification of the choroid that is not present in humans is found in the cow eye? _____

    _____

    What is its function? _____

    _____

2. What is the anatomic appearance of the retina? _____

    _____

    At what point is it attached to the posterior aspect of the eyeball? _____

Microscopic Anatomy of the Retina

1. The two major layers of the retina are the epithelial and nervous layers. In the nervous layer, the neuron populations are arranged as follows from the epithelial layer to the vitreous humor. (Circle all proper responses.)

    bipolar cells, ganglion cells, photoreceptors   photoreceptors, ganglion cells, bipolar cells

    ganglion cells, bipolar cells, photoreceptors   photoreceptors, bipolar cells, ganglion cells

2. The axons of the _____ cells form the optic nerve, which exits from the eyeball.

3. The following statements may be completed by inserting either RODS or CONES. Complete the following sentences:

    The dim light receptors are the _____. Only _____ are found in the fovea centralis, while

    mostly _____ are found in the periphery of the retina. _____ are the photoreceptors that

    operate best in bright light and allow for color vision.

Visual Pathways to the Brain

1. The visual pathway to the occipital lobe of the brain consists most simply of a chain of five neurons. Beginning with the photoreceptor cell of the retina, name them and note their location in the pathway.

    1   Photoreceptor cell, retina

    2 _____

    3 _____

    4 _____

    5 _____

2. Visual field tests are done to reveal destruction along the visual pathway from the retina to the optic region of the brain. Note where the lesion is likely to be in the following cases:

    normal vision in left eye visual field; absence of vision of right eye visual field: _____

    normal vision in both eyes for right half of the visual field; absence of vision in both eyes for left half of

    the visual field: _____

3. How is the right optic *tract* anatomically different from the right optic *nerve?* _____

_____

_____

What does this difference result from? _____

Visual Tests and Experiments

1. Match the terms in column B with the descriptions in column A.

Column A	Column B
_____ light bending	a. accommodation
_____ ability to focus for close (under 20 ft.) vision	b. astigmatism
_____ normal vision	c. convergence
_____ inability to focus well on close objects (farsightedness)	d. emmetropia
_____ nearsightedness	e. hyperopia
_____ blurred vision due to unequal curvatures of the lens or cornea	f. myopia
_____ medial movement of the eyes during focusing on close objects	g. refraction

2. Complete the following statements:

In farsightedness, the light is focused __a__ the retina. The lens required to treat myopia is a __b__ lens. The "near point" increases with age because the __c__ of the lens decreases as we get older. A convex lens, like that of the eye, produces an image that is upside down and reversed from left to right. Such an image is called a __d__ image.

a. _____
b. _____
c. _____
d. _____

3. Check the appropriate column under the vertical headings to characterize events occurring within the eye during close and distant vision.

	Ciliary muscle		Suspensory ligaments		Lens convexity		Degree of light refraction	
close vision	relaxed	contracted	relaxed	contracted	increased	decreased	increased	decreased
distant vision	relaxed	contracted	relaxed	contracted	increased	decreased	increased	decreased

4. Explain why vision is lost when light hits the blind spot. _____

5. What is meant by the term *negative afterimage* and what does this phenomenon indicate? _____

_____

6. Record your Snellen eye test results below:

Left eye (without glasses) _____ (with glasses) _____

Right eye (without glasses) _____ (with glasses) _____

Is your visual acuity normal, less than normal, or better than normal? _____

Explain. _____

_____

Explain why each eye is tested separately when using the Snellen eye chart. _____

_____

Explain 20/40 vision: _____

_____

Explain 20/10 vision: _____

_____

7. Define *astigmatism:* _____

_____

How can it be corrected? _____

8. Record the distance of your near-point of accommodation as tested in the laboratory:

right eye _____   left eye _____

Is your near-point within the normal range for your age? _____

9. Define *presbyopia:* _____

_____

What causes it? _____

10. To which wavelengths of light do the three cone types of the retina respond maximally?

_____   _____   _____

11. Since only three cone types exist, how can you explain the fact that we see a much greater range of colors?

_____

_____

12. From what condition does color blindness result? _____

_____

13. Record the results of the demonstration of the relative positioning of rods and cones in the circle below (use appropriately colored pencils).

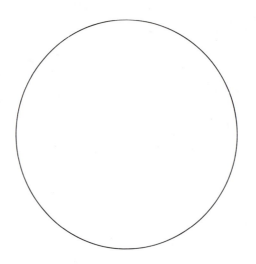

14. Explain the difference between binocular and panoramic vision. _____

_____

_____

_____

What is the advantage of binocular vision? _____

What factor(s) are responsible for binocular vision? _____

_____

_____

_____

15. Why is the ophthalmoscopic examination an important diagnostic tool? _____

_____

16. In the experiment on the convergence reflex, what happened to the position of the eyeballs as the object was moved closer to the subject's eyes? _____

_____

What extrinsic eye muscles control the movement of the eyes during this reflex? _____

_____

What is the value of this reflex? _____

_____

What would be the visual result of an inability of these muscles to function? _____

_____

17. In the experiment on the photopupillary reflex, what happened to the eye pupil exposed to light? _____

_____

What happened to the pupil of the nonilluminated eye? _____

_____ Explanation? _____

_____

_____

18. Many college students struggling through mountainous reading assignments are told that they need glasses for "eyestrain." Why is it more of a strain on the extrinsic and intrinsic eye muscles to look at close objects

than at far objects? _____

_____

Review Sheet

EXERCISE 20

Special Senses: Hearing and Equilibrium

Anatomy of the Ear

1. Select the terms from column B that apply to the column A descriptions. Some terms are used more than once.

Column A

Column B

_____, _____, _____ structures comprising the outer or external ear

_____, _____, _____ structures composing the bony or osseous labyrinth

_____, _____, _____ collectively called the ossicles

_____, _____ ear structures not involved with hearing

_____ allows pressure in the middle ear to be equalized with atmospheric pressure

_____ transmits the vibrations to the ossicles

_____ contains the organ of Corti

_____, _____ contain receptors for the sense of equilibrium

_____ transmits the vibratory motion of the stirrup to the fluid in the Scala vestibuli of the inner ear

_____ acts as a pressure relief valve for the increased fluid pressure in the Scala tympani; bulges into the tympanic cavity

_____ connects the nasopharynx and the middle ear

_____ fluid contained within the membranous labyrinth

_____ fluid contained within the osseous labyrinth and bathing the membranous labyrinth

a.   anvil

b.   cochlea

c.   endolymph

d.   eustachian tube

e.   external auditory canal

f.   hammer

g.   oval window

h.   perilymph

i.   pinna

j.   round window

k.   semicircular canals

l.   stirrup

m.   tympanic membrane

n.   vestibule

2. Sound waves hitting the eardrum initiate its vibratory motion. Trace the pathway through which vibrations and fluid currents are transmitted to finally stimulate the hair cells in the organ of Corti. (Name the

appropriate ear structures in their correct sequence.) Eardrum → _____

_____

3. Match the membranous labyrinth structures listed in Column B with the descriptive statements in Column A.

Column A

_____, _____ found within the vestibule

_____ contains the organ of Corti

_____, _____ sites of the maculae

_____ positioned in all spatial planes

_____ hair cells of organ of Corti rest on this membrane

_____ gelatinous membrane overlying the hair cells of the organ of Corti

_____ contains the crista ampullaris

_____, _____, _____, _____ function in static equilibrium

_____, _____, _____, _____ function in dynamic equilibrium

_____ carries auditory information to the brain

_____ gelatinous cap overlying hair cells of the crista ampullaris

_____ grains of calcium carbonate in the maculae

Column B

a. ampulla

b. basilar membrane

c. cochlear duct

d. cochlear nerve

e. cupula

f. membranous semicircular canals (ducts)

g. otoliths

h. saccule

i. tectorial membrane

j. utricle

k. vestibular nerve

4. Identify all indicated structures and ear regions in the following diagram.

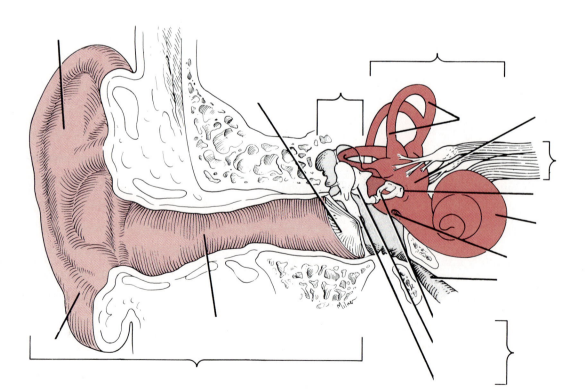

5. Describe how sounds of different frequency (pitch) are differentiated in the cochlea. _____

_____

_____

6. Explain the role of the endolymph of the semicircular canals in activating the receptors during angular

motion. _____

_____

_____

_____

7. Explain the role of the otoliths in perception of static equilibrium (head position). _____

_____

_____

_____

Laboratory Tests

1. Was the auditory acuity measurement made during the experiment on page 198 the same or different for

both ears? _____ What factors might account for a difference in the

acuity of the two ears? _____

_____

2. During the sound localization experiment on page 199, in which position(s) was the sound least easily

located? _____

How can this phenomenon be explained? _____

_____

3. In the experiment on page 199, which tuning fork was the most difficult to hear? _____ cps.

What conclusion can you draw? _____

_____

4. When the tuning fork handle was pressed to your forehead during the Weber test, where did the sound

seem to originate? _____

Where did it seem to originate when one ear was plugged with cotton? _____

_____ How do sound waves reach the cochlea when conduction deafness is present?

_____

5. Indicate whether the following conditions described relate to conduction deafness (C) or sensorineural (central) deafness (S):

_____ can result from the fusion of the ossicles

_____ can result from a lesion on the cochlear nerve

_____ sound heard in one ear but not in the other during bone and air conduction

_____ can result from otitis media

_____ can result from impacted cerumen or a perforated eardrum

_____ can result from a blood clot in the auditory cortex

6. The Rinne test evaluates an individual's ability to hear by air- or bone-conducted sound. Which is more indicative of normal hearing? _____

7. Define *nystagmus:* _____

_____

_____

_____

8. The Barany test investigated the effect of rotatory acceleration on the semicircular canals. Explain *why* the subject still had the sensation of rotation immediately after being stopped. _____

_____

9. What is the usual reason for conducting the Romberg test? _____

_____

_____

Was the degree of sway observed greater with the eyes open or closed? _____

Why? _____

10. Normal balance or equilibrium is dependent on input from a number of sensory receptors. Name them.

_____

_____

Review Sheet

EXERCISE 21

Special Senses: Taste and Olfaction

Localization and Anatomy of Taste Buds

1. Name three sites where receptors for taste are found, and circle the predominant site.

   _____, _____ and _____

2. Describe the cellular makeup and arrangement of a taste bud. (Use a diagram if helpful.) _____

   _____

   _____

Localization and Anatomy of Olfactory Receptors

1. Describe the cellular composition and location of the olfactory epithelium. _____

   _____

   _____

2. How and why does sniffing improve your sense of smell? _____

   _____

Laboratory Experiments

1. Taste and smell receptors are both classified as _____ because they both respond

   to _____

2. Why is it impossible to taste substances with a dry tongue? _____

   _____

3. State the most important sites of the taste-specific receptors as determined during the plotting exercise in the laboratory.

   salt _____   sour _____

   bitter _____   sweet _____

4. The basic taste sensations are elicited by specific chemical substances or groups. Name them.

   salt _____   sour _____

   bitter _____   sweet _____

5. Name three factors that influence our appreciation of foods. Substantiate each choice with an example from the laboratory experience.

_____   Substantiation _____

_____

_____

_____   Substantiation _____

_____

_____

_____   Substantiation _____

_____

_____

Which of the factors chosen is most important? _____

Substantiate your choice with an example from everyday life. _____

_____

Expand on your explanation and choices by explaining why a cold, greasy hamburger is unappetizing to

most people. _____

_____

6. Babies tend to favor bland foods, whereas adults tend to like highly seasoned foods. What is the basis for

this phenomenon? _____

_____

_____

7. How palatable is food when you have a cold? _____

Explain. _____

_____

8. What is the mechanism of olfactory adaptation? _____

_____

_____

In your opinion, is olfactory adaptation desirable? _____ Explain your answer.

_____

_____

Review Sheet

EXERCISE 22

Anatomy and Basic Function of the Endocrine Glands

Gross Anatomy and Basic Function of the Endocrine Glands

1. Both the endocrine and nervous systems are major regulating systems of the body; however, the nervous system has been compared to an airmail delivery system and the endocrine system to the pony express.

   Briefly explain this comparison. _____

   _____

   _____

   _____

2. Define *hormone*. _____

   _____

   _____

3. Chemically, hormones belong chiefly to two molecular groups, the _____

   and the _____.

4. What do all hormones have in common? _____

   _____

5. Define *target organ:* _____

   _____

6. Why don't all tissues respond to all hormones? _____

   _____

7. Identify the endocrine organ described by the following statements:

   _____ located in the throat; bilobed gland connected by an isthmus

   _____ found close to the kidney

   _____ a mixed gland, located in the mesentery close to the stomach and small intestine

   _____ paired glands suspended in the scrotum

_____ ride "horseback" on the thyroid gland

_____ found in the pelvic cavity of the female, concerned with ova and female
hormone production

_____ found in the upper thorax overlying the heart; short-lived endocrine function

_____ found in the roof of the third ventricle

8. For each statement describing hormonal effects, identify the hormone(s) involved by choosing a number
   from key A, and note the hormone's site of production with a letter from key B. More than one hormone
   may be involved in some cases.

   Key A   Key B

   1. ACTH   13. MSH   a. adrenal cortex
   2. ADH   14. oxytocin   b. adrenal medulla
   3. aldosterone   15. progesterone   c. anterior pituitary
   4. cortisone   16. PTH   d. hypothalamus
   5. epinephrine   17. serotonin   e. ovaries
   6. estrogen   18. STH (GH)   f. pancreas
   7. FSH   19. testosterone   g. parathyroid glands
   8. glucagon   20. thymosin   h. pineal gland
   9. insulin   21. thyrocalcitonin   i. posterior pituitary
   10. LH   22. thyroxine   j. testes
   11. LTH (prolactin) 23. TSH   k. thymus gland
   12. melatonin   l. thyroid gland

   _____, _____ basal metabolism hormone

   _____, _____ programming of T-lymphocytes

   _____, _____ and _____, _____ regulation of blood calcium levels

   _____, _____ and _____, _____ released in response to stressors

   _____, _____ and _____, _____ development of secondary sexual characteristics

   _____, _____; _____, _____; _____, _____; and _____, _____ regulate the function
   of another endocrine gland

   _____, _____ mimics the sympathetic nervous system

   _____, _____ and _____, _____ regulate blood glucose levels; produced by the same "mixed"
   gland

   _____, _____ and _____, _____ directly responsible for regulation of the menstrual cycle

   _____, _____ and _____, _____ regulation of the ovarian cycle

   _____, _____ and _____, _____ maintenance of salt and water balance in the ECF

   _____, _____ and _____, _____ directly involved in milk production and ejection

   _____, _____ questionable function; may stimulate the melanocytes of the skin

9. Although the pituitary gland is often referred to as the master gland of the body, recent studies show
   that the hypothalamus exerts some control over the pituitary gland. How does the hypothalamus control

   both anterior and posterior pituitary functioning? _____

   _____

_____

_____

10. Indicate whether the release of the hormones listed below is stimulated by A, another hormone, B, the nervous system (neurotransmitters, or releasing factors); or C, humoral factors (the concentration of specific substances in the blood or extracellular fluid):

_____ thyroxin   _____ parathyroid hormone   _____ ADH

_____ insulin   _____ testosterone   _____ TSH, FSH

_____ estrogen   _____ epinephrine   _____ aldosterone

11. Name the hormone that would be produced in *inadequate* amounts under the following conditions:

_____ sexual immaturity

_____ tetany

_____ excessive diuresis without high blood glucose levels

_____ excessive thirst, high blood glucose levels

_____ abnormally small stature, normal proportions

_____ miscarriage

_____ lethargy, hair loss, low BMR, obesity

12. Name the hormone that is produced in *excessive* amounts in the following conditions:

_____ lantern jaw and large hands and feet in the adult

_____ bulging eyeballs, nervousness, increased pulse rate

_____ demineralization of bones, spontaneous fractures

Microscopic Anatomy of Selected Endocrine Glands (optional)

1. Choose a response from the key below to name the hormone(s) produced by the cell types listed:

Key: a. insulin   d. thyrocalcitonin   g. glucagon
      b. GH, prolactin   e. TSH, ACTH, FSH, LH   h. PTH
      c. thyroxine   f. mineralocorticoids   i. glucocorticoids

_____ parafollicular cells of the thyroid   _____ zona reticularis cells

_____ follicular epithelial cells of the thyroid   _____ zona glomerulosa cells

_____ beta cells of the islets of Langerhans   _____ oxyphil cells

_____ alpha cells of the islets of Langerhans   _____ acidophil cells of the anterior pituitary

_____ basophil cells of the anterior pituitary

2. Five diagrams of the microscopic structures of the
   endocrine glands are presented here. Identify each
   and name all indicated structures.

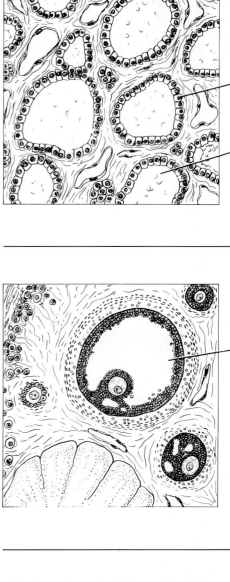

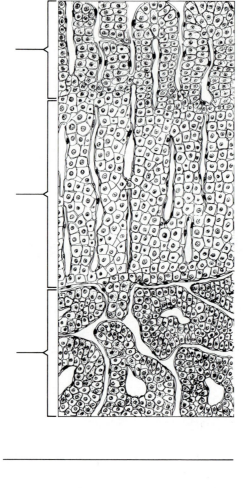

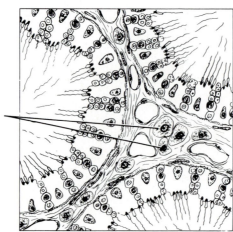

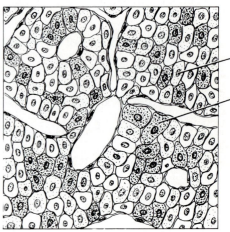

Review Sheet

EXERCISE 23

Blood

Composition of Blood

1. What is the blood volume of an averaged-size adult? _____ liters

2. Explain why blood is classified as a connective tissue. _____

   _____

   _____

3. What determines whether blood is bright red or a dull brick-red? _____

   _____

4. Use the key to identify the cell type(s) or blood elements that fit the following descriptive statements:

   _____ most numerous leukocyte

   _____, _____, _____ granular leukocytes

   _____ also called an erythrocyte, anucleate

   _____, _____ actively phagocytic leukocytes

   _____, _____ agranular leukocytes

   _____ fragments to form platelets

   _____ (a) through (g) are all examples of these

   _____ increases during allergy attacks

   _____ releases histamine during inflammatory reactions

   _____, _____ formed in lymphoid tissue

   _____ contains hemoglobin; therefore involved in oxygen transport

   _____ primarily water, noncellular; the fluid matrix of blood

   _____ increases in number during prolonged infections

   _____, _____, _____, _____, _____ also called white blood cells

   Key: a. red blood cell

   b. megakaryocyte

   c. eosinophil

   d. basophil

   e. monocyte

   f. neutrophil

   g. lymphocyte

   h. formed elements

   i. plasma

5. Supply the following information based on what you know about substances normally found dissolved in

   plasma. Name three nutrient substances. _____, _____ and

   _____

Name two gases. _____ and _____

Name three ions. _____ , _____ , and _____

6. Describe the consistency and color of the plasma you observed in the laboratory. _____
_____

7. What is the average life span of a red blood cell? How does its anucleate condition affect this life span?
_____
_____

8. From memory, describe the structural characteristics of each of the following blood cell types as accurately as possible, and note the percentage of each in the total white blood cell population.

eosinophils _____
_____

neutrophils _____
_____

lymphocytes _____
_____

basophils _____
_____

monocytes _____
_____

Hematologic Tests

1. Broadly speaking, why are hematologic studies of blood so important in the diagnosis of disease?
_____
_____

2. Record information from the blood tests you conducted in the chart below. Complete the chart by recording values for healthy male adults, and indicating the significance of high or low values for each test.

Test	Student Test results	Normal values (healthy male adults)	Significance	
			High values	Low values
total WBC count				

Test	Student Test results	Normal values (healthy male adults)	Significance High values	Low values
total RBC count				
hematocrit				
hemoglobin determination				
sedimentation rate				
coagulation time				

3. Why is a differential WBC count more valuable when trying to pin down the specific source of pathology than a total WBC count? _____

_____

_____

4. Explain the reasons for using different diluents for the RBC and WBC counts. _____

_____

_____

_____

5. What name is given to the process of RBC production? _____

What acts as a stimulus for this process? _____

What organ provides this stimulus and under what conditions? _____

_____

_____

6. Discuss the effect of each of the following factors on RBC count. Consult an appropriate reference as necessary, and explain your reasoning.

athletic training (for example, running 4 to 5 miles per day over a period of 6 to 9 months) _____

_____

_____

a permanent move from sea level to a high-altitude area _____

_____

_____

lack of iron-containing foods in your diet _____

_____

_____

7. Define *hematocrit:* _____

8. If you had a high hematocrit, would you expect your hemoglobin determination to be high or low?

_____ Why? _____

_____

9. What is an anticoagulant? _____

Name two anticoagulants used in conducting the hematologic tests. _____

and _____

What is the body's natural anticoagulant? _____

10. If your blood clumped with both anti-A and anti-B sera, your ABO blood type would be _____

To what ABO blood groups could you give blood? _____ From which ABO donor

types could you receive blood? _____ Which ABO blood type is most common? _____

Least common? _____

11. Explain why an Rh-negative person does not have a transfusion reaction on the first exposure to Rh-

positive blood, but *does* have a reaction on the second exposure. _____

_____

What happens when an ABO blood type is mismatched for the first time? _____

_____

12. Correctly identify the blood pathologies described in column A by choosing a response from column B.

Column A	Column B
_____ abnormal increase in the number of WBCs	a. anemia
_____ abnormal increase in the number of RBCs	b. leukocytosis
_____ condition of too few RBCs or RBCs with hemoglobin deficiencies	c. leukopenia
_____ abnormal decrease in the number of WBCs	d. polycythemia

the _____ of the tissues to the systemic veins to the _____ and

_____ entering the right atrium of the heart.

11. If the mitral valve does not close properly, which circulation is affected? _____

_____

Dissection of the Sheep Heart

1. During the sheep heart dissection, you were asked initially to identify the right and left ventricles without cutting into the heart. During this procedure, what differences did you observe between the two chambers?

_____

_____

Since structure and function are related, how does this structural difference reflect the relative functions

of these two heart chambers? _____

_____

_____

2. Semilunar valves prevent backflow into the _____; AV valves prevent backflow

into the _____. Using your own observations, explain how the operation of

the semilunar valves differs from that of the AV valves. _____

_____

_____

_____

3. Differentiate clearly between the location and appearance of pectinate muscle and trabeculae carneae.

_____

_____

_____

4. Two remnants of fetal structures are observable in the heart—the ligamentum arteriosum and the fossa ovalis. Where is each located, and what common purpose did they serve as functioning fetal structures?

_____

_____

_____

Microscopic Anatomy of Cardiac Muscle

1. How would you distinguish cardiac muscle from skeletal muscle? _____

    _____

    _____

    _____

2. What role does the unique structure of cardiac muscle play in its function? (Note: Before attempting a response, describe the unique anatomy.) _____

    _____

    _____

    _____

Review Sheet

EXERCISE 25

Anatomy of Blood Vessels

Microscopic Structure of Blood Vessels

1. Use key choices to identify the blood vessel tunic described.

   Key: a. tunica intima   b. tunica media   c. tunica externa

   _____ single thin layer of endothelium

   _____ bulky middle coat containing smooth muscle and elastin

   _____ provides a smooth surface to decrease resistance to blood flow

   _____ the only tunic of capillaries

   _____ also called the adventitia

   _____ the only tunic that plays an active role in blood pressure regulation

   _____ supporting, protective coat

2. Servicing the capillaries is the essential function of the organs of the circulatory system. Explain this

   statement. _____

   _____

   _____

3. Cross sectional views of an artery and a vein are shown here. Identify each and on the lines beneath, note the structural details that allowed you to make these identifications.

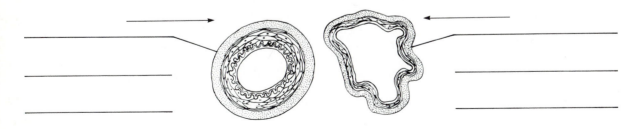

4. Why are valves present in veins but not in arteries? _____

   _____

   _____

5. Name two events occurring within the body that aid in venous return.

   _____ and _____

6. Why are the walls of arteries proportionately thicker than those of the corresponding veins? _____

_____

Major Systemic Arteries and Veins of the Body

1. Use the key on the right to identify the arteries described on the left.

_____ and _____ two arteries formed by the bifurcation of the brachiocephalic artery

_____ first branches off the ascending aorta

_____ and _____ two paired arteries serving the brain

_____ largest artery of the body

_____ artery on the dorsum of the foot checked after leg surgery

_____ serves the posterior thigh

_____ supplies the diaphragm

_____ bifurcates to form the radial and ulnar arteries

_____ artery generally auscultated to determine the blood pressure in the arm

_____ artery serving the kidney

_____ artery serving the liver

_____ artery that supplies the last half of the large intestine

_____ serves the pelvic organs

_____ the external iliac artery becomes the (what?) on entry into the thigh

_____ major artery serving the arm

_____ supplies most of the small intestine

_____ the terminal branches of the dorsal or descending aorta

_____ an arterial trunk that has three major branches, which run to the liver, spleen, and stomach

_____ major artery serving the tissues external to the skull

_____ _____, and _____ three arteries serving the leg inferior to the knee

_____ artery generally used to take the pulse at the wrist

Key:

a. anterior tibial

b. aorta

c. brachial

d. brachiocephalic

e. celiac trunk

f. common carotid

g. common iliac

h. coronary

i. deep femoral

j. dorsalis pedis

k. external carotid

l. femoral

m. hepatic

n. inferior mesenteric

o. intercostals

p. internal carotid

q. internal iliac

r. peroneal

s. phrenic

t. posterior tibial

u. radial

v. renal

w. subclavian

x. superior mesenteric

y. vertebral

z. ulnar

2. The human arterial and venous systems are diagrammed on the next two pages. Identify all indicated blood vessels.

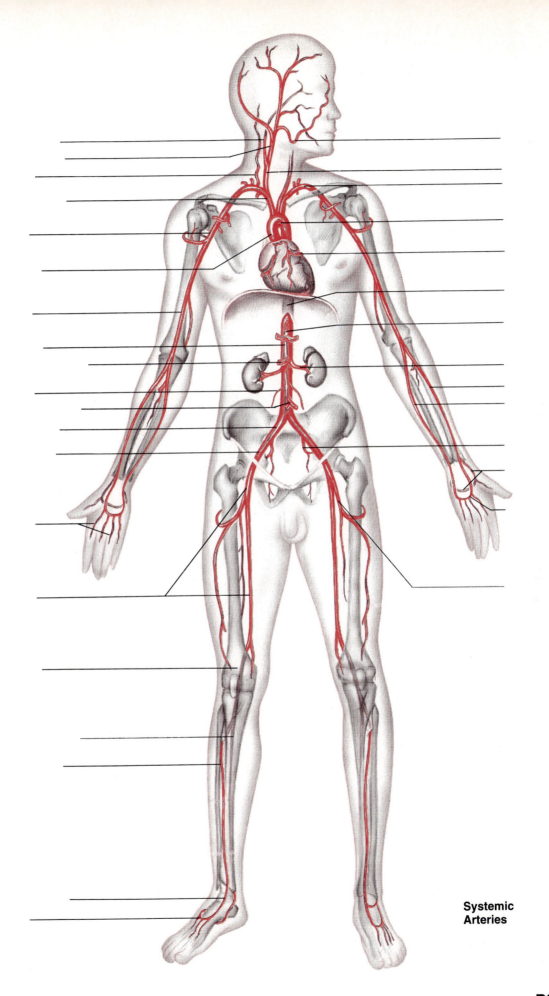

**Systemic
Arteries**

RS101

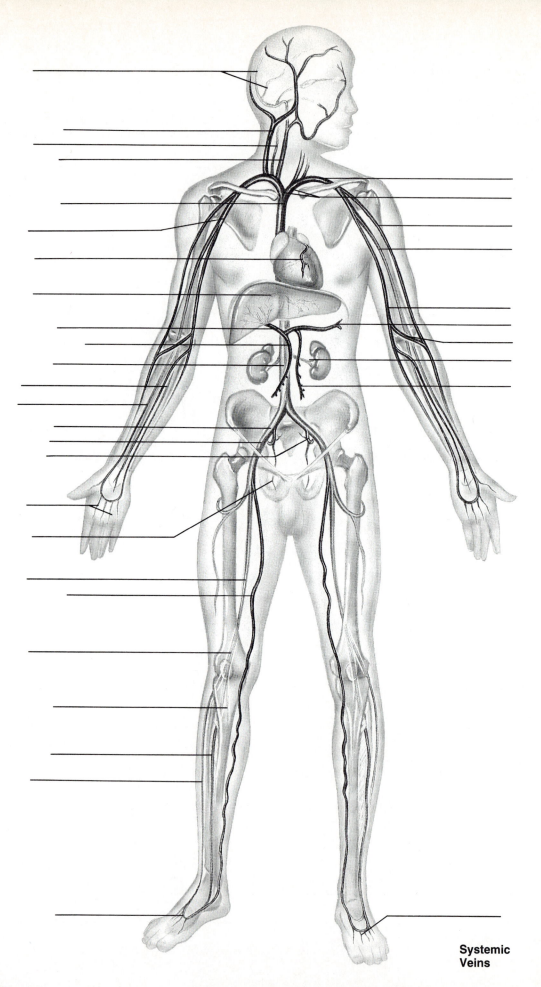

**Systemic
Veins**

Special Circulations

Pulmonary circulation:

1. Trace the pathway of a carbon dioxide gas molecule in the blood in the inferior vena cava until it leaves the bloodstream. Name all structures (vessels, heart chambers, and others) passed through en route.

   _____

   _____

2. Trace the pathway of an oxygen gas molecule from an alveolus of the lung to the right atrium of the heart. Name all structures through which it passes. _____

   _____

   _____

3. Most arteries of the adult body carry oxygen-rich blood, and the veins carry oxygen-depleted, carbon dioxide-rich blood. What is different about the pulmonary arteries and veins? _____

   _____

   _____

4. How do the arteries of the pulmonary circulation differ structurally from the systemic arteries? What condition results from this anatomic difference? _____

   _____

   _____

Arterial supply of the brain:

1. What two paired arteries enter the skull to supply the brain?

   _____ and _____

2. The paired arteries just named cooperate to form a ring of blood vessels encircling the pituitary gland at the base of the brain. What name is given to this communication network? _____

   What is its function? _____

   _____

3. What portion of the brain is served by the anterior and middle cerebral arteries? _____

   _____

   Both the anterior and middle cerebral arteries arise from the _____ arteries.

4. Trace the pathway of a drop of blood from the aorta to the occipital lobe of the brain, noting all structures through which it flows. _____

   _____

Hepatic portal circulation:

1. What is the source of blood in the hepatic portal system? _____

_____

2. Why is this blood carried to the liver before it enters the systemic circulation? _____

_____

3. The hepatic portal vein is formed by the union of the _____, which drains

the _____, _____, _____

_____, and the _____, which drains the _____

and _____. The _____ vein, which drains the lesser
curvature of the stomach, empties directly into the hepatic portal vein.

4. Trace the flow of a drop of blood from the small intestine to the right atrium of the heart, noting all

structures encountered or passed through on the way. _____

_____

_____

Fetal circulation:

1. The failure of two of the fetal bypass structures to close after birth can cause congenital heart disease, in
which the youngster would have improperly oxygenated blood. Which two structures are these?

_____ and _____

2. For each of the following structures, first note its function in the fetus, and then note what happens to it
or what it is converted to after birth. Circle the blood vessel that carries the most oxygen-rich blood.

Structure	Function in fetus	Fate
umbilical artery		
umbilical vein		
ductus venosus		
ductus arteriosus		
foramen ovale		

3. What organ serves as a respiratory/digestive/excretory organ for the fetus? _____

Dissection of the Blood Vessels of the Cat

1. What differences did you observe between the origin of the left common carotid artery in the cat and in the human? _____

   _____

   _____

   _____

   Between the origin of the internal and external iliac arteries? _____

   _____

   _____

   _____

   _____

2. How do the relative sizes of the external and internal jugular veins differ in the human and the cat?

   _____

   _____

3. In the cat the inferior vena cava is called the _____;

   the superior vena cava is referred to as the _____.

Review Sheet

EXERCISE 26   # The Lymphatic System

The Lymphatic System

1. Explain why the lymphatic system is a one-way system, whereas the blood vascular system is a two-way

   system. _____

   _____

   _____

2. How do lymphatic vessels resemble veins? _____

   How do lymphatic capillaries differ from blood capillaries? _____

   _____

3. What is the function of the lymphatic vessels? _____

   _____

4. What is lymph? _____

5. What factors are involved in the flow of lymphatic fluid? _____

   _____

6. What name is given to the terminal duct draining most of the body? _____

7. What is the cisterna chyli? _____

   _____

   How does the composition of lymph in the cisterna chyli differ from that in the general lymphatic stream?

   _____

8. Which portion of the body is drained by the right lymphatic duct? _____

9. Note three areas where lymph nodes are densely clustered: _____,

   _____, and _____

10. What are the two major functions of the lymph nodes? _____

    _____

11. The radical mastectomy is an operation in which a cancerous breast, surrounding tissues, and the under-
    lying muscles of the anterior thoracic wall, plus the axillary lymph nodes, are removed. After such an
    operation, the arm usually swells, or becomes edematous, and is very uncomfortable—sometimes for

months. Why? _____

_____

Microscopic Anatomy of a Lymph Node

1. In the space below, make a rough drawing of the structure of a lymph node. Identify the cortex area, germinal centers, and medulla. For each identified area, note the cell type (T cell, B cell, or macrophage) most likely to be found there.

2. What structural characteristic ensures a slow flow of lymph through a lymph node? _____

_____

Why is this desirable? _____

_____

Review Sheet

EXERCISE 27

Anatomy of the Respiratory System

Upper and Lower Respiratory Structures

1. Complete the labeling of the diagram of the upper respiratory structures (sagittal section).

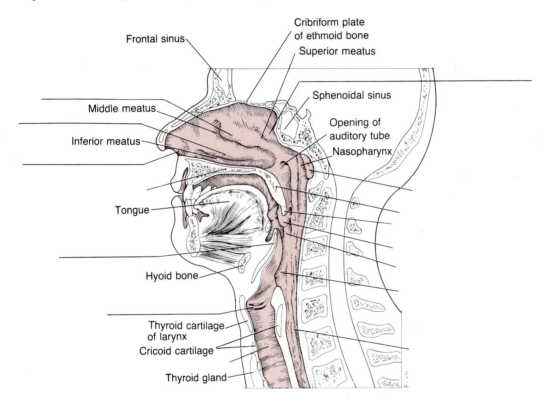

2. Two pairs of vocal folds are found in the larynx. Which pair are the true vocal cords? (Superior or inferior?) _____

3. What is the significance of the fact that the human trachea is reinforced with cartilage rings? _____

_____

Of the fact that the rings are incomplete posteriorly? _____

_____

4. Name the specific cartilages in the larynx that correspond to the following descriptions:

forms the Adam's apple _____ shaped like a signet ring _____

a "lid" for the larynx _____ vocal cord attachment _____

RS109

5. Trace a molecule of oxygen from the external nares to the pulmonary capillaries of the lungs: External

   nares → _____

   _____

   _____

   _____

6. What is the function of the pleural membranes? _____

   _____

7. Name two functions of the nasal cavity mucosa: _____

   _____

8. The following questions refer to the primary bronchi:

   Which is longer? _____ Larger in diameter? _____

   More horizontal? _____
   The most common site for lodging of a foreign object that had entered the respiratory passageways?

   _____

9. Match the terms in column B to those in column A.

Column A	Column B
_____ nerve that serves the diaphragm during inspiration	a. alveoli
_____ separates the oral and nasal cavities	b. bronchioles
_____ food passageway posterior to the trachea	c. conchae
_____ flaps over the glottis during swallowing of food	d. epiglottis
_____ voice box	e. esophagus
_____ windpipe	f. glottis
_____ pleural layer covering the walls of the thorax	g. larynx
_____ actual site of gas exchange	h. palate
_____ autonomic nervous system nerve serving the thoracic region	i. parietal pleura
_____ opening between the vocal folds	j. phrenic nerve
_____ fleshy lobes in the nasal cavity	k. primary bronchi
	l. trachea
	m. vagus nerve
	n. visceral pleura

10. What portions of the respiratory system are referred to as anatomic dead space? _____

_____

Why? _____

11. Define *external respiration:* _____

_____

 internal respiration: _____

_____

12. On the diagram below identify alveolar epithelium, capillary endothelium, alveoli, and red blood cells, and bracket the respiratory membrane.

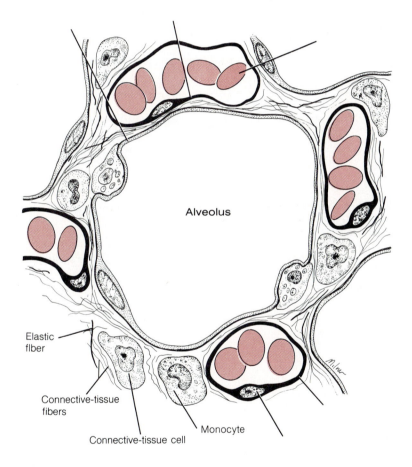

Alveolus

Elastic
fiber

Connective-tissue
fibers

Connective-tissue cell

Monocyte

Sheep Pluck Demonstration

1. Does the lung inflate part-by-part or as a whole, like a balloon? _____ _____

   What happened when the pressure was released? _____

   What type of tissue insures this phenomenon? _____

Examination of Prepared Slides of Lung and Trachea Tissue

1. The tracheal epithelium is ciliated and has goblet cells. What is the function of each of these modifications?

   Cilia? _____

   Goblet cells? _____

2. The tracheal epithelium is said to be pseudostratified. Why? _____

   _____

   _____

3. What structural characteristics of the alveoli make them an ideal site for the diffusion of gases? _____

   _____

   Why does oxygen move from the alveoli into the pulmonary capillary blood? _____

   _____

   _____

4. If you observed pathologic lung sections, what were the conditions responsible and how did the tissue differ from normal lung tissue?

   Slide type   **Observations**

Dissection of the Respiratory System of the Cat

1. Are the cartilaginous rings in the cat trachea complete or incomplete? _____

2. How does the number of lung lobes in the cat compare with the number in humans? _____

   _____

3. Describe the appearance of lung tissue under the dissection microscope. _____

   _____

Review Sheet

EXERCISE 28

Anatomy of the Digestive System

Gross Anatomy of the Human Digestive System

1. Match the key items on the right with the descriptive statements on the left.

_____ structure that suspends the small intestine from the posterior body wall

_____ fingerlike extensions of the intestinal mucosa that increase the surface area for absorption

_____ large collections of lymphatic tissue found in the submucosa of the small intestine

_____ deep folds of the mucosa and submucosa that extend completely or partially around the small intestine circumference

_____, _____ two anatomic regions involved in the mechanical breakdown of foodstuffs

_____ organ that manipulates food in the mouth and initiates swallowing

_____ common passageway for air and food

_____, _____, _____ three structures continuous with and representing modifications of the peritoneum

_____ literally a food chute; no digestive/absorptive function

_____ folds of the gastric mucosa

_____ sacculations of the large intestine

_____ projections of the plasma membrane of a mucosal epithelial cell

_____ prevents food from moving back into the small intestine once it has entered the large intestine

_____ primary region of food and water absorption

_____ membrane securing the tongue to the floor of the mouth

_____ region of the digestive tract primarily involved in water absorption and the formation of feces

_____ area between the teeth and lips/cheeks

_____ blind sac that outpockets from the cecum

_____ region in which protein digestion begins

Key:

a. anus

b. appendix

c. esophagus

d. frenulum

e. greater omentum

f. hard palate

g. haustra

h. ileocecal valve

i. large intestine

j. lesser omentum

k. mesentery

l. microvilli

m. oral cavity

n. parietal peritoneum

o. Peyer's patches

p. pharynx

q. plicae circulares

r. pyloric valve

s. rugae

t. small intestine

u. soft palate

v. stomach

(*key continues*)

RS113

_____ structure attached to the lesser curvature of the stomach

_____ organ into which the stomach empties

_____ sphincter controlling the movement of food from the stomach into the duodenum

_____ posterior superior boundary of the oral cavity; uvula forms its terminus

_____ organ that receives pancreatic and bile secretions

_____ serous lining of the abdominal cavity wall

_____ principal site for the synthesis of vitamin K by microorganisms

_____ region containing two sphincters through which feces are expelled from the body

_____ anteriosuperior boundary of the oral cavity, underlain by bone

w. tongue

x. vestibule

y. villi

z. visceral peritoneum

2. The tubelike digestive system canal that extends from the mouth to the anus is the _____ canal.

3. The general anatomic features of the digestive tube have been presented. Fill in the table below to complete the information listed.

Wall layer	Subdivisions of the layer	Major functions
mucosa		
submucosa	(not applicable)	
muscularis externa		
serosa or adventitia	(not applicable)	

4. How is the muscularis externa of the stomach modified? _____

_____

How does this modification relate to the function of the stomach? _____

_____

5. Correctly identify all organs indicated on the diagram on the following page.

RS114

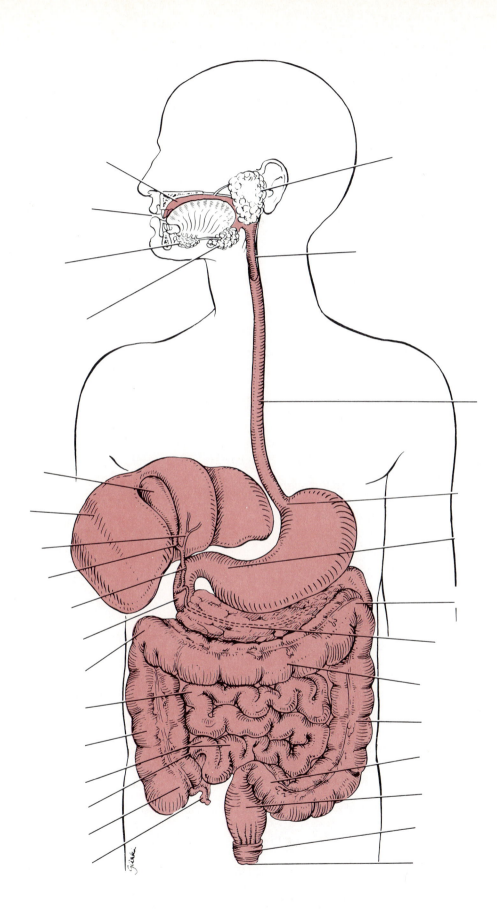

Accessory Digestive Organs

1. Various types of glands form a part of the alimentary tube wall or duct their secretions into it. Match the glands listed in column B with the function/locations described in column A.

Column A

_____ produce mucus; found in the submucosa of the small intestine

_____ produce a product containing amylase that begins starch breakdown in the mouth

_____ produces a whole spectrum of enzymes and an alkaline fluid that is secreted into the duodenum

_____ produces bile that it secretes into the duodenum via the common bile duct

_____ produces HCl and pepsinogen

_____ found in the mucosa of the small intestine; produce intestinal juice

Column B

a. Brunner's glands

b. crypts of Lieberkühn

c. gastric glands

d. liver

e. pancreas

f. salivary glands

2. Which of the salivary glands produces a secretion that is mainly serous? _____

3. What is the role of the gall bladder? _____

_____

4. Use the key to identify each tooth area described below.

_____ visible portion of the tooth in situ

_____ material covering the tooth root

_____ hardest substance in the body

_____ attaches the tooth to bone and surrounding alveolar structures

_____ portion of the tooth embedded in bone

_____ forms the major portion of tooth structure; similar to bone

_____ form the dentin

_____ site of blood vessels, nerves, and lymphatics

_____ entire portion of the tooth covered with enamel

Key: a. anatomic crown

b. cementum

c. clinical crown

d. dentin

e. enamel

f. gingiva

g. odontoblasts

h. periodontal membrane

i. pulp

j. root

5. In the human, the number of deciduous teeth is _____; the number of permanent teeth is _____.

6. The dental formula for deciduous teeth is: $\dfrac{2,1,0,2}{2,1,0,2}$

Explain what this means: _____

_____

RS116

What is the dental formula for the permanent teeth? _____

7. What teeth are the "wisdom teeth"? _____

Microscopic Anatomy of the Digestive System and Accessory Organs

1. You have studied the histologic structure of a number of organs in this laboratory. Four of these are diagrammed below. Identify each.

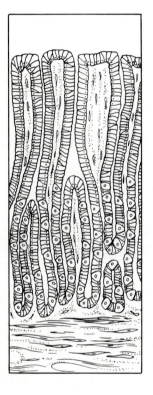

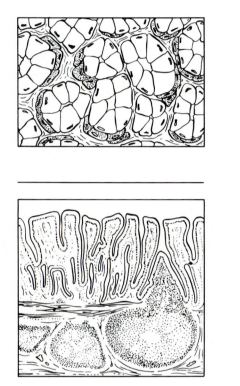

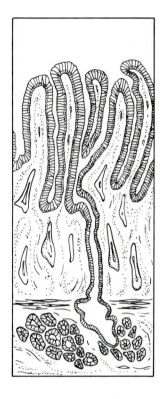

_____   _____   _____

2. What transition in epithelium type exists at the cardiac-esophageal junction? _____

_____

How do the epithelia of these two organs relate to their specific function? _____

_____

_____

3. What cells of the stomach produce HCl? _____

Pepsinogen? _____

4. Name three structures always found in the triad regions of the liver. _____,

_____, and _____.

5. Where would you expect to find the van Kupffer cells of the liver? What is their function? _____

_____

6. Why is the liver so dark red in the living animal? _____

_____

Dissection of the Digestive System of the Cat

1. Several differences between cat and human digestive anatomy should have become apparent during the dissection experience. Note the pertinent differences between the human and the cat relative to the following structures:

Structure	Cat	Human
tongue papillae		
number of liver lobes		
appendix		
means of attachment of the large intestine to body wall		

Review Sheet

EXERCISE 29

Anatomy of the Urinary System

Gross Anatomy of the Human Urinary System

1. Complete the following statements:

   The kidney is referred to as an excretory organ because it excretes __a__ wastes. It is also a major homeostatic organ because it maintains the electrolyte, __b__, and __c__ balance of the blood.

   Urine is continuously formed by the __d__ and is routed down the __e__ by the mechanism of __f__ to a storage organ called the __g__. Eventually, the urine is conducted to the body __h__ by the urethra. In the male, the urethra is __i__ inches long and transports both urine and __j__. The female urethra is __k__ inches long and transports only urine.

   Voiding or emptying the bladder is called __l__. Voiding has both voluntary and involuntary components. The voluntary sphincter is the __m__ sphincter. An inability to control this sphincter is referred to as __n__.

   a. _____

   b. _____

   c. _____

   d. _____

   e. _____

   f. _____

   g. _____

   h. _____

   i. _____

   j. _____

   k. _____

   l. _____

   m. _____

   n. _____

2. What is the function of the fat cushion that surrounds the kidneys in life? _____

_____

3. Define *ptosis*. _____

_____

4. Why is incontinence a normal phenomenon in the child under 1½ to 2 years old? _____

_____

What events may lead to its occurrence in the adult? _____

_____

_____

5. Complete the labeling of the diagram to correctly identify the urinary system organs.

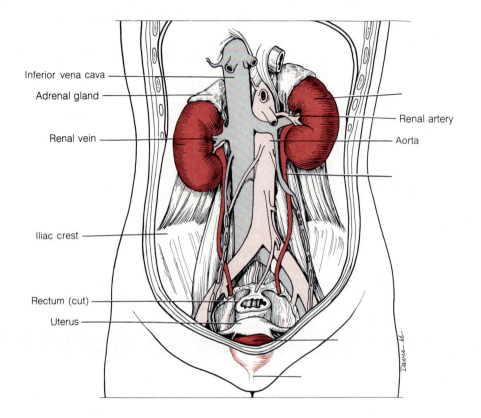

Inferior vena cava

Adrenal gland

Renal vein

Renal artery

Aorta

Iliac crest

Rectum (cut)

Uterus

Gross Internal Anatomy of the Pig or Sheep Kidney

1. Match the appropriate structure in column B to its description in column A.

Column A

_____ smooth membrane, tightly adherent to the kidney surface

_____ portion of the kidney containing mostly collecting ducts

_____ portion of the kidney containing the bulk of the nephron structures

_____ superficial region of kidney tissue

_____ basinlike area of the kidney, continuous with the ureter

_____ a cup-shaped extension of the pelvis that encircles the apex of a pyramid

_____ areas of cortical tissue found between the medullary pyramids

Column B

a. cortex

b. medulla

c. minor calyx

d. renal capsule

e. renal columns

f. renal pelvis

Microscopic Anatomy of the Kidney and Bladder

1. Match each of the lettered structures on the diagram of the nephron (and associated renal blood supply) on the left with the terms on the right.

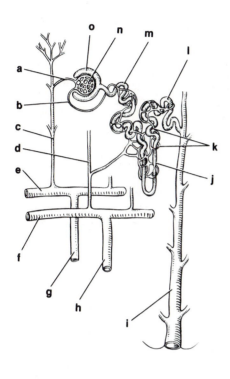

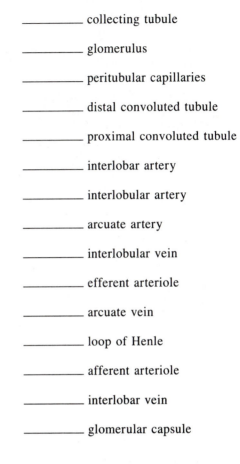

_____ collecting tubule

_____ glomerulus

_____ peritubular capillaries

_____ distal convoluted tubule

_____ proximal convoluted tubule

_____ interlobar artery

_____ interlobular artery

_____ arcuate artery

_____ interlobular vein

_____ efferent arteriole

_____ arcuate vein

_____ loop of Henle

_____ afferent arteriole

_____ interlobar vein

_____ glomerular capsule

2. Using the terms provided in item 1, identify the following:

_____ site of filtrate formation

_____ primary site of tubular reabsorption

_____ secondarily important site of tubular reabsorption

_____ structure that conveys the processed filtrate (urine) to the renal pelvis

_____ blood supply that directly receives substances from the tubular cells

_____ its inner (visceral) membrane forms part of the filtration membrane

3. Complete the following statements:

   The glomerulus is a unique high-pressure capillary bed because the __a__ arteriole feeding the glomerulus is larger in diameter than the __b__ arteriole draining the bed. Glomerular filtrate is very similar to __c__, but it has fewer proteins.

   Three reabsorption mechanisms that occur in the renal tubule are __d__, __e__, and __f__. As an aid for the reabsorption process, the cells of the proximal convoluted tubule have dense __g__ on their luminal surface, which increases the surface area dramatically. Other than reabsorption, an important tubular function is __h__.

   a. _____

   b. _____

   c. _____

   d. _____

   e. _____

   f. _____

   g. _____

   h. _____

4. Trace a drop of blood from the time it enters the kidney in the renal artery until it leaves the kidney through the renal vein.   renal artery → _____

   _____

   _____

   _____ → renal vein

5. Trace the anatomic pathway of a molecule of creatinine (metabolic waste) from Bowman's capsule to the urethra. Note each microscopic and/or gross structure it passes through in its travels. Name the subdivisions of the renal tube.   Bowman's capsule → _____

   _____

   _____ → urethra

6. What is important functionally about the specialized epithelium (transitional epithelium) in the bladder?

   _____

   _____

Dissection of the Cat Urinary System

1. How does the position of the kidneys in the cat differ from their position in humans? _____

   _____

   _____

2. How does the site of urethral emptying in the female cat differ from its termination point in the human female? _____

   _____

   _____

3. What gland encircles the neck of the bladder in the male? _____ Is this part of the urinary system? _____ What is its function? _____

   _____

   _____

Review Sheet

EXERCISE 30

Anatomy of the Reproductive System

Gross Anatomy of the Human Male Reproductive System

1. List the two principal functions of the testis:

    _____

2. Identify all indicated structures or portions of structures on the diagrammatic view of the male reproductive system below.

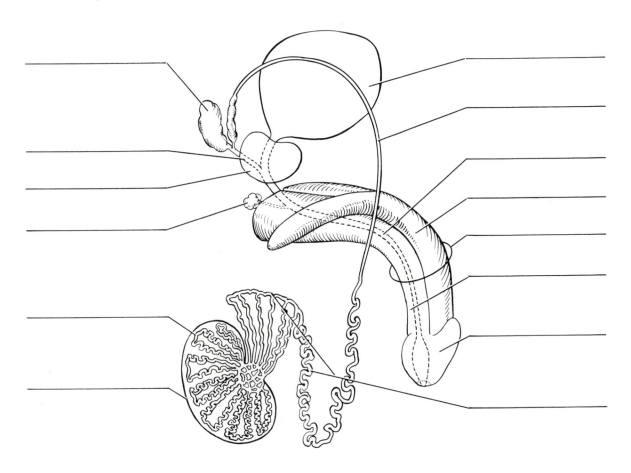

3. A common part of any physical examination of the male is palpation of the prostate gland. How is this

    accomplished? (Think!) _____

    _____

4. How might enlargement of the prostate gland interfere with urination or the reproductive ability of the male? _____

_____

5. Match the terms in column B to the descriptive statements in column A.

Column A

_____ copulatory organ/penetrating device

_____ site of sperm/androgen production

_____ passageway conveying sperm from the epididymis to the ejaculatory duct; in the spermatic cord

_____ conveys both sperm and urine down the length of the penis

_____ tubular storage and maturation site for sperm; hugs the lateral aspect of the testis

_____ extra-abdominal sac, which houses the testis

_____ cuff of skin encircling the glans penis

_____ portion of the urethra between the prostate gland and the penis

_____ empties a secretion into the prostatic urethra

_____ empties a secretion into the membranous urethra

Column B

a. Cowper's glands

b. epididymis

c. glans penis

d. membranous urethra

e. penile urethra

f. penis

g. prepuce

h. prostate gland

i. prostatic urethra

j. seminal vesicles

k. scrotum

l. testes

m. vas deferens

6. Why are the testes located in the scrotum? _____

_____

7. Describe the composition of semen and name all structures contributing to its formation. _____

_____

_____

8. Of what importance is the fact that semen is alkaline? _____

_____

9. What structures comprise the spermatic cord? _____

_____

Where is it located? _____

_____

10. Using the following terms, trace the pathway of sperm from the testes to the urethra: rete testis, epididymis, seminiferous tubule, ductus deferens.

_____ → _____ → _____ → _____

11. Using an appropriate reference, define cryptorchidism and discuss its significance.

_____

_____

Gross Anatomy of the Human Female Reproductive System

1. On the diagram of a frontal section of a portion of the female reproductive system seen below, identify all indicated structures.

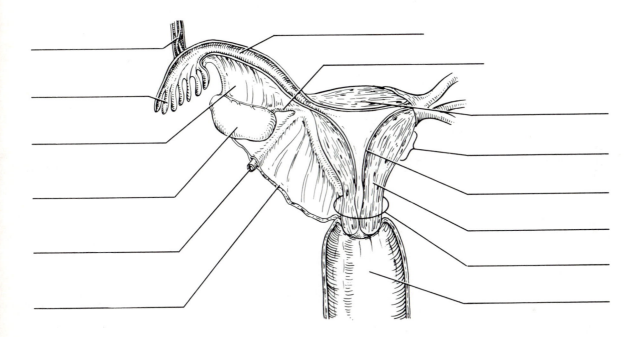

2. Identify the female reproductive system structures described below:

_____ chamber that houses the developing fetus

_____ copulatory canal

_____ usual site of fertilization

_____ becomes erectile during sexual stimulation

_____ duct conveying the ovum to the uterus

_____ partially closes the vaginal canal; a membrane

_____ primary reproductive organ

_____ undulate to create fluid currents to draw the ovulated egg into the fallopian tube

3. Do any sperm enter the pelvic cavity of the female? Why or why not? _____

_____

4. What is an ectopic pregnancy, and how can it happen? _____

_____

_____

5. Name the structures composing the external genitalia, or vulva, of the female. _____

_____

_____

6. Put the following vestibular-perineal structures in their proper order from the anterior to the posterior aspect: vaginal orifice, anus, urethral opening, and clitoris.

   Anterior limit: _____ → _____ → _____ → _____

7. Name the male structure that is homologous to the female structures named below.

   labia majora _____   clitoris _____

8. Assume a couple has just consummated the sex act and the male's sperm have been deposited in the woman's vagina. Trace the pathway of the sperm through the female reproductive tract.

_____

_____

_____

9. Define *ovulation:* _____

_____

_____

10. To describe breast function, complete the following sentences: Milk is formed by _____

   within the _____ of the breast. Milk is then excreted into enlarged storage regions called

   _____ and then finally through the _____.

11. Describe the procedure for self-examination of the breasts. (Men are not exempt from breast cancer, you

   know!) _____

_____

_____

_____

_____

_____

_____

Microscopic Anatomy of Selected Reproductive Organs

1. What is the function of the cavernous bodies seen in the male penis? _____

   _____

2. Name the three layers of the uterine wall from the inside out.

   _____ , _____ , _____

   Which of these is sloughed off during menses? _____

   Which contracts during childbirth? _____

3. The testis is divided into a number of lobes by connective tissue. Each of these lobes contains one to four

   _____ , which converge on a tubular region at the testis hilus called

   the _____ .

4. On the diagram showing the sagittal section of the human testis, correctly identify all indicated structures.

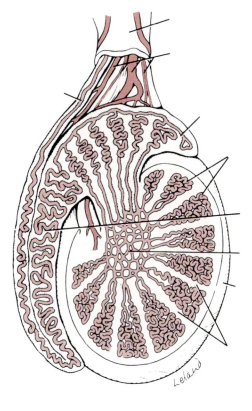

Dissection of the Reproductive System of the Cat

1. The female cat has a _____ uterus; that of the human female is _____.

    Explain the difference in structure of these two uterine types. _____

    _____

    _____

2. What reproductive advantage is conferred by the feline uterine type? _____

    _____

3. Cite differences noted between the cat and the human relative to the following structures:

    uterine tubes or oviducts _____

    site of entry of ductus deferens into the urethra _____

    _____

    location of the prostate gland _____

    _____

    seminal vesicles _____

    urethral and vaginal openings in the female _____

    _____

    _____

Review Sheet

EXERCISE 31

Survey of Embryonic Development

Early Embryology of the Sea Urchin and the Human

1. Use the key choices to identify the embryonic stage or process described below.

   Key: a. cleavage   c. zygote   e. blastula
           b. morula   d. fertilization f. gastrulation

   _____ fusion of egg and sperm nuclei

   _____ solid ball of embryonic cells

   _____ process of rapid mitotic cell division without intervening growth periods

   _____ the fertilized egg

   _____ process involving cell migrations and rearrangements to form the three embryonic germ layers

   _____ embryonic stage in which the embryo consists of a hollow ball of cells

2. What is the importance of cleavage in embryonic development? _____

   _____

   How is cleavage different from mitotic cell division, which occurs later in life? _____

   _____

   _____

3. Explain the process of gastrulation. _____

   _____

   _____

4. Name the primary germ layers, and describe their relative positions in the embryo.

   _____ , _____

   _____ , _____

   _____ , _____

5. The cells of the human blastula (more commonly called the blastocyst or chorionic vesicle) have various
   fates. Which blastocyst structures have the following fates?

   _____ produces the embryonic body

_____ becomes the chorion and cooperates with uterine tissues to form the placenta

_____ produces the amnion, yolk sac, and allantois

_____ produces the primordial germ cells (an embryonic membrane)

_____ an embryonic membrane that provides the structural basis for the body stalk or umbilical cord

6. What is the function of the amnion and the amniotic fluid? _____

_____

_____

_____

7. Describe the process of implantation, noting the role of the trophoblast cells. _____

_____

_____

_____

8. How many days after fertilization is implantation generally completed? _____ What event in the

female menstrual cycle ordinarily occurs just about this time if implantation does not occur? _____

_____

9. What name is given to the part of the uterine wall directly under the implanting embryo? _____

_____ That surrounding the rest of the embryonic structure? _____

_____

10. Using an appropriate reference, find out what *decidua* means and state the definition. _____

_____

How is this terminology applicable to the deciduas of pregnancy? _____

_____

11. Referring to the illustrations and text of "Life Before Birth," answer the following:

Which two organ systems are extensively developed in the *very young* embryo?

_____ and _____

Describe the direction of development by circling the correct descriptions below:

proximal-distal   distal-proximal   caudal-rostral   rostral-caudal

Does bodily control during infancy develop in the same directions? Think! Can an infant pick up a common pin (pincer grasp) or wave his arms earlier? Is arm-hand or leg-foot control achieved earlier? _____

_____

12. Note whether each of the following organs or organ systems develop from the (a) ectoderm, (b) endoderm, or (c) mesoderm. Use an appropriate reference as necessary.

_____ skeletal muscle   _____ respiratory mucosa   _____ nervous system

_____ skeleton   _____ circulatory system   _____ serosa membrane

_____ lining of gut   _____ epidermis of skin   _____ liver, pancreas

In Utero Development

1. Make the following comparisons between a human and the pregnant dissected animal structures.

Comparison object	Human	Dissected animal
shape of the placenta		
shape of the uterus		

2. Where in the human uterus do implantation and placentation ordinarily occur? _____

_____

3. Describe the function(s) of the placenta. _____

_____

_____

What embryonic membranes has it more or less "put out of business"? _____

_____

4. When does the human embryo come to be called a fetus? _____

5. What is the usual and most desirable fetal position in utero? _____

Why is this the most desirable position? _____

_____

_____

_____

Gross and Microscopic Anatomy of the Placenta

1. Describe fully the gross structure of the human placenta as observed in the laboratory. _____

    _____

    _____

    _____

2. What is the tissue origin of the placenta: fetal, maternal, or both? _____

3. What are the placental barriers that must be crossed to exchange materials? _____

    _____

Index

Note: Page numbers in *italic* refer to illustrations.

Photo Credits

Figure Number	Credit
Inside covers	© Ed Reschke
2.5	© Tom Tracy/The Stock Shop
3.1	AO Scientific Instruments
3.5	Marian Rice
4.1c	R. Roldewald, University of Virginia/BPS
6.3a–c	Marian Rice
6.3d–f	© Ed Reschke
6.4a	Marian Rice
6.4b–e	© Ed Reschke
6.5a	© Eric Grave/Photo Researchers, Inc.
6.5b	Marian Rice
6.5c	University of California, San Francisco
6.6	© Ed Reschke
7.2	Marian Rice
9.3	© Ed Reschke
14.1	© Ed Reschke
14.2	Marian Rice
14.4	© Ed Reschke
15.3c, 15.4c, 15.5c, 15.6c, 15.8f, 15.10b, 15.11c, 15.12c	From *A Stereoscopic Atlas of Human Anatomy* by David L. Bassett, M.D.
15.14–15.25	Paul Waring, BioMed Arts
16.2	© Manfred Kage/Peter Arnold, Inc.
16.3	G. L. Scott, J. A. Feilbach, and T. A. Duff
16.6	From *Tissues and Organs: A Text-Atlas of Scanning Electron Microscopy* by Richard G. Kessel and Randy H. Kardon, W. H. Freeman and Company, Copyright © 1979. "A Closer Look": © M.P.L. Fogden/Bruce Coleman, Inc.
17.2b	From *A Stereoscopic Atlas of Human Anatomy* by David L. Bassett, M.D.
17.8–17.10	Ann Allworth
18.9b, 18.10b	Paul Waring, BioMed Arts
19.5b	From *Tissues and Organs: A Text-Atlas of Scanning Electron Microscopy* by Richard G. Kessel and Randy H. Kardon. W. H. Freeman and Company, Copyright © 1979.
23.5	Ann Allworth
23.6	Marian Rice
24.3b	Ann Allworth
24.6	Marian Rice
25.1	Marian Rice
25.9b; 25.11b	Paul Waring, BioMed Arts
27.3c	Marian Rice
28.11b; 28.12b	Paul Waring, BioMed Arts
28.13a–c	Marian Rice
29.3b; 29.4b	Paul Waring, BioMed Arts
29.5b	Ann Allworth
29.7	From *Tissues and Organs: A Text-Atlas of Scanning Electron Microscopy* by Richard G. Kessel and Randy H. Kardon. W. H. Freeman and Company, Copyright © 1979.
30.6b; 30.7b	Paul Waring, BioMed Arts
Color Plates A, B, C, D, E	Paul Waring, BioMed Arts
Color Plate F	© Manfred Kage/Peter Arnold, Inc.
Color Plates G, H, I	Marian Rice